P. Neumann
HFH LAB.

D1503355

A GUIDE
TO HUMAN
FACTORS AND
ERGONOMICS

Second Edition

A GUIDE TO HUMAN FACTORS AND ERGONOMICS

Second Edition

Martin Helander

Nanyang Technological University
Singapore

Taylor & Francis
Taylor & Francis Group
Boca Raton   London   New York

A CRC title, part of the Taylor & Francis imprint, a member of the
Taylor & Francis Group, the academic division of T&F Informa plc.

Published in 2006 by
CRC Press
Taylor & Francis Group
6000 Broken Sound Parkway NW, Suite 300
Boca Raton, FL 33487-2742

International Standard Book Number-10: 0-415-28248-9 (Hardcover)
International Standard Book Number-13: 978-0-415-28248-2 (Hardcover)
Library of Congress Card Number 2005047023

Library of Congress Cataloging-in-Publication Data

Helander, Martin, 1943-
  A Guide to human factors and ergonomics / Martin Helander.--2nd ed.
    p. cm.
    Rev. ed. of: A guide to ergonomics of manufacturing. London ;Bristol, PA : Taylor & Francis c1995.
  Includes bibliographical references and index.
  ISBN 0-415-28248-9 (alk. paper)
  1. Human engineering. 2. Work environment. 3. Industrial hygiene. I. Helander, Martin, 1943- Guide to ergonomics of manufacturing. II. Title.

T59.7.H46 2005
620.8'2--dc22   2005047023

Taylor & Francis Group
is the Academic Division of Informa plc.

Visit the Taylor & Francis Web site at
http://www.taylorandfrancis.com

and the CRC Press Web site at
http://www.crcpress.com

To Mahtun

Foreword

This book is based on some of my teaching and research experiences at university and in industry. It is intended as a basic text for a first course in Human Factors/Ergonomics (HFE) at the undergraduate and the graduate levels. Some of the text is adapted from a book that I wrote a few years ago: *A Guide to the Ergonomics in Manufacturing*, but most is new and was written at Nanyang Technological University in Singapore.

Human Factors is an interdisciplinary science, and the book is suitable for many types of students and professionals: engineers, computer scientists, behavioral scientists as well as medical doctors and physiotherapists.

I have portrayed HFE as a systems science. To design an artifact or an interface the human factors expert must consider the interactions between three important systems components: The Operator/User, the Environment, and the Machine/Computer, and these interactions can be modeled using a systems approach.

Some HFE design problems have obvious and immediate solutions, and the benefit-cost ratio of implementing the solutions may be high. But in some cases design problems are complex, and it is difficult to propose good design solutions. One may have to consider expertise from various professions, and there may be several alternative design solutions. To decide which alternative is best, it is common to evaluate the design alternatives using test persons or users. One can then measure how well the user is interacting with the system. A system that is well designed allows the operator to finish a task quickly and is well designed; a system that makes the user commit many errors is not well designed. There are often trade-offs in design—one particular design A may lead to a reduction in user performance time but increase in user errors. For another design B performance time may increase, but errors are reduced. To decide between A and B one will then have to examine other task related factors. Good design solutions are hard to find, but the process of arriving at the solutions is rewarding and exciting.

Another aspect that is important to many human factors professionals is compassion with fellow workers and users. We want to put things right—enhance performance and reduce errors, accidents and injuries. Operators and users have a right to work with well designed systems.

The following quote of C.N. Anadurai, a former Chief Minister of Tamil Nadu, India sums up both our methodology and compassion:

Dear Friends: Go to the people,
Live among them, Learn from them,
Love them, Serve them,
Plan with them, Start with what they know,
Build on what they have.

I have worked in different parts of the world as a teacher and as a consultant: Luleå University, Human Factors Research Inc. in Santa Barbara, Virginia Tech, The University at Buffalo, Linköping University, Hong Kong University of Science and Technology, and currently at Nanyang Technical University in Singapore. I learned much from my university colleagues in research and from industrial partners in implementing HFE in the real world and I am grateful for the experience.

My wife Mahtun gives me everyday honest and constructive criticism. She read the book and helped with editing. It is a true pleasure to dedicate the book to her.

The cover of the book is from a copper print from 1785 with the title: A General Display of the Arts and Sciences. It is a frontispiece to the Royal Encyclopedia and the artist's name is Grignon. He selected an environment from the antique Rome. All types of work activities are illustrated: design, planning, team collaboration and physical work, thereby illustrating the issues of concern in this book.

There are several examples and exercises in the book. Solutions and discussions of these are provided at www.ntu.edu/martin/guidebook.

The Author

Martin G. Helander is Professor at the School of Mechanical and Aerospace Engineering at Nanyang Technological University in Singapore. He received a Ph.D. from Chalmers University of Technology in Göteborg, Sweden, and became Docent of Engineering Psychology at Luleå University. He has held faculty positions at Luleå University, the State University of New York at Buffalo, and Linköping University, and visiting appointments at Virginia Tech, MIT, and Hong Kong University of Science and Technology. His primary research interests are in human factors engineering and ergonomics. In 1996, he established the Graduate School of Human–Machine Interaction in Linköping, Sweden. He is currently the director of the graduate program in human factors engineering at NTU. Dr. Helander has authored 300 publications, including 8 books. He is a Fellow of the International Ergonomics Association, of the Human Factors and Ergonomics Society in the U.S., and of the Ergonomics Society in the U.K. and the Institute of Engineers, Singapore. He is a former president of the International Ergonomics Association.

Contents

PART I   Information-Centered Human Factors

PART II Human-Body-Centered Ergonomics

PART III   Organization/Management-Centered Human Factors

Part I

Information-Centered Human Factors

1 Introduction to Human Factors and Ergonomics

Science never appears so beautiful as when applied to the uses of human life.

Thomas Jefferson

1.1 INTRODUCTION

The purpose of this chapter is to give an overview of human factors and ergonomics (HFE) and to show how these two sciences developed—ergonomics in Europe and HFE in the U.S.

The word *ergonomics* is derived from the Greek words *ergo* (work) and *nomos* (laws). It was used for the first time by Wojciech Jastrzebowski in a Polish newspaper in 1857 (Karwowski, 1991). In the U.S., *human factors engineering* and *human factors* have been close synonyms. European ergonomics has its roots in work physiology, biomechanics, and workstation design. Human factors, on the other hand, originated from research in experimental psychology, where the focus was on human performance and systems design (Chapanis, 1971).

But there are several other names, such as engineering psychology, and more recently cognitive engineering and cognitive systems engineering. The latter emphasizes the importance of human information processing for our science (Hollnagel and Woods, 2005).

Despite the differences between human factors and ergonomics in the type of knowledge and design philosophy, the two approaches are coming closer. This is partly due to the introduction of computers in the workplace. Design of computer workplaces draws from a variety of human factors and ergonomics knowledge (see Table 1.1). We can illustrate the problem as shown in Figure 1.1. Here a user of a computer is perceiving information on a display. The information is interpreted and an appropriate action is selected. The action is executed manually as a control input, which in turn affects the information status on the display. A new display is generated.

To solve a problem that is related to computer workplaces, an ergonomist must be able to identify the problem, analyse it, and suggest improvements in the form of design solutions. This leads to our first maxim:

The primary purpose of human factors and ergonomics is design.

In designing a workplace, the existing situation must first be analysed, new design solutions must be synthesized, and these design solutions must be analyzed again. The design process may be described using a control loop, as shown in Figure 1.2. Through successive design iterations, sometimes over extended periods of time,

TABLE 1.1
Design Problems and Corresponding Knowledge Arising from the Introduction of Computers in the Workplace

Problem	Knowledge Required to Solve Problem
Work posture and keying	Biomechanics
Size of screen characters, contrast, colors	Vision research, perception
Environmental factors	Noise, environmental stress
Layout of screen information	Cognitive psychology, cognitive engineering
Design of new systems	Systems design and cybernetics
Collaboration on the net	Psychology, cognitive psychology, anthropology
Problem solving at work	Cognitive work analysis, task analysis

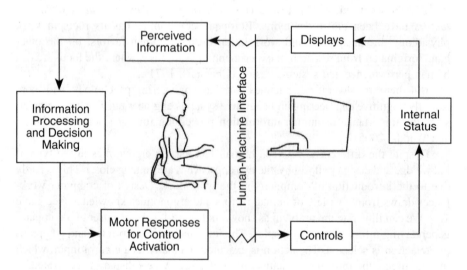

FIGURE 1.1 Analysis of the human-machine interface requires interdisciplinary knowledge of biomechanics, cognitive psychology, and systems design methodology.

design is improved. In fact, even simple designs, such as that of the paper clip, took more than 120 years to mature, from the first patent in 1814, which was for a paper pin, to the present paper clip, which was patented in 1934 (Petroski, 1992). We can understand from this example that the design of complex systems such as computers is still in its infancy.

It follows that interdisciplinary knowledge is required in ergonomics design for the following reasons: (1) to formulate systems goals; (2) to understand functional requirements; (3) to design a new system; (4) to analyze a system; and (5) to implement a system. From the feedback loops in Figure 1.2 it also follows that

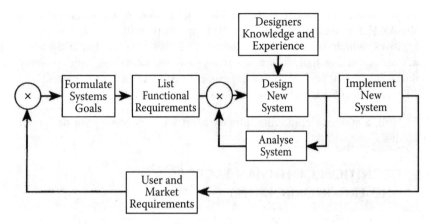

FIGURE 1.2 Procedure for design and redesign of a system.

design is a never-ending activity. There are always opportunities for improvements or modifications.

The interdisciplinary nature of ergonomics is obvious when one notes the mixed professional background of ergonomists. They come from a variety of professions—engineering, psychology, and medical professions, to mention a few.

Many HFE experts are designers. A common design scenario may be as follows. Imagine that the system in Figure 1.1 was redesigned with three displays rather than one, and that some of the operator's decisions were taken over by a decision support system, and that the user's input to the computer system was made by voice recognition technology rather than manual keying. This type of system is currently used in fighter aircrafts. In designing such a system the HFE specialist would have to consider many issues:

- Should the user always be in charge, or are there situations where the computer should take over and fly the aircraft automatically? If so, define in what situations the computer should take over.
- How should the information on the three displays be laid out? What type of information should go on what display? Do we gain anything from using color displays (which are expensive)?
- Can a voice recognition system understand the pilot despite the noisy background? How can one avoid misrecognitions by the computer? We had better make sure that critical commands, such as "fire," are correctly understood.
- Are there design constraints, such as economic and organizational constraints, and possibly constraints from a labor union? Some important constraints are dictated by training requirements. Pilots may be confused if the new model of an aircraft is very different from the old model, and they may revert to the control behavior for the old aircraft, particularly under stressful conditions.

The HFE specialist will analyze the situation and obtain information from users and management. To come up with a good design it will be necessary to get information about the design of similar systems. Many good design ideas can be found in textbooks and scientific articles. The HFE specialist will generate a few design alternatives, which will then be evaluated. The HFE specialist will then have to select an evaluation tool. There are many options, including rapid prototyping, usability studies, or performing an experiment with users as test subjects.

This scenario leads to our second maxim:

In HFE, a systematic, interdisciplinary approach is necessary for design and analysis.

1.2   DEFINITION OF HUMAN FACTORS AND ERGONOMICS (HFE)

There are many definitions in the HFE literature. The following is from Helander (1997):

- Considering environmental and organizational constraints,
- Use knowledge of human abilities and limitations
- To design the system, organization, job, machine, tool, or consumer product
- So that it is safe, efficient, and comfortable to use.

Note again that the main purpose is design (Chapanis, 1995). Ergonomics is thereby different from most of the bodies of knowledge that are used to support HFE. Ergonomics is different from anthropology, cognitive science, psychology, sociology, and medical sciences, since their primary purpose is to understand and model human behavior—but not to design.

The International Ergonomics Association (2000) provides the following definition:

"Ergonomics (or human factors) is the scientific discipline concerned with the understanding of interactions among humans and other elements of a system, and the profession that applies theory, principles, data and methods to design in order to optimize human well-being and overall system performance."

"Ergonomists contribute to the design and evaluation of tasks, jobs, products, environments and systems in order to make them compatible with the needs, abilities and limitations of people."

Throughout the book, I will use the terms ergonomics, human factors, and HFE interchangeably. I will assume that there are no differences between these words, although, as we shall see below, the histories of human factors and ergonomics are quite different.

1.3   THE EARLY DEVELOPMENT OF HUMAN FACTORS AND ERGONOMICS (HFE)

One may argue that designing for human use is nothing new. Hand tools, for example, have been used since the beginning of mankind, and ergonomics was always a concern. If hand tools are appropriately designed they can concentrate and deliver power, and aid the human in tasks such as cutting, smashing, scraping, and piercing. Various hand tools have been developed since the Stone Age, and the interest in ergonomic design can be traced back in history (Childe, 1944, Braidwood, 1951).

Bernardino Ramazzini was a professor of medicine at Padua and Modena in Italy. In 1717 he published a book called *The Diseases of Workers*, which documented links between many occupational hazards and the type of work performed. He described, for example, the development of cumulative trauma disorder, which he believed was caused by repetitive motions of the hand, by constrained body posture, and by excessive mental stress. Although he did not have the present tools of science to support his findings (such as statistical testing), he proposed many innovative solutions to improve the work place.

The Frenchman LaMettrie's controversial book *L'homme Machine* (*Man, the Machine*) was published in 1748, at the beginning of the Industrial Revolution (Christensen, 1962). He examined the analogies between humans and machines and concluded that people are … quite similar. Two things can be learned from LaMettrie's writings. First, the comparison of human capabilities and machine capabilities was a sensitive issue already in the 18th century. Second, by considering how machines operate, one can learn much about human behavior. Both issues remain debated in ergonomics in our day. For example, an industrial robot has many constraints. Some assembly tasks are really difficult for a robot to do, such as putting a washer on a screw before the screw is tightened. Therefore, in the design of new products one must consider "design for assembleability." There can be no washers in the design, there can be no precision tasks that are difficult to program, and so forth. It turns out that the same design features are also very helpful for human assembly operators; we just did not think about this issue until we had to deal with robots. It seems a great irony that only with the introduction of robots did designers start to consider the requirements of the human operator (Helander, 1995).

Rosenbrock (1983) pointed out that during the Industrial Revolution in England there were efforts to apply the concepts of a "human-centered design" to tools such as the Spinning Jenny and the Spinning Mule. The concern was to allocate interesting tasks to the human operator, but let the machine do repetitive tasks. This is another common reason for robotics in our days: to make the work more interesting we must remove repetitive and uninteresting tasks.

At the beginning of the 20th century, Frederick Taylor introduced the scientific study of work. This was followed by Frank and Lillian Gilbreth, who developed the time and motion study and the concept of dividing ordinary jobs into several small elements called "therbligs" (Konz and Johnson, 2004). Today, there are objections against Taylorism, which has been seen as a tool for exploiting workers. This is because there are behavioral aspects of work simplification: give a person a repetitive and mindless job, and there is a great risk that the person will turn mindless. Nonetheless, time and motion study remains useful for measuring and predicting work activities, such as the time it will take to perform a task (Helander, 1997). These are valuable tools if used for the right purpose!

1.4 THE CURRENT DEVELOPMENT OF HUMAN FACTORS IN THE U.S. AND EUROPE

In the beginning of the 20th century, industrial psychology did much research to find principles for selecting operators who were the most suitable to perform a task. The research on accident proneness is typical for the 1920s. Accident proneness assumes that certain individuals have certain enduring personality traits, which make them more prone to have accidents than others. This is because they have a "bad personality." If one can understand how these individuals differ from "normal" people, one can exclude them from activities where they will incur accidents. This approach, which dominated research for about 40 years, was not fruitful. It turns out that accident proneness and many personality features are not stable features, but change with age and experience (Shaw and Sichel, 1971). A person may have many accidents in his young age, but 10 years later he is a different person with no accidents. In current ergonomics there is a realization that human error is mostly caused by poor design, and one should not blame operators for accidents. Instead the goal should be to design environments and artifacts that are safe for all users.

In Europe, ergonomics started seriously with industrial applications in the 1950s, and used information from work physiology, biomechanics, and anthropometry for the design of workstations and industrial processes. The focus was on the well-being of workers as well as on improved manufacturing productivity. Ergonomics was well established in the 1960s, particularly in the U.K., France, Germany, Holland, Italy, and the Scandinavian countries. In many European countries, labor unions took an early interest in promoting ergonomics as being important for worker safety, health, comfort, and convenience. The labor unions are particularly strong in the Scandinavian countries, in France, and in Germany, where they can often dictate what type of production equipment a company should purchase. Good ergonomics design is now taken for granted. As a result, even heavy equipment, such as construction machines, is designed to be very comfortable and convenient to operate.

In the U.S. human factors emerged as a discipline after World War II. Many problems were encountered in the use of sophisticated equipment such as airplanes, radar and sonar stations, and tanks. Sometimes these problems caused human errors with grave consequences. For example, during the Korean War, more pilots were killed during training than in actual war activities (Nichols, 1976). This surprising finding led to a review of the design of airplanes as well as procedures and strategies in operation. Several new design issues were brought up:

- How can information be better displayed so that pilots can quickly under-stand what the situation is. (In our present day lingo we would refer to increased "situation awareness.")
- How can controls be integrated with the task so that they were intuitive and easier to handle? (Control-display compatibility is a useful design concept; see Chapter 13.)

Much research was done in HFE to support new designs and many improvements were implemented, such as a pilot's control stick which combined several control functions and made it easier to handle the airplane and auxiliary combat functions (Wiener and Nagel, 1988). As a result of these improvements and new pilot training programs, the number of fatalities in pilot training decreased to 5% of what they had been previously. Ever since this happened, much research in human factors has been sponsored by the U.S. Department of Defense. Consequently, the information in human factors textbooks is often more influenced by military than by civilian applications of ergonomics. However, with the introduction of computers in the work place in the early 1980s this situation has changed. The workplace has become as high tech as the military scenario, and presently much more funding is channeled to solve these problems.

Several government agencies have sponsored research on civilian applications of HFE. In the U.S. there are many examples: the Federal Highway Administration (design of highways and road signs), NASA (human capabilities and limitations in space; design of space stations), the National Highway Traffic Safety Administration (design of cars, including crash worthiness and effects of drugs and alcohol on driving), the Department of the Interior (ergonomics in underground mining), the National Institute of Standards and Technology (safe design of consumer products), the National Institute of Occupational Safety and Health (ergonomic injuries at work, industrial safety, work stress), the Nuclear Regulatory Commission (design requirements for nuclear power plants), and the Federal Aviation Administration (aviation safety).

In the U.S., ergonomics applications in manufacturing are fairly recent. Eastman-Kodak in Rochester, New York, was probably the first company to implement a substantial program around 1965. Their approach is well documented in an excellent book (Eastman Kodak Company, 2004). At IBM Corporation, interest in manufacturing ergonomics started around 1980. At that time IBM had many human factors experts, but they worked on consumer product design, computers, and software systems. Most manufacturing ergonomics was undertaken by industrial engineers and company nurses (Helander and Burri, 1994).

As I have noted, human factors developed from military problems, and has its origin in experimental psychology and systems engineering. During the last ten years there has been an upsurge of interest in workplace ergonomics, and it seems that the two traditions of human factors and ergonomics have fused. The name change of the Human Factors Society in the USA to the Human Factors and Ergonomics Society is indicative of the changing times.

Ergonomics and human factors have proliferated since the 1950s in Asia, Africa, Latin America, and Australia (Luczak, 1995). In many industrially developing countries (IDCs) ergonomic problems have manifested themselves, and have become more obvious in this era of rapid industrialization. The fast transformation from a rural-agrarian to an urban-industrialized life has come at a cost, and workers are "paying" in terms of a tremendous increase of industrial injuries and increased stress at work. Many of these problems remain hidden, because official statistics which can illuminate the true state of affairs are usually not available. For example, workers in Asian countries do not like to complain about ergonomics problems, which hence go unnoticed.

TABLE 1.2
Important Emerging Areas in Ergonomics around the World

Methodology to change work organization and design
Work-related musculoskeletal disorders
Usability testing for consumer electronic goods
Human–computer interface: software
Organizational design and psychosocial work organization
Ergonomic design of physical work environment
Control room design of nuclear power plants
Training of ergonomists
Interface design with high technology
Human reliability research
Measurement of mental workload
Workforce cost calculation
Product liability
Road safety and car design
Transfer of technology to developing countries

In the transition to industrial status, IDCs have bypassed several stages of development and are now totally immersed in the computerized global environment. What took the Western world 200 years has been accomplished in just 20 years by IDCs. Associated with this development are new HFE problems in education and training personnel. There is a tremendous need for training local employees to understand ergonomics, so that they can themselves monitor potential hazards. Human factors specialists, who understand training problems, could have a significant role to play.

A world-wide survey of HFE professional societies was undertaken by the International Ergonomics Association (Brown, Hendrick, Noy, and Robertson, 1996). The results are given in Table 1.2.

Table 1.2 clearly illustrates that, while traditional physical ergonomics and biomechanics remain important, there is a need for cognitive ergonomics and organizational issues. This finding came as a surprise, since IDCs have in the past only sponsored research in physical ergonomics. The three dominating occupations are in agriculture, construction, and manufacturing—all requiring heavy physical work.

1.5 A SYSTEMS DESCRIPTION

The purpose of this section is to describe HFE in a *systems context*. Most HFE problems are well described by a systems approach. In Figure 1.3 we consider an environment-operator-machine system (Helander, 1997). The operator is the central

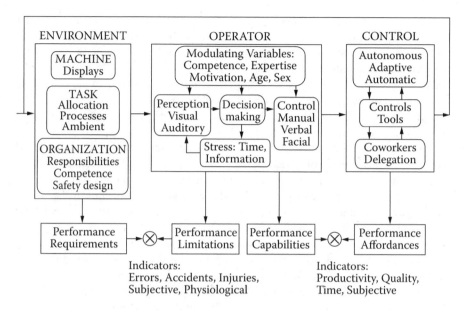

FIGURE 1.3 Ergonomics systems model for measurement of safety and productivity.

focus in ergonomics and should be described in an organizational context, which is the purpose of Figure 1.3. The figure illustrates the most important operator concepts. In reality, human perception, information processing, and response are much more complex, with many feedback loops and variables that are not dealt with here.

In HFE research studies, a classification of independent and dependent variables is used to analyze a problem. Chapter 3 describes the methodology used in HFE research. For example, we may study the effect of environmental factors on operator performance. The characteristics of the environment (such as noise and heat) are independent variables that we can manipulate through design, and the measures of operator performance are the dependent variables. One common operator performance measure is the time it takes to assemble a widget. Another measure is the number of errors committed by the operator per hour. One can also ask the operator what he or she thinks about the new design—how good it is and how satisfying it is to work with. Satisfaction—rated on a seven-point scale—would be a third type of dependent variable.

The operator perceives the environment mainly through the visual and auditory senses, then considers the information, makes a decision, and, finally, produces a control response. Perception is guided by the operator's attention. From the millions of bits of information available, the operator will attend to the information that would seem to be the most relevant to the task. Some attentional processes are automatic and subconscious (pre-attentive) and are executed instantaneously (Neisser, 1967). They are so automatic that the operator will not be able to tell what happened. Some other processes become automatic only after training, while some require deep thinking and pondering of alternatives and hence require more time to analyze.

For new or unusual tasks, decision making can be time consuming. The operator will have to interpret the information, consider alternatives for action, and assess if the actions are relevant to achieve the goals of the task. For routine tasks, decisions are more or less automatic and it takes much less time to decide. One may question whether "decision making" is an appropriate term; it is more of an automatic reaction, and operators usually do not reflect on the decision. Klein (1993) used the term Recognition-Primed Decision (RPD) making. An operator recognizes a scenario and he knows immediately what to do.

The purpose of the operator's response is to convey information through manual response, such as control of a machine (e.g., a computer) or a tool (e.g., a hammer) or an artifact (e.g., a football), or through verbal response to a coworker. For some technology verbal response may also be used in controlling the machine by speaking.

There are several *modulating variables* that affect task performance, including operator needs, attitudes, competence, expertise, motivation, age, gender, body size, and strength. These are idiosyncratic variables and they differ between individuals. For example, an experienced, competent operator will perceive a task differently than a novice operator. The experienced operator will focus on details of importance, filter out irrelevant information, and "chunk" the information into larger units, so that it is possible to make faster and more efficient decisions. A novice operator, on the other hand, may not know where to look for important information, and may think the work is very stressful. Another modulating variable is body size. Different body dimensions have consequences for the design of workstations. These issues are dealt with in anthropometry.

Stress is an important variable that affects perception, decision making, and response selection. High psychological stress levels are common when the time to perform a task is limited, or when there is too much information to process. Under such conditions the bandwidth of attention may narrow, and operators develop "tunnel vision." Thereby the probability of operator error increases. In general, high stress levels lead to increased physiological arousal, which can be measured using various physiological measures (e.g., heart rate, EEG, blink rate, and excretion of catecholamines). These are dependent variables for monitoring of stress.

The Sub-system Environment is used to conceptualize the task as well as the context in which it is performed. It could be a steel worker monitoring an oven. Here the organization of work determines the task allocation: some tasks may be allocated to fellow workers, or supervisors, or computers. Task allocation is a central problem in ergonomics. How can one best allocate work tasks among machines and operators so as to realize both company goals and individual goals? Task allocation affects how information is communicated between employees and computers, and it also affects system performance.

The operator receives various forms of feedback from his or her actions. There may be feedback from task performance, from coworkers, from management, and so forth. To enhance task performance, communication, and job satisfaction, such feedback must be informative. This means that individuals must receive feedback on how well or how poorly they are doing, as well as feedback through communication.

The *ambient environment* describes the influence of environmental variables on the operator. For example, a steel worker is exposed to high levels of noise and heat. This increases physiological arousal and stress, thereby affecting task performance, safety, and satisfaction.

The importance of the *organizational environment* has been increasingly emphasized during the last few years. This movement in ergonomics is referred to as *macroergonomics* (Hendrick, 1995; Hendrick, 2001). Work is undertaken in an organizational context, which deeply affects the appropriateness of alternative design measures. Company policies with respect to communication patterns, decentralization of responsibilities, and task allocation have an impact on ergonomics design. One should first decide who should do what and how people should communicate. Following this activity, individual tasks, machines, displays, and controls can be designed.

Macroergonomics is a much neglected area, and until recently there had not been considerable research. One exception is the socio-technical research developed in the U.K. in the 1950s (e.g., the Tavistock group). Perhaps because human factors research in the military setting was quite dominant in the U.S., the importance of organizational context was not emphasized.

Organizational considerations are important in the work context, but are less important for design of leisure systems and consumer products. These are typically used by individuals who do not have to consider collaboration and task delegation.

The *Machine* sub-system is broadly conceptualized in Figure 1.3. The term "machine" is in a sense misleading. Here it stands for any controlled artifact. The "machine" could be a computer, a video cassette recorder (VCR), or a football. The term *controls* denotes machine controls that are used by the operator. Note that in some systems, machine control may be taken over fully or partially by automation and computers.

As a result of machine control, there is a changing state which is displayed. It can be seen or heard: a pocket calculator will show the results of a calculation; the melting iron in a steel plant will change temperature and color; a computer will produce a sound: or a toaster will pop up the bread. All of these are examples of displays. They convey visual or auditory information, and they can be designed to optimize systems performance. Streitz (2004) presented several examples of human–artifact interaction where displays need not be necessarily CRT displays but can be artifacts, such as toast.

It is important to note that the system in Figure 1.3 has *feedback*. Machine information is fed back to the environment subsystem and becomes integrated with the task. Ergonomics is concerned with *dynamic systems*—it is necessary to go around the loop and incorporate the effect of feedback. Ergonomics, in this sense, is different from other disciplines. In experimental psychology, for example, there is no requirement for studying dynamic systems.

With the system as a basis, we will now discuss three major systems goals in HFE that were mentioned in the definition above: safety, productivity, and operator satisfaction.

1.6 THE GOAL OF SAFETY

Ergonomics is rarely a goal in itself. Safety, operator satisfaction, and productivity are common goals. Ergonomics is a design methodology that is used to arrive at safety, productivity, and satisfaction.

The safety status of a system may be assessed by comparing the performance requirements of the environment with the performance limitations of the operator (see Figure 1.3). Let's take the example of driving a car. The driving task imposes a demand for operator attention, but this demand varies over time. Sometimes a car driver must look constantly at the traffic, and at other times the traffic situation is less demanding. At the same time, operator attention varies over time. A sleepy driver has a low level of attention, while an alert driver has a high level of attention. If the task demands are greater than the available attention, there is an increased risk for accidents or errors. Hence it is important to understand how the limitations imposed by operator perception, decision making, and control action can be taken into consideration in design, so as to create systems with low and stable performance requirements.

Injuries and accidents are relatively rare in the workplace. Rather than waiting for accidents to happen it may be necessary to predict safety problems by analyzing other indicators (or dependent variables) such as operator errors, subjective assessments, and physiological response variables. These measures are indicated in Figure 1.3 under the heading "Measures of Negative Outcome."

If a system must be redesigned to make it safer, there are several things one can do:

1. Examine the allocation of tasks between workers and machines or computers. Workers may be moved from a hazardous area and automation could take over their job.
2. Poor work posture leads to fatigue and poor work quality. Redesign work processes and workstation to improve worker posture, comfort, and convenience.
3. The ambient environment—illumination, noise, vibration, and heat or cold—can be stressful. For example, inadequate illumination makes it difficult to see safety hazards, and therefore the low illumination imposes stress.
4. Organizational factors, such as allocation of responsibility and autonomy, as well as policies for communication, can be changed. Sometimes operators are not in charge of their own processes. Valuable time is lost if they must contact supervisors to get permission to shut down a process.
5. Design features of a machine can be improved, including changes of controls and displays.

1.7 THE GOAL OF PRODUCTIVITY

System design has three goals: safety, productivity, and operator satisfaction. Their relative importance varies depending on the system. In a nuclear power plant, safety and production of electricity are two self-evident goals, and together they determine the design of the plant.

To enhance system performance, one can design a system which improves performance affordances. This means that through efficient design of the system the operator can excel in exercising his or her skills. Such system design makes it possible to perceive quickly, make fast decisions, and exercise efficient control.

To improve systems performance an ergonomist could, for example, design systems affordances so that they enhance important skill parameters: handling of machine controls becomes intuitive (e.g., through control-response compatibility), and interpretation of displays becomes instantaneous (e.g., through use of ecological displays).

In Figure 1.3, several measures of positive outcome are indicated. One can measure productivity, quality, and time to perform a task, and one can ask the operator how well the system works (subjective assessment). These measures are the common dependent variables used to measure the productivity of a system.

1.8 THE TRADE-OFF BETWEEN PRODUCTIVITY AND SAFETY

Industrial managers often expect employees to work more quickly with fewer quality errors. However, research has shown that people cannot simultaneously reduce errors and increase speed. In general, the greater the speed (of vehicles, production machinery, etc.) the less time an operator has to react, and as a result he or she will make more errors. Shorter work cycles improve productivity but compromise safety. Operators hence have a choice between increased speed *or* increased accuracy. This is referred to as the speed–accuracy trade-off or SATO (Wickens and Hollands, 2000). Managers' expectations are often contrary to SATO and hence impossible to achieve.

It is, however, possible to improve safety and quality of production at the same time. A reduction in the number of operator errors will typically lead to improved safety as well as improved production quality. An emphasis on quality of production is therefore more appropriate and more effective than the traditional approach in industry, which stresses speed and quantity of production.

1.9 THE GOAL OF OPERATOR SATISFACTION

We discuss satisfaction in a broad sense—work satisfaction as well as user satisfaction. Various aspects of dissatisfaction are also considered, such as job dissatisfaction and consumer dissatisfaction. It is important to note that satisfaction as well as dissatisfaction may be understood only if the operator's or user's needs are clearly understood. Different people have different needs and different expectations, and these vary substantially between countries and cultures. What are considered essential workers' rights in Sweden are sometimes less important in the U.S. or Asia. For example, in Sweden there is a law that requires that office workers must have an office with a window. As a result office workers have "acquired" a need for a window, and lack of a window would cause great dissatisfaction. Office workers in the U.S.

or Asia may not think twice about the lack of a window. It is not expected and therefore they are happy anyway.

Job satisfaction does not influence productivity or safety. One would think that a satisfied worker would produce more and a dissatisfied worker would produce less. One would also think that a satisfied worker would be safer and a dissatisfied worker not so safe. But extensive research on these issues has demonstrated that there is no relationship between satisfaction and productivity, safety, or quality.

1.10 CONCLUSION

Since the beginning of the history of ergonomics around 1950, society and technology have developed tremendously, and HFE has followed along. The following can characterize the development over the last 50 years. Different issues have driven the development of our science from 1950 to the present.

- 1950s: Military ergonomics
- 1960s: Industrial ergonomics
- 1970s: Consumer products ergonomics
- 1980s: Human-computer interaction and software ergonomics
- 1990s: Cognitive ergonomics and organization ergonomics
- 2000s: Global communication, internet, and virtual collaboration

As I have emphasized, HFE is a systems science and a science of design. The systems approach in Figure 1.3 is useful for conceptualizing problems and suggesting design solutions. Our profession is driven by design requirements from users, markets, industries, organizations, and governments. We must be able to respond quickly to the changing needs of society. HFE is therefore at the forefront of technological development.

Ergonomics will continue to evolve and professional ergonomists must extend their knowledge to deal with a rapidly changing scenario. I believe that this will require increasing interaction with other disciplines to solve problems. Most problems in this world are of an interdisciplinary nature.

In the design of complex systems it is necessary to apply many design criteria simultaneously. All these criteria must be at least partially satisfied—or, to use Simon's (1996) terminology, multiple criteria must be "satisficed." In other words, there are many goals that drive a design. In manufacturing there are goals related to quality, productivity, and worker satisfaction. One can probably not find a design solution that can fully satisfy all criteria. The problem is then to identify a design solution that is good enough—where all assessment criteria have reached an acceptable level. Multiple criteria are thereby satisfyced.

In Chapter 2, we discuss the benefits and costs of HFE improvements in two areas: manufacturing and human–computer interaction. We will note that design changes can improve all aspects of system performance, productivity as well as satisfaction—a win-win situation, as they say.

2 Cost-Benefit Analysis of Improvements in the Human Factors Design

2.1 INTRODUCTION

Principles of human factors engineering can be used to redesign a system to make it more productive and safer. Such a redesign will cost money, and in order to justify the expense we must be certain that there will be benefits associated with the improvements. The question is, are improvements worthwhile? Do they pay off? In this chapter two case studies are presented. The first study concerns a manufacturing plant. In manufacturing there can be several types of benefits: improved productivity and quality, reduced injury rate, and improved worker comfort. In this case, the economic benefits from improved productivity were substantial, and much larger than the other benefits.

The second case study deals with improvements in human–computer interaction. There are many ways to measure the benefits, such as reduced task performance time, reduced number of key strokes, or number of user errors. For example, one can measure the time it takes to complete filling in credit card information on an e-commerce web page.

2.2 ERGONOMICS IMPROVEMENTS IN CARD ASSEMBLY

At IBM in Austin, Texas, printed circuit boards for computers were manufactured. The boards consisted of multiple layers of copper sheeting and fiber glass with etched circuitry. Holes were drilled through the circuit board for insertion of components. Much of the component insertion was automated using special machines— so-called card-stacking machines. However, there were many tasks which could not be automated, including quality control and inspection of component parts and finished products.

One important measure of quality in the manufacturing of boards is the percentage production yield. One may set a target yield such that, say, 95% of the boards must pass the quality control test at the end of the manufacturing process. In this case, plant management complained that the yield was consistently 5–10% below target (Burri and Helander, 1991a). Most of the quality problems were defects inside a circuit board, which could have occurred in a department called "Core Circuitize." This was located just prior to the determination of the percentage yield, about halfway through the manufacturing process. Figure 2.1 shows an overview of this area of the plant.

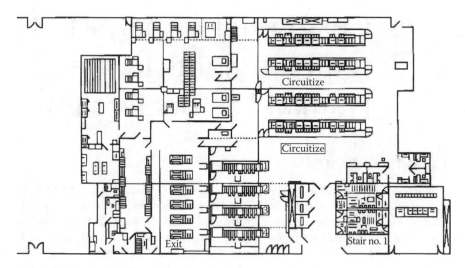

FIGURE 2.1 Overview of the plant.

Altogether, 132 individuals, mostly operators, worked at this location, which had 59 workstations. To evaluate the manufacturing scenario, information was collected from five different sources:

1. *Discussion with management.* We asked them what the problem was and what should be the focus of our study. These questions brought up new issues to pursue. In addition to repetitive motion injuries, we also learned about the problem in yield rate.
2. *Plant walk-through, inspection, and note taking.* How is the manufacturing and material flow organized?
3. *Discussion with operators.* How do they perform their tasks? How long does it take to learn a new task? What are the problems that newly employed workers have?
4. *Discussion with first-line supervisors.* Often these are able workers who have been promoted. They are a great source of information.
5. *Measurements in the plant of illumination, noise, and the design of the workstation.*

These measurement values should be recorded and documented in a systematic fashion, so that comparisons can be made between different workstations. For example, what are the illumination and noise levels at the different work places?

On the basis of these discussions, opinions, and measurement data, we derived a comprehensive assessment of both the manufacturing system and the operators' tasks. In addition, the findings revealed significant opportunities for improvement.

Most of the 59 workstations were different, and it is not relevant to summarize the data here. Instead we focus on the recommendations. Based on the information that was collected, we identified 14 design improvements (see Table 2.1). Some of these

---

TABLE 2.1
Ergonomic Improvements at the IBM Plant in Austin, Texas

1. Uniform illumination level at 1000 lux
2. Installation of special lighting for inspection
3. Job rotation to avoid monotony
4. Personal music was distracting and was discontinued
5. Ergonomic chairs certified for clean rooms
6. Improved communication
7. Materials-handling guidelines
8. Automation of monotonous jobs
9. Metric to decimal conversion charts
10. Housekeeping improved
11. Noise reduction
12. Ergonomics training
13. Continuous flow manufacturing
14. Use of protective gloves

---

were rather conventional ergonomic measures, and some required redesign of the manufacturing process.

2.3 DESIGN IMPROVEMENTS

ILLUMINATION LEVEL

Improved illumination turned out to be the most important of all the measures. Several operators were performing a relatively simple task. They placed circuit boards into card-stacking machines for automatic insertion of components. The managers thought of the operators as supervisors of the automatic machines. However, interviews with the operators disclosed that they regarded themselves more as quality inspectors rather than as machine tenders. They would inspect cards and components that were placed in the machine, and they inspected the finished product as it was removed from the machine. One of the most critical aspects of this task was to inspect the magazines containing the electronic components that were put into the card-stacking machines. A common problem was that the components were turned in the wrong direction in the magazines and would therefore be inserted in the wrong direction into the board.

SPECIAL LIGHTING FOR INSPECTION

The illumination was measured in the plant. The average level was about 500 lux, which is inadequate for inspection work. In some areas the illumination was as low as 120 lux. It was decided to increase the illumination to 1000 lux throughout. This was achieved by installing fluorescent light tubes, switching on lights that had been turned off for energy-conservation reasons, and lowering light fixtures from high ceilings to a location closer to the workstations.

In addition to these measures, some polarized lights were installed to make it easier to see imperfections and quality defects. Many examples of special illumination systems for inspection are presented in Chapter 4.

Job Rotation and Shift Overlap

Visual inspection is often monotonous, and the operators had problems in sustaining their attention throughout an entire work shift. To break the monotony, job rotation was incorporated. Operators could then split their time between two jobs (Grandjean, 1985). Existing rest-break patterns were evaluated, but it did not seem necessary to increase the length of the rest break. The time overlap between shifts was reduced from 30 to 12 minutes. The shift overlap made it possible for the outgoing shift to inform the incoming shift about potential problems, such as problems with machines and processes. However, the existing overlap of 30 minutes was found to be excessive and unproductive.

Personal Music

An experiment was performed to introduce personal music into the work place. However, the music was distracting to the work and it was therefore discontinued. A common problem is to find music that everybody likes. Some prefer hard rock and would be irritated to listen to country and western, and vice versa.

Ergonomic Chairs

New ergonomic chairs increased comfort, and at the same time increased productivity as operators could remain seated during inspection. The chairs were manufactured to be used in a clean-room environment. There were several adjustability functions, including seat height, back-rest angle, and seat-pan angle. For some operators, sit/stand types of chairs were also provided for occasional use.

Operator Communication and Feedback

In order to enhance verbal communication and feedback between operators, the enclosures of some of the workstations were removed. The open access to coworkers improved communication significantly, and was helpful, particularly with respect to quality control (Bailey, 1996).

Materials Handling

Manufacturing parts and finished products were stored on racks. The lowest shelves were taken away. This made it impossible to store materials at a low height, which in turn reduced the amount of bending and back injuries. In addition, guidelines for a maximum weight of parts were established.

AUTOMATION OF MONOTONOUS JOBS

Some operations were converted from manual work to automation. One of the jobs involved a task where a protective tape was removed from a board. This was a highly monotonous and repetitive task and did not provide any job satisfaction. Therefore it was automated. The operator now supervises several pieces of automation, a situation which provided a more varied and interesting job.

METRIC TO DECIMAL CONVERSION

The conversions between metric and decimal units of measurements were confusing to several operators, and a conversion chart was put up at each workstation.

HOUSEKEEPING

Through collaboration with management, an example of good housekeeping was set up in a part of the plant. The area was cleaned up and organized. This inspired operators in other areas as well, and housekeeping improved. As part of the house-keeping effort, the manufacturing facility was converted to a 10,000-type clean-room facility. Clean-room clothing and smocks were evaluated and their use recommended.

NOISE REDUCTION

The noise levels were well within the 85 dBA stipulated in the regulations by National Institute of Occupational Safety and Health (NIOSH). However, to enhance verbal communication, sound insulating covers were installed for several processes. The ambient noise level at these workstations was then reduced from about 75 to 60 dBA, which made it possible for operators to have an undisturbed conversation.

ERGONOMICS TRAINING

An ergonomics training and awareness program was provided for operators. This was a 4-hour program which addressed a variety of problems. The motto is to increase self-vigilance through informing the operators of ergonomic hazards.

CONTINUOUS FLOW MANUFACTURING

Continuous flow manufacturing was implemented at several locations in the plant. The main purpose was to reduce the amount of space required for manufacturing, rather than to enhance ergonomics. There were important benefits accomplished in that the distance between adjacent operators decreased so that it became possible to talk to other operators.

EVALUATION OF PROTECTIVE GLOVES

Many operators used protective gloves to avoid cuts from the sharp edges of the boards. However, some types of gloves reduced tactile sensation, so that it was difficult to manipulate components. Several different gloves were tested by operators. The selected glove was comfortable and at the same time enhanced tactile sensation.

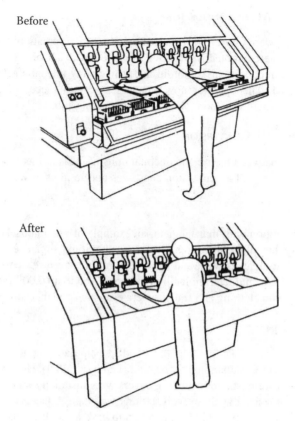

FIGURE 2.2 The location of the drill bits forced excessive reaching and required great caution. After the modification, drill bits were located closer to drills, thus reducing reach.

SPECIFIC PROBLEMS

Besides the general problems identified, there were specific problems at several workstations. At one work station, operators supervised a machine which was used to drill holes in boards. To replace the drill bits the operators had to bend over the machine (see Figure 2.2). They also had to bend very carefully to prevent the drill bits from sticking into their stomachs. The machine was changed. On the new machine the drill bits were relocated, which made them easier to reach, besides improving work posture. Equally important was that they made the work easier to do; it saved about 1.5 minutes per changing operation. Since there were many machines and many changes of drill bits, the yearly saving was $270,000.

COST EFFICIENCY OF IMPROVEMENTS

Based on the experience gained in previous field studies, we had projected a 20% improvement in process yield, a 25% improvement in operator productivity, and a 20% reduction in injuries. The actual improvements were close to our predictions

TABLE 2.2
Improvements and Cost Reduction in Dollars

	Improvement (%)		Cost Reduction ($)	
	Projected	Actual	Projected	Actual
Yield improvement	20	18	2,268,800	2,094,000
Operator productivity	25	23	5,647,500	5,213,000
Injury reduction	20	19	73,400	68,000
Total			7,989,700	7,375,000

and resulted in a cost reduction of $7,375,000 (see Table 2.2). The cost of materials for ergonomic improvements (such as improved illumination) was $66, 400. The labor cost for the implementation was about $120,000. The benefit/cost ratio for these improvements was approximately 40:1 for the first year—put another way, the payback time was about 1 week (Helander and Burri, 1994).

Reductions in injury costs were fairly minor compared with the improvements in productivity and yield. This case study demonstrates that improvements in productivity can sometimes be extraordinary, and ergonomics can play a large role in productivity improvement. The management was impressed by the results and hired two ergonomists with an industrial engineering background to continue with the improvement work.

There were also improvements in operator comfort, convenience, and job satisfaction. Informal interviews were conducted among a large number of operators and with management. There were no negative effects of the new system. Operators generally appreciated what had been done and were happy with the new system. These types of improvements are more tangible and difficult to quantify in terms of cost savings than are improvements in productivity and safety.

CONCLUSIONS

From a scientific point of view the study was unsatisfactory. There was no control group that can be used to compare the results. Therefore we cannot easily substantiate the claim that the benefits were due to ergonomic improvements. Such is often the case in industry, since it is usually very difficult to find an identical control group. The improvements could possibly have been due to other factors, such as continuous flow manufacturing, which was also implemented. Improvements could also be due to "uncontrolled factors" such as changes in leadership style. Since this was not recorded, we do not know if there was a real change.

In an effort to validate the results we interviewed 26 managers and engineers at the plant. They agreed that approximately half of the savings could be attributed to ergonomics, while the remaining was attributed to other improvements including continuous flow manufacturing. The management was extremely positive about the ergonomic improvements, particularly the increased illumination levels for visual inspection. This turned out to be a critical change that improved both quality and productivity.

This case study also demonstrates that ergonomic improvements cannot be undertaken in isolation of the manufacturing process. There must be a clear understanding of technological alternatives for improving productivity and of how ergonomics is affected by the choice of technical system, process layout, equipment, and communication patterns between employees.

2.4 BENEFIT–COST IMPROVEMENTS IN HUMAN–COMPUTER INTERACTION

Benefit–cost improvements are easier to demonstrate in human–computer interaction than in industrial ergonomics. Tullis (1981) was among the first to demonstrate economic benefits from improving human computer interaction. He investigated a very specific task, that of identifying and correcting faults in telephone networks. This is a common task in telephone companies, and his findings had a significant economic impact.

Figure 2.3 shows the design of the interface for analyzing the telephone line before (to the left) and after the redesign (to the right). The interface to the left is a narrative format for presenting the results, while the interface to the right presents a structured, symbolic format. In the first case it took the average experienced user 8.3 seconds to interpret the display, while in the second case it took 5.0 seconds. The outcome was a 40% reduction in performance time. Considering that telephone maintenance is a common task in the telephone industry, the redesign saved several

```
TEST RESULTS     SUMMARY: GROUND

  GROUND, FAULT T-G
  3 TERMINAL DC RESISTANCE
    > 3500.00 K OHMS T-R
    =   14.21 K OHMS T-G
    > 3500.00 K OHMS R-G
  3 TERMINAL DC VOLTAGE
    =    0.00 VOLTS   T-G
    =    0.00 VOLTS   R-G
  VALID AC SIGNATURE
  3 TERMINAL AC RESISTANCE
    =    8.82 K OHMS   T-R
    =   14.17 K OHMS   T-G
    =  628.52 K OHMS   R-G
  LOGITUDINAL BALANCE POOR
    =   39   DB
  COULD NOT COUNT RINGERS DUE TO
    LOW RESISTANCE
  VALID LINE CKT CONFIGURATION
  CAN DRAW AND BREAK DIAL TONE
```

```
  *********************************
  *                               *
  *   TIP GROUND    14 K    *
  *********************************
```

DC RESISTANCE	DC VOLTAGE	AC SIGNATURE
3500 K T-R		9 K T-R
14 K T-G	0 V T-G	14 K T-G
3500 K R-G	0 V R-G	629 K R-G
BALANCE		CENTRAL
		OFFICE
39 DB		VALID LINE CKT
		DIAL TONE OK

FIGURE 2.3 To the left, an example of a "narrative" format for presenting results of tests on a telephone line. The average time for experienced users was 8.3 sec. To the right, the "structured" format for presenting results of tests on a telephone line. The average time for experienced users was 5.0 sec—a reduction of 40% compared to the narrative format.

TABLE 2.3
The Value of User-Centered Design in Human–Computer Interaction

Application and Measure	Improvement in Productivity (%)
A variety of products at Digital Equipment Corp.	30
Mainframe installation at IBM	400
Transaction system for data entry at NCR	67
Text browser for finding information at Bellcore	100
Security log-on procedures for network at IBM	720

Adapted from Landauer (1995).

---

hundred thousand dollars. Tullis' example is easy to illustrate. Many other usability problems are more abstract and cannot be captured by a figure so easily.

Landauer (1995) presented the outcome of several usability studies performed by computer companies. As illustrated in Table 2.3, improved human–computer interaction can produce significant productivity effects.

The problem with the present evaluation methods for computer systems and software is that they are typically viewed from the engineering perspective: how fast they compute, how much data they store, how flawlessly they run, the quality of the graphics, and the impressiveness of the tricks (Nickerson and Landauer, 1997). This perspective often neglects usability.

The good news is that there are many techniques for improving usability, and on average the efficiency of a work task can be improved by approximately 50%. Nielsen (1994) showed that the average computer interface has about 40 usability bugs. This should not be surprising since almost 60% of the computer code deals with the user interface. Nielsen was of the opinion that five to six users can find most of the usability bugs. In this case users will try to use the system and at the same time analyze the interface and comment on design features that they think should be removed or replaced. Different users will find different problems. Testing with two users may identify half of the flaws, while six users will find almost 90% of the usability bugs. The testing will take about 2 days. After a single day of usability testing, the work efficiency can be expected to improve by around 50%.

Usability testing is therefore a successful method for reducing the difficulty and time for performing a task on a computer. Usability testing can dramatically improve the quality of the work with respect to productivity and also with respect to job satisfaction. A task that has an easy and smooth work flow is simply more interesting and more satisfying to perform.

Karat (1997) elaborated further on the importance of usability. She noted that usability engineering has many direct and indirect positive effects for an organization. It is like the quality movement in manufacturing. Because of the increased

---

TABLE 2.4
Increased Usability Has Many Beneficial Effects on Systems Development (Karat, 1992)

Product design and product performance will improve.

User satisfaction will increase.

Since usability errors can be detected much earlier in the systems development cycle, the development time and the development cost for the interface will be reduced.

As a result of the improved software design, sales and revenue will increase.

Because interface is easier to handle it will take less time to train employees.

There will also be reduced maintenance costs, reduced personnel costs, and improved user productivity.

---

focus on the user, there is a better understanding of user needs. As an added benefit, the definition of the product or the software improves (see Table 2.4).

Take, for example, software maintenance. Studies have shown that 80% of maintenance costs are spent on unforeseen user requirements; only 20% are due to software bugs. If usability engineering can identify and resolve a majority of user requirements prior to product release, the organization will accrue substantial benefits and avoid future costs. It is best to conduct usability studies early in the product development life cycle in order to provide feedback on product design and performance.

An example: Karat (1990) reported on the design of a system that was used to complete 1,308,000 tasks per year. She studied the software for this task and managed to improve the usability, so that the task performance time was reduced by 9.6 minutes per task. The first-year benefits due to increased productivity could then be calculated: $1,308,000 \times 9.6$ minutes = 209,280 hours, which corresponded to a cost savings of $6,800,000. The usability testing and the related costs amounted to $68,000. Therefore the cost–benefit ratio could be calculated as 1:100 for the first year of application.

2.5 DISCUSSION OF COSTS AND BENEFITS

The two case studies have illustrated the costs and benefits of ergonomics improvement in industry and usability improvements in human–computer interaction. In the industrial case it was difficult to attribute the improvements in productivity to ergonomics alone. There were many other simultaneous changes in the manufacturing plant, and it can therefore be argued that industrial implementations of ergonomics are not so well controlled. Nevertheless, most of the interviewed employees agreed that the improvements in productivity were primarily due to the improved illumination system. As a result of the improvements, operators could see what they were doing, and many quality errors could be avoided.

For usability in human–computer interaction, it is much easier to account for the benefits by means of a simple comparison before and after the improvements are implemented. This can be performed as a "laboratory study" with full control of external factors. It is also much less expensive to perform a study on usability of a computer interface than on a manufacturing plant. Such arguments justify the need for usability analysis of the user interface. As a result, the number of usability engineering professionals has grown tremendously in the last ten years. Companies such as Microsoft and IBM now employ a large number of usability professionals.

The issues concerning human–computer interaction will be treated in greater detail in Chapter 12.

RECOMMENDED READING

Hendrick, H.W., 1996, Good ergonomics is good economics, *Proceedings of Human Factors and Ergonomics Society 40th Annual Meeting*, Santa Monica, CA: Human Factors and Ergonomics Society, pp. 1-10. Also available for free at www://hfes.org/publications/goodergo.pdf.

3 Conducting a Human Factors Investigation

3.1 INTRODUCTION

Human factors professionals are often involved in the design of new equipment and interfaces. Many design decisions have to be made, and the new design must be evaluated. Handbooks and articles may offer some help, but the design issues are often quite specific, and it is difficult to find published data that can be applied directly. The HFE professional may then undertake her own investigation. It can be a formal research study or it can be a quick collection of data to illuminate critical aspects of the research problem. This chapter gives an overview of common investigative methods in HFE.

There are three different types of studies in HFE:

1. Descriptive studies, which are used to characterize a population of users
2. Experimental studies, which test the effect of some design feature on human performance
3. Evaluation studies, which test the effect of a system on human behavior

These different approaches will be explained below.

3.2 DESCRIPTIVE STUDIES

The objective of descriptive studies is to describe user characteristics, such as anthropometric sizes, hearing capabilities, visual performance, age, and so forth. The main purpose is to collect data with the aim of understanding a pattern, trend, or characteristic. The data can be classified as independent or modulating variables, and can include information such as age, gender, size, education, and experience, as shown in the center of Figure 3.1 under the heading Modulating Variables.

For example, to set up an anthropometric database for Singapore, three major population segments must be considered: Chinese, Malays, and Indians. Indians are slightly taller than the rest of the population, Malays are relatively shorter, and the Chinese in between. In measuring the size of people, participants are selected randomly within each of the three populations. The randomization process makes it possible to generalize the results to each population type. To adequately describe the three populations as well as the entire Singaporean population one would need to measure about 5000 individuals.

Anthropometric data are needed for the design of many products. A chair manufacturer, for example, may want to understand the size of the users so as to design

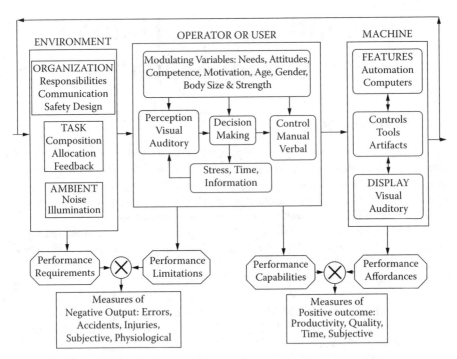

FIGURE 3.1 A systems approach to measurement of human performance and preferences.

the adjustability of the seat correctly. The office desk should also be at a comfortable height. In the U.S. the desk height is usually 29 in (74 cm) above the floor. This is far too high for an Asian population. Unfortunately most countries in Asia have adopted U.S./European standards without reflecting on these issues. Access to an anthropometric database for the local population would make it possible to design for local users.

Descriptive data can sometimes be collected using questionnaires. Unfortunately many questionnaires are not returned. Let's assume that you ask people to rate back pain on a scale from 1 to 7, with 1 being "No Pain" and 7 "Extreme Pain." Let's say that 30% of the questionnaires were returned and the average back pain was rated 6. This raises an important question: did the persons who returned the questionnaire have any particular reason for responding? Maybe they all had back pain, and maybe those who had not responded did not have back pain, and therefore were not motivated to return the questionnaire. If their responses had been taken into consideration, the average back pain could have been around 2 or 3. Hence, one must be cautious in the interpretation of the data, particularly when the response rate is low.

3.3 EXPERIMENTAL RESEARCH

Experiments are done to investigate specific issues. They are sometimes performed in a laboratory environment and sometimes in the real field environment. One major benefit of an experiment is that the results can be used to draw firm conclusions with respect to cause and effect. In Figure 3.1 there are several types of dependent variables. They are listed as indicators at the bottom of the figure and include: number of errors, number of injuries, time on task, and quality of work. There are also subjective variables—that is, information that comes directly from asking people what they think of a specific task or design.

To summarize, three different types of dependent variables can be evaluated in experiments:

1. Performance variables, such as task performance time and errors
2. Physiological variables, such as heart rate, electromyography (EMG), and galvanic skin response (GSR)
3. Subjective variables—information that comes from asking people, as with questionnaires or rating scales

The independent variables, as shown in Figure 3.1, are the design variables in the environment and the machine subsystems. Examples are temperature and noise in the environment subsystem and design of controls and displays in the machine subsystem.

There is a relationship between the variables so that

$$y = f(x)$$

where y is the dependent variable (performance, physiological, or subjective data) and x is the independent variable, the design that is under investigation.

Assume that you would like to measure the extent of drivers' mental stress reactions (y) in traffic situations as a function of road and traffic conditions (x). To determine stress (y), there are several measures that can be taken, including heart rate, heart rate variability, GSR, blink rate, pupil parameter, and excretion of stress hormones. First, the investigator must decide what variable(s) to measure.

- Heart rate, although it is easy to measure, may not be appropriate since it is affected by physical work as well as mental work.
- Heart rate variability has often been used in traffic studies. However, there are many different ways of defining and measuring heart rate variability, which leads to uncertainty in collecting the data.
- Galvanic skin response (GSR) is a measure of the electrical conductivity of a particular type of sweat gland. A good location to measure GSR is in the palm of the hand. It there is a sudden increase in task demand, or if there is an unexpected event that startles the driver, there will be a GSR response. Therefore, it is an appropriate measure for measuring human stress in traffic.
- Blink rate is a very good measure of stress for specific tasks, such as fighter pilots in the cockpit. Under very high stress, pilots stop blinking! The question is if this measure would also work in traffic, where the stress level is lower.

- Pupil diameter is sensitive to mental workload, such as mental arithmetic. The pupil increases in size as mental workload increases. But the changing illumination conditions inside the car would also affect the pupil size, so it would not be a valid measure in traffic.
- Subjective evaluation can also be used. One can ask the driver to use a scale of, say, 1 to 5 to continuously rate the difficulty or stress of the traffic conditions as they keep driving. This measure usually provides a reasonably good assessment of stress.
- Excretion of stress hormones is more useful for assessing long term stress than for temporary stress in traffic. For example, this measurement can show the cumulative effects of stress from a day of hard work.

Several other potential measures may be considered, but typically heart rate variability, galvanic skin response, and subjective evaluation are the most appropriate in the traffic scenario.

As mentioned above, the independent variables in the traffic environment are design-related, such as road illumination, the width of the road, and traffic density. These variables matter most to traffic engineers in designing roads. For example, to increase traffic capacity, a traffic engineer must decide between options, such as making the road wider or building more traffic lanes. The latter option is more expensive, so there are several trade-off decisions to be made.

EXAMPLE

Helander (1978) measured galvanic skin response (GSR) during different types of traffic events. Fifty test persons drove an experimental vehicle on a rural narrow road which was 24 km long. An experimenter was sitting in the passenger seat next to the driver, marking every traffic event on a keyboard as it occurred. The traffic events were recorded on a digital recorder together with galvanic skin response and variables describing vehicle behavior, such as brake pressure and steering wheel angle.

The average brake pressure and the average GSR were calculated for 15 traffic events (see Table 3.1). The traffic events were then ranked in order, so that the event with the greatest average brake pressure obtained rank 1 for brake, the traffic event with the greatest average GSR obtained rank 1 for GSR, and so forth. From Table 3.1 it is clear that in most cases the rank orders of brake pressure and GSR follow each other perfectly. Spearman rank order correlations were calculated between brake and GSR (Siegel and Castellan, 1988). Before performing the statistical analysis, we noted that for events 15 and 13 the driver was passing or being passed, and there was no reason why he should brake. If these events were excluded, the rank correlation coefficient between brake pressure and GSR is $R_S = 0.95$, which is statistically significant with $p < 0.0001$. The significant p value means that the correlation coefficient could have been obtained by chance only in one case out of 10,000. The finding is therefore conclusive.

We can conclude from this study that traffic events which require drivers to brake will also be perceived as stressful. To design a less stressful traffic environment one could try to construct traffic environments where the traffic flow is smooth and reduces the need for braking. The traffic flow on freeways and highways with several lanes and good illumination is usually much smoother than on rural country roads and requires less braking. Freeways are also about four times as safe as rural country roads

TABLE 3.1
Rank Orders for Average Brake Pressure and GSR for Fifteen Different Traffic Events while Driving on a Rural Road

Traffic Event	Number of Events	Brake	GSR
1. Cyclist or pedestrian + MOC*	28	1	1
2. Other car merges in front	47	2	2
3. Multiple events	163	3	3
4. Leading car diverges	207	4	5
5. Cyclist or pedestrian	839	5	7
6. Own car passes other car + CF**	126	6	6
7. Cyclist or pedestrian + CF**	65	7	10
8. Car following + MOC*	353	8	12
9. MOC*	1,535	9	9
10. CF**	13,049	10	11
11. Parked car	742	11	15
12. No event	112,630	12	13
13. Other car passes own car	157	13	8
14. Parked car + car following	64	14	14
15. Own car passes other car	3,590	15	4

* MOC = meeting other car.
** CF = car following.
If events 13 and 15 are excluded, the rank order correlation coefficient $r_s = 0.95$, $p < 0.001$.

3.4 EVALUATION RESEARCH

Evaluation research is more global and less specific in its purpose than experimental studies. It typically involves the study of human performance in real-world settings, such as operators in a manufacturing plant, medical doctors in the operating theater, or firemen fighting a fire. These studies can be exploratory in nature, which means that there may not be a direct purpose, but the investigator would like to educate himself and understand how workers perform their tasks, the effect of the environment, and so forth.

The manufacturing case study in Chapter 2 is a good example of evaluation research. The purpose was to identify reasons for musculoskeletal disorders among factory workers. Later we ran into a more important problem in quality control. It would not be possible to set up an experimental study to investigate this problem. Instead we collected information by talking to people who worked in the plant.

Beyer and Holtzblatt (1988) developed a method called contextual inquiry. The purpose of this method is to go to the operator, observe real work, interview people while they are working, and understand what operators look for and how they make decisions. In fact, our manufacturing case study in Chapter 2 was performed along these lines.

EXAMPLE

The Hawthorne studies constitute a good example of evaluation research. These studies, which were undertaken around 1930, are famous since they supposedly illustrate that a caring supervisor can increase workers' job motivation and productivity at work.

There were seven female workers in a plant. They manufactured transformers. Researchers thought that the productivity could be increased if the illumination level at the workstations was improved. The illumination level was increased in several steps over some time. Each time the productivity increased, and the conclusion was that increased illumination improves productivity. To check the reliability of the findings the investigators lowered the illumination level. The surprising result was that productivity improved even more. The investigators therefore arrived at a new conclusion: the reason for the improved productivity was not the illumination level, but rather the concern that the supervisors had for the well-being of the workers. This increased job motivation and thereby productivity. This group of workers was indeed treated differently from other workers in the plant and there were many conversations concerning the effect of the illumination and how the workers liked it.

Many years later Parsons (1986) reanalyzed the findings. He pointed out that the women were paid piece rate for each transformer. Behind the workstation was a counter showing the number of finished transformers, so that each worker knew exactly how many she manufactured each day. There was also a competitive spirit among the workers and the payment system also provided incentives for the women to work fast. Some of the improved productivity was probably a training effect; workers learn a job and they work faster (see Chapter 15). Neither illumination nor job motivation were very important in this case—although they could have played a minor role. The feedback from the counter and from the salary increases were more important.

This example demonstrates the difficulties in drawing firm conclusions from evaluation studies. An experimental study, which compares different groups of workers who received different "treatments," would have been more informative.

There could, for example, have been three illumination levels and three groups of workers. In real life it is very difficult to arrange these studies, as there is usually little incentive for the owners of a plant to participate.

3.5 SELECTION OF DEPENDENT VARIABLES

In selecting a dependent variable, there are three main criteria that must be fulfilled: it must be reliable, valid, and sensitive.

RELIABILITY

Here we refer to the consistency or stability of a measure over time. For example, to get entrance to a university a student ay need to take the Scholastic Aptitude Test (SAT). If a student takes the test many times, the test score should be about the same each time; SAT must be a reliable measure. To measure the reliability one can correlate the test results for two occasions and calculate the test–retest reliability coefficient. There is indeed a training effect in SAT; however, the improvement in SAT scores from one occasion to another is fairly small. A test–retest reliability coefficient of 0.8 is often considered adequate (1.0 indicates perfect agreement).

VALIDITY

Does the instrument measure what we intend to measure? An instrument can be reliable but not valid. For example, body temperature is easy to measure and very reliable. But it would not be a good measure of mental workload. Heart rate variability is also reliable and easy to measure and would be a good indicator of mental workload. There are three main ways to evaluate validity: face validity, content validity, and construct validity.

Face Validity

In this case we can ask an expert if he or she believes that the measure is valid. The expert could say, "Looks good to me," or "It has worked before." Such an assertion indicates that face validity is achieved.

Content Validity

A measure must include all aspects that we want to measure. For example, the Federal Aviation Administration may want to design a test for selection of persons who will become air traffic controllers. Such a test must measure the variety of skills that good air traffic controllers possess: the ability to perform several tasks at the same time (dual-task performance capability), the ability to communicate effectively with other air traffic controllers (team communication), the ability to keep track of all airplanes (attention span and memory capability), and the ability to predict how the present scenario will develop (spatial and predictive capability). A selection test that incorporates all these capabilities is said to have content validity.

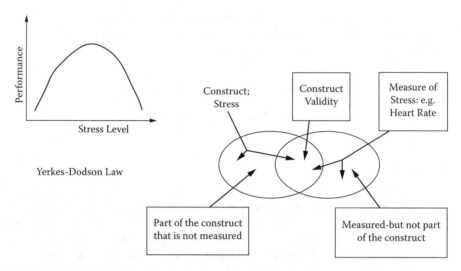

FIGURE 3.2 The right part of the figure illustrates that heart rate measures some of the stress in traffic but not all. In addition to reacting to traffic stress, heart rate also responds to other parameters, such as physical work in steering the car and many other stressors in life that the driver happens to think about.

Construct Validity

In this case we need to know if the measure addresses an underlying construct or theory. Many research studies build upon a theory, and develop assumptions and hypotheses about possible events. Let's assume that we would like to measure the physiological stress of operators working in a call center. This is considered a stressful job, as the operators are selling services to potential customers who may not be interested; instead they can become annoyed at the operators. To measure the stress level, one can opt to analyze the concentration of stress hormones, such as adrenaline and noradrenalin in samples of saliva, urine, or blood. The question, then, is to what extent stress hormones can measure stress at work and to what extent they may be related to a theory of stress at work. This is illustrated in Figure 3.2.

Let's assume that to enhance traffic safety, it is important to identify the high peaks of driver stress and then to redesign parts of the roadway so that they are less demanding and reduce driver stress. The theory or construct is based on Yerkes-Dodson Law. This law implies that if there are situations where most people have a high stress level, then there is also a chance that the performance level is reduced. There is an optimal stress level where there is a chance that performance is greater.

Sensitivity

In addition to reliability and validity, a dependent variable must also have sensitivity, so that one can distinguish between different levels of the independent variables. Assume that we would like to find a good measure of mental workload in driving. It should have the sensitivity to distinguish between situations such as driving on a freeway, entering the freeway, and exiting from a freeway. One type of measure is the capacity a driver has to perform two tasks at the same time, or dual-task capacity. To investigate these types of problems Michon (1967) used a tapping task. The idea is that the driver can establish an even rhythm in verbal tapping—saying "ta-ta-ta" at a rate of one "ta" per second. But when traffic conditions become difficult the rhythm becomes irregular, because there is not enough mental capacity for both driving and tapping. If this particular measure can distinguish between the levels of difficulty in of driving on different road or traffic environments, then it is a sensitive measure.

EXERCISE

1. The purpose of this exercise is to design a study and select dependent and independent variables as discussed in this chapter. The research problem can be an environment–driver vehicle system, similar to the system in Figure 3.1. Or it can investigate a manufacturing system, or the use of consumer products, with a focus on human, system, or product performance.

    Decide on a study that you want to do.
    • Why is this study interesting to investigate, and what do you want to achieve? Formulate the objectives.

2. Select the independent (design) variables.
    • What are the design alternatives that you want to investigate?

3. Select the dependent variables.
    • In relation to Figure 3.1, discuss possible measures of negative outcome as well as positive outcome. Then select one or several dependent variables that you want to measure.

4. Decide how you want to collect data
    • What environment will you use for testing?
    • Who are the test subjects?
    • How long will the test take?
    • How will you instruct the test subjects?

5. Draw a system.
    • Draw a system where all independent and dependent variables are indicated. Use Figure 3.1 as a model for your system. Remove the modulating variables that you do not need; then add your own. Likewise remove boxes (subsystems) and independent and dependent variables that you don't need. This means that the contents of all the small boxes in the system should be modified to fit the purpose of your study. However, the sequence of operator information processing, perception, decision-making, control, and stress remains the same.

6. Discussion.
    • Discuss why you selected the dependent variable(s). Are they reliable and valid? Will they have the sensitivity to distinguish between different independent (design) variables?

3.6 METHODS

In HFE there are many investigative methods (see Table 3.2). Some of the measures we have talked about in this chapter are listed in the table, including performance measures, stress measurement, psychophysiology methods (GSR), and questionnaires. Other methods are inferred from Figure 3.1: accident analysis and error analysis. Several of the methods in the table are for the design of new systems. The main purpose in presenting the table is to emphasize that there is a wealth of methods, most of which are unique to HFE. Some of the methods support analysis and some support design. Some of them are discussed in later chapters where their usage is more relevant. A comprehensive overview of methods is given by Brookhuis, Hendrick, Hedge, Salas, and Stanton (2004).

TABLE 3.2

One Hundred Methods and Techniques for Collecting and Analyzing Data in Human Factors and Ergonomics for Design of New Systems

Accident Analysis	Fatigue Measurement
Activity Analysis	Fault Tree Analysis
Anthropometric Analysis/Design	Function Allocation
Biomechanical Analysis	Goals/Means Task Analysis
Body Rhythms and Shift Work Design	GOMS Analysis
Checklist Analysis	Hazard Analysis
Climate Analysis	Human Reliability Assessment
Cognitive Abilities Testing	Human–Computer Interaction
Cognitive Systems Design	Illumination Measurement
Cognitive Task Analysis	Information Analysis
Cognitive Walkthrough	Information Visualization
Cognitive Work Analysis	Injury Analysis
Comfort Rating	Intervention Studies
Communication Analysis	Interview Technique
Cost/Benefit Analysis	Job Motivation Assessment
Critical Incident Technique	Job Satisfaction Measurement
Decision Support System Design	Kansei Engineering
Decision/Action Analysis	Link Analysis
Design Reviews	Macroergonomics
Direct Observation/Activity Analysis	Manual Materials Handling Assessment
Discomfort Rating	Mental Model Assessment
Environmental Sampling	Mental Workload Assessment
Error Analysis	Menu Design
Error Classification	Mockup Design/Analysis
Experimental Design	Multimedia Design
	Natural Language Interface Design
Failure Mode and Effects Analysis	Operational Sequence Analysis

TABLE 3.2 (continued)
One Hundred Methods and Techniques for Collecting and Analyzing Data in Human Factors and Ergonomics for Design of New Systems

Operator Performance Assessment	Systems Safety
Organizational Analysis	Task Analysis
Organizational Design	Task Performance Measures
Performance Measures (Time and Error)	The Human Error Rate Prediction (THERP)
Performance Ratings	Thermal Stress Measurement
Physical Work Load Assessment	Time and Motion Study
Predetermined Time Analysis	Time Lapse Photography
Production Systems Analysis	Training Needs Assessment
Psycho-Motor Proficiency Testing	Usability Analysis
Psychophysics and Scaling	Usability Engineering
Psychophysiological Measurements	Usability Testing
Questionnaire	User Log Books
Rapid Prototyping	User Population Definition
Repetitive Motion Injury Assessment	Verbal Protocol Analysis
Scenario-Based Design	Vibration Measurement
Screen Design	Video Recording
Selection Testing	Virtual Environment Design
Simulation Design/Evaluation	Visibility/Legibility Analysis
Standards and Guidelines for Design	Vision/Hearing Testing
Stress Measurement	Visual Performance Assessment
Survey Design	Walkthrough Analysis
Systems Analysis	Work Condition Evaluation
	Working Posture Analysis
	Workload Analysis
	Workspace Design

FURTHER READINGS

There are several interesting books dealing with human factors methods.

Wilson, J.R. and Cornett, E.N., 2002, *Evaluation of Human Work*, 2nd ed., London: Taylor & Francis.

Young, M.S. and Stanton, N., 1999, *Guide to Methodology in Ergonomics: Designing for Human Use*, Boca Raton, FL: CRC Press.

Brookhuis, K., Hendrick, H.W., Hedge, A., Salas, E., and Stanton, N., 2004, *Handbook of Human Factors and Ergonomics Methods*, Boca Raton, FL: CRC Press.

4 Vision and Illumination Design

4.1 INTRODUCTION

In this chapter we will first explain the anatomy of the eye and the properties of vision. The second part of the chapter is devoted to design of illumination systems to enhance visibility of text as well as in working environments.

4.2 THE STRUCTURE OF THE EYE

The eye is a slightly irregular sphere with a diameter of about 2.5 cm (see Figure 4.1). In the front of the eye, covering the eye pupil, is the cornea. The cornea protects the eye. It is very tough—even harder than a fingernail (Snyder, 1995). It has a high refractive index, which is helpful for focusing images on the retina. The aqueous humor, between the cornea and the pupil, is a fluid substance that doesn't seem to serve any great function, except perhaps lubricating the iris. The pupil size ranges from a small diameter of 2 mm to a large diameter of 8 mm; therefore the larger opening is 16 times as large as the smaller opening. The pupil is under autonomous nervous control, which means that the pupil size cannot be changed deliberately.

The ciliary muscle is a ring muscle that goes around the lens. It can contract or it can relax, and by doing so it either pushes the lens, so that it bulges, or it pulls the lens, so that it becomes flatter. To look at close objects, the lens has to bulge, thereby increasing the refractive power, while for distant objects it flattens out and reduces the refractive power. However, it is of interest to note than 70–80% of the refractive power is in the cornea, and the rest, 20–30%, is "fine tuning" which is performed by the lens. There are in fact two "lenses" of the eye: the cornea and the lens.

The inside of the eyeball is lined by the retina. This is a paper-thin layer of light-sensitive cells. All the cells are connected to the optic nerve, which transmits the information to the visual cortex, the main location for visual information processing.

The visual axis extends from the cornea to the fovea, which is the central part of the retina. The area of the fovea corresponds to the central vision of the eye. It has a different set of cells than the rest of the retina. In the fovea there are mostly cones, which are responsible for the color vision of the eye. The cones also have very high resolution, which means that they can sense very small details. Figure 4.2 shows the distribution of the rods and the cones along the retina.

The light sensors in the peripheral vision are mostly rods. The rods are not sensitive to color but rather to black and white. The rods have less resolution but greater light sensitivity than the cones. This is because several rods are coupled in

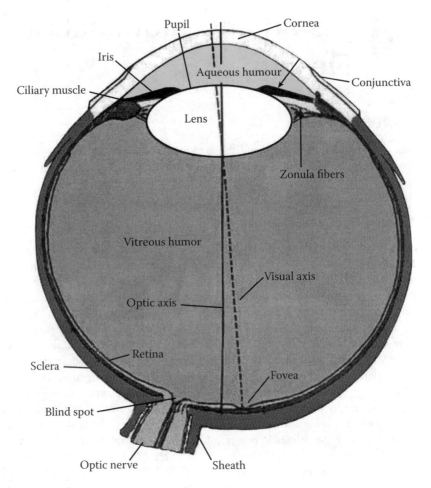

FIGURE 4.1 The structure of the eye.

series so that the incoming light energy is "amplified," but since the receptor area is increased, there is a loss in resolution (see Figure 4.3).

A driver at nighttime can notice many of the effects of the visual system. As it gets darker, the color vision is lost, since the cones, which have low sensitivity, are no longer active. The incoming light energy is not enough to excite the cones so the driver relies on the rods in his peripheral vision. An experienced nighttime driver, such as a truck driver, will sometimes turn his head to the side. In this way he can focus the image directly on the peripheral field of vision (the rods) rather than the central. This will enhance his sensitivity, and he can see the road ahead more clearly.

4.3 ACCOMMODATION OR FOCUSING OF THE EYE

Through accommodation of the eye we can see objects sharply at different distances. Accommodation is a mechanism that changes the shape of the lens in order to bring

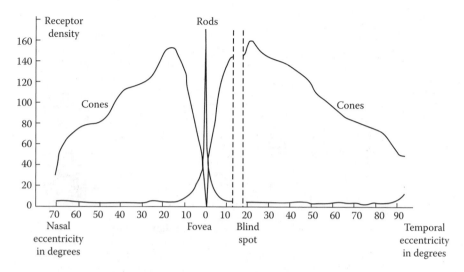

FIGURE 4.2 The distribution of rods and cones in the retina. To remember the function of the cones, think of CCC: cones, central, and color.

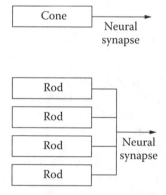

FIGURE 4.3 Each cone is connected one-to-one to a neural synapse. This leads to high resolution, but low light sensitivity. For the rods there is a many-to-one nerve connection. This is increases light sensitivity about 1000 times compared to cones, but at the price of lower resolution.

an image into sharp focus on the retina. This is done by contraction and relaxation of the ciliary muscles, and as a consequence the curvature of the lens increases or decreases, thereby affecting its refractive power. As we will explain below, accommodation ability or the amplitude of accommodation is lost with age (see Figure 4.11). At birth, the amplitude of accommodation is approximately 16 diopters (D), which is reduced to about 7–8 D by the age of 25 and about 1–2 D by the age of 60. These age changes are called presbyopia.

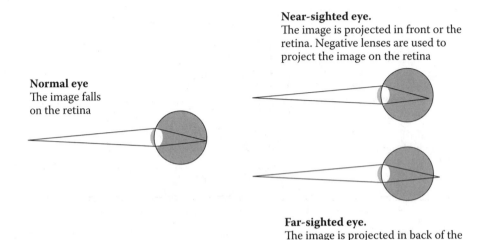

Normal eye
The image falls
on the retina

Near-sighted eye.
The image is projected in front or the
retina. Negative lenses are used to
project the image on the retina

Far-sighted eye.
The image is projected in back of the
retina. Positive lenses are used to
project the image on the retina

FIGURE 4.4 For myopia the image is projected behind the retina (a), and for hyperopia it is projected in front of the retina (b).

4.4   REFRACTIVE ERRORS

Many people have difficulties in focusing on objects due to myopia (near-sightedness), hyperopia (far-sightedness), astigmatism (the vision along vertical axis is different from the horizontal axis), and presbyopia (changes due to age). Refractive errors are caused by the shape of the eye and by the changes in the lens of the eye. For myopia, or near-sightedness, the images are projected in front of the retina rather than on the retina itself (see Figure 4.4). This can happen because of two different phenomena. Either the eyeball is too long, or the refractive power of the cornea and the lens is too strong. The cure is to wear concave or negative lenses. Hyperopia, or far-sightedness, occurs when the images are projected behind the retina. This may happen because the eyeballs are too short, or the refractive powers of the cornea and lens are too weak. Astigmatism may occur because of irregularities in the curvature of the cornea. For example, the cornea may have a sharp curvature in the horizontal direction but a more flattened shape in the vertical direction. Therefore, part of the image may be projected in front of the retina and part of it in the back.

Presbyopia, or changes due to age, also results in blurred images. With increasing age, the lens of the eye gets harder, so that it cannot accommodate, or bulge in and out. For a normal-sighted person (emmetrope), this may have the effect that she develops myopia as well as hyperopia. This means that she can neither see objects in the far distance nor in the close range. I vividly remember my grandfather, who at the age of 75 would pick up his newspaper and move it back and forth so as to find the appropriate focusing range where he could read. We will return to this issue below in 4.11.

4.5 LIGHT ADAPTATION AND DARK ADAPTATION

The human visual system has a tremendous capability to sense light. The dynamic range of the eye goes from 0.0000004 foot-lamberts (fhr L), which is the minimum sensitivity of the eye, to a maximum of 30,000 fhr L. The lower level corresponds to the light from a candle at about 2 km distance. The higher level of light corresponds to a white sand beach under intense sunshine. Beyond this level there is too much glare and dazzle and one cannot see so well. The range of vision corresponds to approximately 12 logarithm units. In this range, the eye can sense a range of about 3 log units without adaptation.

Of the 12 log units, only a small portion of the adaptation place in the pupil. As we have mentioned, the difference in size between the fully contracted and the fully expanded pupil is about 1:16, which corresponds to only 1.5 log units. The rest of the adaptation comes from other sources; retinal adaptation accounts for 4 log units. Nervous system gain adds another 3 log units. It is not all in the eye; there are also changes in the visual cortex. Adding up, we obtain: 3 + 1.5 + 4 + 3 = 11.5 log units.

When we walk into a dark movie theatre, it will take a few minutes before the eyes adapt to the darkness. After that, we can see the surrounding people. This process is called dark adaptation, and it is achieved by photo-chemical processes in the eye.

Both rods and cones contain light-sensitive chemicals called photopigments. The photopigment in the rods is called rhodopsin. The cones contain three different types of photopigments for the sensing of red, blue, and green. When light hits the photopigments, they undergo a chemical reaction that converts the light energy into electrical activity. This chemical reaction is referred to as light adaptation. In this process, the photopigments are decomposed. Intense light will decompose the photopigments rapidly and completely, thus reducing the sensitivity of the eyes so that it becomes difficult to see in dim light. The photopigments are then regenerated during dark adaptation, such as when a person walks into a dark environment and needs more photopigments in order to see. The process of dark adaptation is shown in Figure 4.5.

The figure illustrates the reduction of the light threshold as a function of time. When a person first walks into a dark environment, the threshold intensity is high. This means that the photopigments are not very sensitive, and as a result many photons are necessary in order to produce neural impulses. During the first 10 minutes the cones develop more rhodopsin, so that the eyes become much more sensitive to low-level light; the threshold for vision is reduced. This initial phase is called the photopic phase.

The rods then take over. As we can see in the figure, the threshold curve for the rods goes below the threshold curve of the cones. This means that the rods are more sensitive and can respond to less light energy than the cones. The second, or scotopic, phase results from regeneration of rhodopsin in the rods. There is a thousand-fold increase in sensitivity, and the dark adaptation is finished in about 30 to 40 minutes.

4.6 COLOR VISION

Visible light ranges in wavelengths from 380 nanometers (nm) to 760 nm. This corresponds to the colors violet, indigo, blue, green, orange, and red. An adjacent

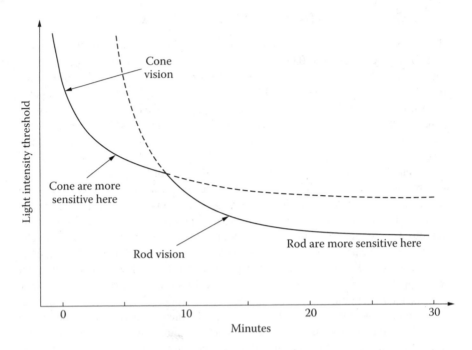

FIGURE 4.5 The dark adaptation curve, depicting the threshold levels of cones and rods.

portion of the spectrum, although not visible, can also affect the eye. Ultraviolet (UV) wavelengths extend from 180 nm to 380 nm. Exposure to UV radiation can produce ocular tissue damage. Infrared (IR) wavelength occurs from 760 nm up to the microwave portion of the spectrum. IR, or thermal radiation, can also damage tissues in the eyes.

The rods and the cones have different sensitivity to wavelengths. The cones are sensitive to the entire light spectrum from 380 to 720 nm. Rods on the other hand are not so sensitive to the upper or red part of the spectrum; their sensitivity is limited to 400 to about 560 nm (see Figure 4.6). The maximum sensitivity for the cones is at a wavelength of about 510 nm, corresponding to yellowish green, and for the rods at 510 nm, corresponding to green (perceived as black/white). Above the cone threshold curve (level 4), colors are visible. Below the rod threshold curve, nothing is visible—it is too dark. In the area between the cone curve and the rod curve there is black and white vision.

As the illumination drops, there is a shift in color, so that the perceived colors become more and more greenish. Above the cones sensitivity curve, at level 4 in the figure, only green is visible. Red is not visible—not even by the rods.

Under the cones' sensitivity curve one can only see grey colors; below level 4 in the figure, there is only grey vision. This is because illumination of the rods will only cause a black and white perception. From this figure we may note again that the rods are more sensitive.

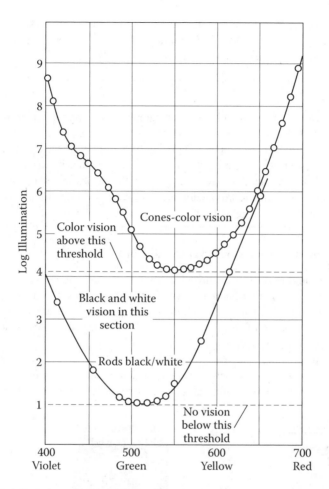

FIGURE 4.6 Relative sensitivity of rods and cones.

There are several design implications that follow from the figure.

1. Cones do not operate in low illumination. To perceive small objects, such as small letters, the cones must be used. Hence objects and lettering presented under low illumination must be much larger than in daylight conditions, so that they can be seen by the cones.
2. Don't use color coding when the illumination is very low, because there will not be discrimination between the colors.
3. For some tasks it is important for the operator to maintain dark adaptation. This is the case for pilots who fly at night in a dark cockpit. The challenge is to be able to see light on the ground and to navigate. Since a dark adaptation takes up to 30 minutes, pilots may dark-adapt prior to a flight by wearing dark red goggles. These provide enough light to get around in daylight but block out the shorter wavelengths (blue–orange). Only the red part of the spectrum can pass through the glasses. The rods are quite

insensitive to red wavelengths, and the rods can therefore maintain adaptation to the dark. Once in the dark cockpit the glasses are taken off and the pilot is perfectly dark-adapted.

EXAMPLE

A few years ago I was involved in designing machines for underground coal mines in the U.S. We wanted to propose warning signs to put on the machines used underground in coal mines. Typically, warning signs have red text, so we brought a few prototypes, but we quickly found out that it was so dark in the mine that it was difficult to see red. Color coding does not work out in a low illumination environment. We decided to use black text on white background, to make the contrast as large as possible.

4.7 MEASUREMENT OF VISUAL ACUITY

The most common way to measure visual acuity is by using a chart with letters of different sizes, a so-called Snellen chart. This was developed by Hermann Snellen in the 1860s. The Snellen charts, although commonly used, are a bit inaccurate, because the letters are of different width, and therefore some letters are easier to distinguish than others (see Figure 4.7). Landolt rings and checkerboard patterns produce more reliable results.

A common way of expressing visual acuity is by using numbers such as 20/20. 20/20 means that a person can read at a distance of 20 feet what a normal sighted

Snellen chart

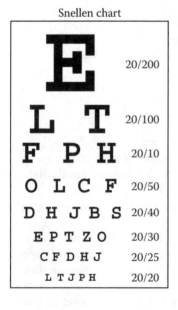

Landolt C ring

E-Chart. The character extends a visual angle of 5 min of arc. Each detail is 1 min of arc.

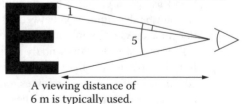

A viewing distance of 6 m is typically used.

FIGURE 4.7 Several different types of visual stimuli are used to test visual acuity.

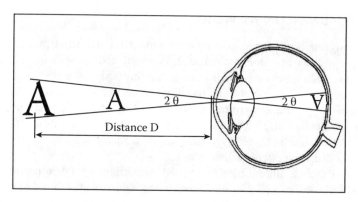

FIGURE 4.8 Standards for minimum size of text on visual displays.

person should be able to read at 20 feet. 20/40 means that the person must stand at 20 feet to read the same letter that a person with 20/20 vision can read at 40 feet and so on. For young persons, an acuity of 20/10 is normal, corresponding to 0.5 minutes of arc of resolution. A better analogy is that 20/20 corresponds to a visual acuity of 1 minute of arc or 1/60 of a degree. Visual angle is defined as the angle that an object subtends from the eye (see Figures 4.7 and 4.8).

A problem with the Snellen method is that the results depend on reading ability, and therefore Snellen is not a useful method for children. In addition the different letters have different sizes and some therefore easier to see than others.

To overcome this problem, one can use Landolt C-rings or so-called E-charts, as illustrated in Figure 4.7. To conduct the vision test C-rings of different sizes are rotated in four different directions, and the respondent has to identify whether the direction of the gap in the C is up, down, left, or right. Likewise for the E-chart, different sizes of E are presented, and the test person has to report in what direction the open side is pointing—up, down, left, or right.

The standards are formulated in terms of visual angle (see Figure 4.8). For regular text the recommendations are that the size must be about 20 arc minutes (arcmin), that is, about 20/60, or one third of a degree. Letters that are smaller than 10 arcmin are difficult to read, and letters that are larger than 25 arcmin are too large, because the larger size makes it difficult to scan several words with just one glance of the eye. Moving the eye's point of regard back and forth takes a longer time and reading is therefore slower. To calculate the readability of a traffic sign, use the following formula:

$$\text{Tan } \theta = (A/2)/D$$

where A is the height of the character, D is the distance to the character from the eye, and 2 is the visual angle of the character. For small values of θ, tan $\theta = \theta$.

EXAMPLE

Assume that the letters on a traffic sign are 0.2 m. Calculate the maximum distance from which the driver can read the sign, assuming that the critical size of the sign is 30 arcmin.

4.8 ILLUMINATION AT WORK

A well-designed illumination system is important for industrial productivity and quality, as well as operator performance, comfort, and convenience (Hopkinson and Collins, 1970). Below we will explain how to design illumination. Improved illumination is not just a matter of installing more lights, but also of how this is done. There are several ways of improving the quality of illumination; for example, by using indirect lighting. Such lighting can be important since it reduces the amount of glare. As we will note, older persons are particularly sensitive to glare, which may have a disabling effect on their vision.

We also discuss illumination for visual inspection. Visual inspection can be enhanced by using special-purpose illumination, which makes flaws more visible. Illumination for computer workstations is discussed in Chapter 14.

4.9 MEASUREMENT OF ILLUMINANCE AND LUMINANCE

The distinction between illuminance (also called illumination) and luminance is important. Illuminance is the light falling on a surface. After it has fallen on the surface, it is reflected as luminance. Luminance is therefore a measure of light reflected from a surface. Luminance is also used to measure light emitted from a computer screen. This may be theoretically incorrect, but for all practical purposes, light from a computer screen can be treated as having the same properties as reflected light.

To calculate how much luminance can be generated from a surface, one must know how reflective the surface is. This is specified by measuring reflectance, a number which varies from 0 to 1. It is practically impossible to achieve a perfect reflectance of 1.0; a piece of white paper has a reflectance of about 0.85. A non-reflective black surface has a reflectance of 0.

Measurement units are typically specified in the SI system (the metric system). Illuminance is measured in lux and luminance in candela per square meter (cd/m^2), also called "nits." These are the preferred measurement units (Boyce, 1981b).

In the U.S. the "English system" is still used, although the SI system is gaining ground. According to the English system, illuminance is measured in foot-candles (fc). One foot-candle equals 10.76 lux, but for practical purposes a conversion factor of 10 is sufficient. Thus 1000 lux illumination, which would be appropriate for an industrial workstation, corresponds to 100 fc. In the English system, luminance is measured in foot-lamberts (fL). One foot-lambert is equivalent to 3.4 cd/m^2 (or 3.4 nits). The measurement units are illustrated in Table 4.1.

There are simple formulas for converting illuminance to luminance.

For the SI system:

$$\text{Luminance (cd/m}^2) = \text{Illuminance (lux)} \times \text{Reflectance})/\pi$$

For the English system:

$$\text{Luminance (fhr L)} = \text{Illuminance (fc)} \times \text{Reflectance}$$

TABLE 4.1
Units for Measuring Illuminance and Luminance (SI Units are Preferred)

	English	SI
Amount of Light Falling on a Surface		
Illuminance	1 foot-candle (fc) =	10 lux (Ix)
(or illumination)	(or lumen/ft²)	(or lumen/m²)
of Light Reflected from a Surface		
Luminance	1 foot-lambert (fL) =	3.4 candela/m²
	(or candela/hr ft²)	(cd/m² or 3.4 nits)

4.10 MEASUREMENT OF CONTRAST

Contrast is the difference in luminance between two adjacent objects. It is calculated as a contrast ratio between the luminances of the two areas A and B:

$$\text{Contrast ratio} = \text{Luminance A}/\text{Luminance B}$$

An alternative way of expressing contrast is as modulation contrast:

$$\text{Modulation contrast} = (\text{Lum}_{max} - \text{Lum}_{min})/(\text{Lum}_{max} + \text{Lum}_{min})$$

where Lum_{max} is the greater of the two luminances. Modulation contrast is less than 1.0. Some experts prefer this expression of contrast, since it has properties that better resemble the sensitivity of the human eye (Snyder, 1988).

Both contrast and illuminance are important for visibility. For many items in the working environment contrast is rather high. For example, for black print on white paper the contrast is around 1:40, which provides excellent visibility. However, for characters on a VDT screen a contrast of 1:8 is not unusual, which is somewhat less visible (Shurtleff, 1980). For the screens of current mobile phones the contrast is often only 1:5. We will probably see these values improve in the future.

EXAMPLE: CONTRAST REQUIREMENTS IN MANUFACTURING

In manufacturing assembly, visual contrast may be critical. For one particular assembly it was important to distinguish between gold-colored electrodes, copper-colored electrodes, and copper oxide. This involved very small details in electronic manufacturing. Operators were looking through a microscope and bonding the copper electrodes to the gold electrodes. Work with microscopes is very demanding, and to relieve the postural strain a TV system with a TV camera and a monitor was brought in. Instead of looking into the microscope the operator could now look at the monitor while still performing the bonding operation manually. However, it turned out that the color rendering of the TV system was insufficient to distinguish the rather subtle differences between gold, copper, and copper oxide. The TV system had to be removed and the operators returned to using the microscope.

A very large contrast between large objects can cause discomfort glare. For example, the contrast between a window and an adjacent wall is often as large as 100:1. It is a common recommendation not to locate workstations so that the operator will face a bright window. Discomfort glare may cause oscillations of the eye pupils, but people are usually unaware of this phenomenon. Although discomfort glare is harmless, it is nonetheless annoying and discomforting.

For the same reason one should avoid extreme contrasts in the workplace.

A common recommendation is that the contrast ratio between the task and large items in the workstation should be less than 10:1 (or greater than 1:10). Some recommendations specify that the contrast between the task and the adjacent surroundings should be less than 3:1 (Illuminating Engineering Society, 1982). But 3:1 is too restrictive, and 10:1 is more reasonable (Kokoschka and Haubner, 1985). Grandjean (1988) recommends that the maximum luminance ratio within an office should not exceed 40:1.

Illuminance, luminance, and contrast ratio can be measured with a hand-held photometer. This device is similar to a camera light meter, except that it provides a direct readout in lux (or cd/m²). A photometer is color-corrected so that it simulates the human sensitivity to color. Thus, since the human sensitivity to violet and red (at the opposite ends of the color spectrum) is less than to green and yellow (at the centre of the color spectrum), the photometer will produce lower values for violet and red than for green and yellow. Therefore, in determining the luminance one need not be concerned about color, since the photometer will convert the values to simulate the sensitivity of the human eye.

A good photometer has two different settings: one for measuring illuminance and one for measuring luminance (Figure 4.9). To measure the illuminance that falls on a surface, one must consider contributions from a variety of sources: light sources (luminaries), windows, and wall reflections. The photometer must therefore have a wide angle of acceptance. The photometer must be cosine-corrected to account for contributions which are not perpendicular to the photocell on the photometer.

To measure luminance the photometer must have a narrow angle attachment, for example an attachment with 1 degree of acceptance. This enables precise readings of adjacent areas with different reflectances. To measure the contrast ratio between two adjacent areas, two luminance readings are obtained, and the contrast ratio is calculated.

The contrast ratio between characters on a VDT and the screen background is important for visibility, but is difficult to measure. The characters are composed of a rectangular array of dots (pixels). To measure the luminance of a pixel, a special

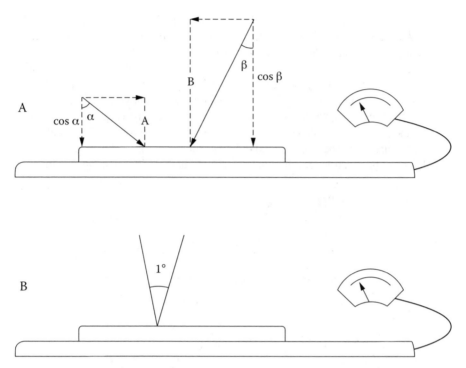

FIGURE 4.9 Use of a photometer for measuring (A) illuminance and (B) luminance. (A) Wide acceptance angle with a cosine correction. (B) Narrow acceptance angle for spot measurement.

photometer with a micro-image slit is required. The procedures are specified in U.S. Standard ANSI/HFS 100 (Human Factors Society, 1988).

Many experiments have been performed to determine the appropriate illumination levels for different tasks. Over the years there has been a succession of recommendations, each claiming to provide adequate illumination. The recommended levels, however, continually increase. Current recommended levels are about 5 times greater than the levels recommended 30 years ago for the same tasks (Sanders and McCormick, 1993).

One method for determining the required illumination is based on laboratory research by Blackwell (Blackwell, 1964,1967; Blackwell and Blackwell, 1971). The experimental task was to detect the presence of a uniformly luminous disk subtending a visual angle of 4′ (four minutes, which equals about 1.1 mm at a distance of 1 m). Blackwell found that when the background luminance decreased, the contrast of the just barely visible disk had to be increased to make it just barely visible again. Laboratory studies are not without problems. Blackwell's studies can be criticized for being overly artificial, since there are few real-life situations that resemble his experimental setup. In addition, subjects in laboratory studies know that they are participating in an experiment and they are usually motivated to perform well— much better than a person would do at work.

Field studies also have their problems, since there are many simultaneous independent variables that affect the outcome. These independent variables are usually not manipulated; sometimes they are not even measured. It can therefore be equally difficult to draw conclusions from field studies. As an example of a successful field study, Bennett et al. (1977) measured the task completion time for several industrial tasks, while varying the illuminance. The general conclusion was that increasing the illumination beyond 1000 lux seems to have limited benefits (Figure 4.10).

Bennett's research is also verified in the present illumination standards, published some 25 years later. There are several independent standards, and they seem to agree on the recommendation of illumination at workplaces. Illuminating Engineering

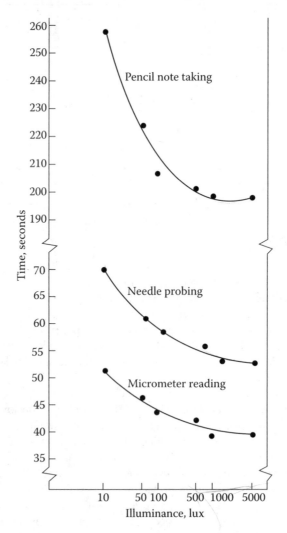

FIGURE 4.10 Relationship between the amount of illuminance and task completion time (Bennet et al., 1977).

TABLE 4.2
Illuminance Recommended by the IESNA for Industrial Tasks

Type of Task	Range of Illuminance (lux)
Workplaces where visual tasks are only occasionally performed	100–200
Visual tasks of high contrast or large size: printed material, rough bench and machine work, ordinary inspection	200–500
Work at visual display terminals for extended periods of time*	300–500
Visual tasks of medium contrast or small size; e.g., penciled handwriting, difficult inspection, medium assembly	500–1000
Visual tasks of low contrast or very small size; e.g., handwriting in hard pencil on poor-quality paper, very difficult inspection	1000–2000
Visual tasks with low contrast items and very small size over a prolonged period; e.g., fine assembly, highly difficult inspection	1000
Performance of exacting visual tasks such as extra fine machine work, exacting assembly and manual crafts, precision arc welding	3000

*This recommendation is from ANSI/HFS 100 (Human Factors Society, 1988).

Adapted from Kaufman and Christensen (1984). The upper values in the range are for individuals aged over 55 years and the lower values are for individuals younger than 40 years.

Society (IES) publishes recommended values of illumination (Table 4.2). Depending upon the size of the visual task and the contrast of the task, different levels of illumination are required. These guidelines also take into account the worker's age, the importance of speed and accuracy, and the reflectance of the task background. The upper end of the recommended range in the table should be used to accommodate older workers and the lower values are for younger workers. It is also suggested that local task lighting rather than general ambient illumination be used, particularly if the illumination at the workplace is above 1000 lux.

4.11 THE AGING EYE

For older individuals there are several physical changes in the eye. The most important is the loss of focusing power (accommodation) of the lenses in the eye (Safir, 1980). This is because with increasing age the eye lenses lose some of their elasticity, and therefore cannot bulge or flatten as much as before.

Figure 4.11 illustrates that the average accommodation for a 25-year-old is about 11 diopters, but for a 50-year-old it is only 2 diopters and for a 65-year-old it is 1 diopter. The number of diopters translates into a range of clear vision that is defined by its far point and its near point. Assume that for the 25-year-old the far point is

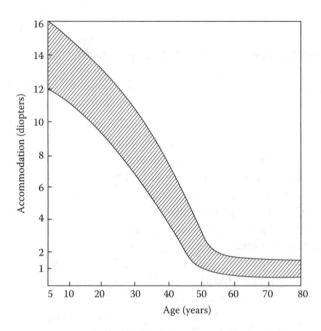

FIGURE 4.11 Changes in accommodation of the eye with age. The shaded area indicates that there is a large variability between individuals (Handbuch für Beleuchtung, 1975).

at infinity. The near point of accommodation is then 9 cm, which can be calculated using the equation:

$$f = 1/D$$

where f is the focusing distance (in meters), and D is the number of diopters of accommodation.

Likewise, if the far point for a 50-year-old with 2 diopters of accommodation is at infinity, then the near point is 50 cm (Figure 4.12). But assuming that the same 50-year-old has 3 diopters of uncorrected short-sightedness (myopia), then the far point (without glasses) is 33 cm and the near point is 20 cm. A person who is myopic at a young age will typically find that with increasing age the far point moves closer and the near point moves further away. For an individual with no refractive errors as a young person, the near point moves further away, while the far point may stay at infinity.

The implication for industrial work is that the different ranges of vision not only affect the visibility of the task but also the work posture. To compensate for poor vision, a myopic (short-sighted) person will move closer, and a hyperopic (far-sighted) individual will move further away. Poor work posture observed in industry is therefore often due to poor vision. If the vision is corrected with eyeglasses, the bad posture may also correct itself automatically. Workers are often not well informed about what kinds of visual corrections are feasible. To help in advising workers, some companies hire optometrists who measure the exact viewing distances

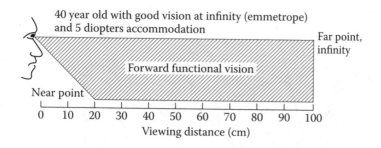

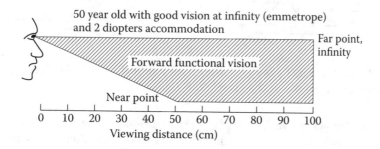

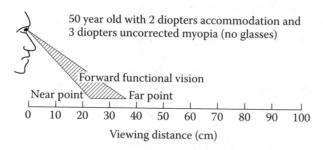

FIGURE 4.12 Calculation of the near point and far point of forward functional vision. The clear range of vision depends on the range of accommodation (in diopters) of the lens in the eye, and the refractive error.

from the eyes to the various task elements. Eyeglasses that are tailored to the conditions at work can then be prescribed.

The limited range of clear vision makes it necessary that items in a workplace are put at a distance where they can be seen clearly. In the same way that forward functional reach limits the physical organization of a workspace, so does forward functional vision.

The second most important effect of age is the clouding of vision. In the vitreous humor, between the lens and the retina, there are particles and impurities. With age these impurities increase in size. They impair clear vision, because they scatter incoming light over the retina. Older persons are therefore particularly sensitive to glare sources or stray illumination, which add a veiling luminance (a cloud) over

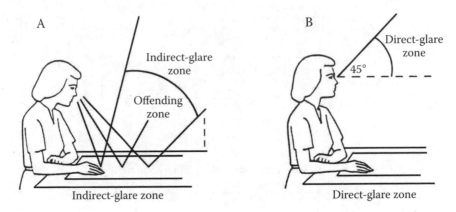

FIGURE 4.13 Indirect glare (A) arises from reflected light, while direct glare (B) arises directly from the light source.

the retina (Wright and Rea, 1984). As a result, the contrast on the retina decreases. For older persons it is therefore important to minimize stray illumination and glare that is not part of the task (Figure 4.13).

Direct glare comes from light sources, such as overhead luminaries, that are shining directly into the operator's eyes. The reflected or indirect glare is from light that is reflected in the workplace from glass or plastic covers, shiny metal, or key caps on a keyboard. One way of solving the problems of both direct and reflected glare is to use task illumination. This involves directing lamps with a restricted light cone towards the visual task. Some examples of task lights are shown in Figure 4.14 (Carlsson, 1979).

4.12 USE OF INDIRECT (REFLECTED) LIGHTING

Many office architects and interior designers prefer to use indirect (reflected) lighting because it creates a pleasant environment (Carlsson, 1979). In this case, about 65% of the illumination is directed upward to the ceiling and then reflected from the ceiling back to the workplace (Figure 4.15). The use of indirect lighting minimizes both direct glare and indirect glare. It minimizes direct glare because the light is directed towards the ceiling rather than the operator's eyes, and it minimizes reflected glare because the light reflected from the ceiling is not directional, and will therefore generate so-called diffuse reflection.

There is one disadvantage of indirect lighting, namely, the loss of light when it is reflected from the ceiling. It is preferable to use a white ceiling with a high reflectance value. Indirect lighting is mostly suitable for offices and clean manufacturing workplaces where the ceilings do not become soiled. Indirect lighting would probably not be effective for dirty manufacturing processes, since the light sources and light fixtures become covered with dirt and it is necessary to clean luminaires and paint ceilings at regular intervals.

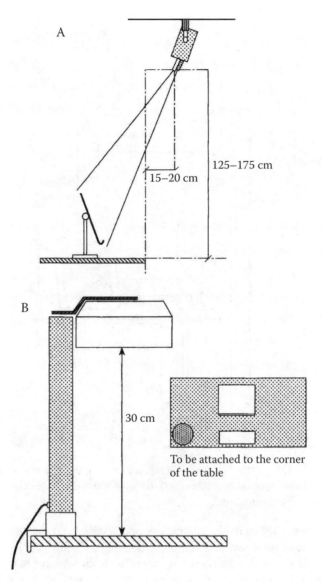

FIGURE 4.14 Examples of task illumination. (A) An overhead task light with a limited light cone is used to illuminate a source document on a document holder. (B) A table-top lamp can be used on a workbench to provide task illumination that does not generate glare.

4.13 COST EFFICIENCY OF ILLUMINATION

Konz (1992b) has provided convincing arguments that the cost of industrial lighting is minimal. In fact, a generous illumination level costs only about 1% of the worker's salary (in the U.S.). As demonstrated in the first case study in Chapter 2, efficient

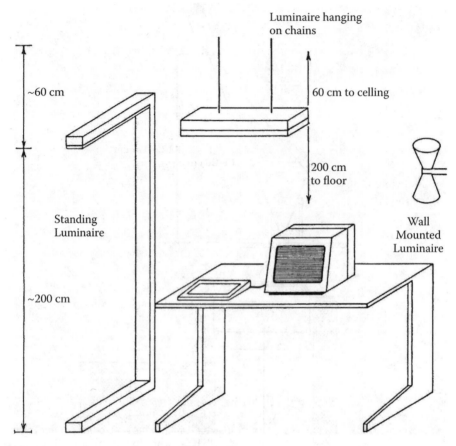

FIGURE 4.15 Three different types of indirect luminaires. The use of indirect light creates a pleasant atmosphere. About 60–65% of the light is directed upward and reflected diffusely downwards.

illumination typically increases quality in manufacturing and manufacturing yield. It is ill advised to cut down on the illumination to save a few pennies.

The efficiency of a light source is measured in lumens per watt (lm/W). As illustrated in Table 4.3, some light sources are very efficient whereas others are less efficient. But there is an important trade-off, namely, the color rendering of the light. The color rendering index (CRI) is a measure of how colors appear under a light source as compared with daylight. A perfect CRI score is 100. The main concern is that the color of the light may distort perception. Low-pressure sodium light, which is intensely yellow, makes faces look grey and should not be used indoors. It is mainly used for outdoor lighting, but even in this situation it is difficult to, say, find a car in the car-park because all colors look similar. Measures of light source efficiency and color rendering are also presented in Table 4.3.

TABLE 4.3
Efficiency of Light Sources and Their Color Rendering Index (CRI)

Type	Efficiency (lm/W)	CRI	Comments
Incandescent	17–23	92	The least effective but most commonly used light source
Fluorescent	50–80	52–89	Efficiency and color rendering vary considerably with type of lamp
Coolwhite Deluxe		89	
Warmwhite Deluxe		73	
Mercury	50–55	45	Very short lamp life
Metal halide	89–90	65	Adequate color rendering
High pressure sodium	85–125	26	Very efficient, but poor color rendering
Low pressure sodium	100–180	20	Most efficient, but extremely poor color rendering; used for roads

The maximum value of the CRI is 100.

Adapted from Wotton (1986).

Incandescent light produces the best color rendering, so that faces look natural, but its efficiency is only 17–23 lm/W, which makes this light expensive to use. Fluorescent lights have fair to good color rendering. The best color rendering is obtained with the Coolwhite Deluxe source, which has more red colors in the spectrum and looks more natural. The light efficiency varies quite a lot (50–80 lm/W).

The other light sources (mercury, metal halide, and high-pressure sodium) have fairly poor color rendering and should not be used in manufacturing plants or offices. They are more appropriate in environments where there are few people (e.g., in warehouses, shipping and receiving, and outdoors) (Wotton, 1986; Boyce, 1988).

4.14 SPECIAL PURPOSE LIGHTING FOR INSPECTION AND QUALITY CONTROL

Special types of illumination can be used to detect faults in manufacturing. For example, to make surface scratches on glass or plastic visible, it is common to use edge lighting that is directed from the side. There are many other special types of light, including polarized lights, cross-polarization, spotlights, convergent lights, and transillumination (Faulkner and Murphy, 1973). The information given in Table 4.4 is adapted from Eastman Kodak (1983), where more complete information is given. The second column in the table describes special purpose lights or other aids, and the last column describes how the techniques work.

Table 4.4
Special-Purpose Lighting for Inspection Tasks

Desired Enhancement in Inspection Task	Special-Purpose Lighting or Other Aids	Technique
Enhance surface scratches	Edge lighting can be used for a glass or plastic plate at least 1.5 mm thick	Internal reflection of light in a transparent product; use of a high-intensity fluorescent or tubular quartz lamp
	Spotlight	Assumes linear scratches of known direction; provides adjustability so that they can be aligned to one side of the scratch direction; uses louvres to reduce glare for the inspector
	Dark-field illumination (e.g., microscopes)	Light is reflected off or projected through the product and focused to a point just beside the eye; scratches diffract light to one side
Enhance surface projections of indentations	Surface grazing or shadowing	Collimated light source with an oval beam
	Moiré patterns (to accentuate surface curvatures)	Project a bright collimated beam through parallel lines a short distance away from the surface, looking for interference patterns (Stengel, 1979); either a flat surface or a known contour is needed

Table 4.4 (continued)
Special-Purpose Lighting for Inspection Tasks

Desired Enhancement in Inspection Task	Special-Purpose Lighting or Other Aids	Technique
	Spotlight	Adjust angle to optimize visualization of these defects
	Polarized light	Reduces subsurface reflections when the axis of is transmission parallel to the product surface
	Brightness patterns	Reflection of a high-contrast symmetrical image on the surface of a specular product; pattern detail should be adjusted to product size, with more detail for a smaller surface
Enhance internal stresses and strains	Cross-polarization	Place two sheets of linear polarizer at 90° to each other, one on each side of the transparent product to be inspected; detect changes in color or pattern with defects
Enhance thickness changes	Cross-polarization	Use in combination with dichroic materials
	Diffuse reflection	Reduce contrast of brightness patterns by reflecting a white diffuse surface on a flat specular product; produces an iridescent rainbow of colors that will be caused by defects in a thin transparent coating
	Moiré patterns	See "Enhance surface projections or indentations" above
Enhance nonspecular defects in a specular surface, such as a mar on a product	Polarized light	A specular non-metallic surface acts, under certain conditions, like a horizontal polarizer and reflects light; non-specular portions such as a mar will depolarize it; project a horizontally polarized light at an angle of 35° to the horizontal

Table 4.4 (continued)
Special-Purpose Lighting for Inspection Tasks

Desired Enhancement in Inspection Task	Special-Purpose Lighting or Other Aids	Technique
Enhance opacity lights changes	Transillumination	For transparent products, such as bottles, adjust to give uniform lighting to the entire surface; use opalized glass as a diffuser over fluorescent tubes for sheet inspection; double transmission transillumination can also be used
Enhance color changes, as in color matching in textiles	Spectrum-balanced lights	Choose lighting type to match the spectrum of lighting conditions expected when the product is used; use 3000 K lights if the product is used indoors, 7000 K lights if it is used outdoors
	Negative filters, as in inspecting layers of color film for defects	These filters transmit light mainly from the end of the spectrum opposite to that from which the product ordinarily transmits or reflects; this reversal makes the product surface appear dark, except for blemishes of a different hue, which are then brighter and more apparent
Enhance fluorescing defects	Black light	Use ultraviolet light to detect cutting oils and other impurities; may be used in clothing industry for pattern marking; fluorescing ink is invisible under white light, but very visible under black light
Enhance hairline of breaks in castings	Coat with fluorescing oils	Use of ultraviolet light inspection will detect pools of oil in the cracks

Adapted from Eastman Kodak Co. (1983), with permission.

EXERCISE: MEASUREMENT OF ILLUMINATION (ILLUMINANCE)

Use a light meter to measure the illumination characteristics in an office.

1. Measure the amount of illumination falling on several office desks. Compare the recorded illumination with the standards in Table 4.2. Make recommendations with respect to the appropriateness of the illumination level.
2. Measure the illumination falling on VDT screens. In this case, the light meter must be held so that the light sensitive surface is parallel to the screen surface. Compare results to Table 4.2.
3. Measure the illumination levels in several other areas, such as corridors, special workplaces for drafting work, etc., and compare to Table 4.2.

FURTHER READINGS

An excellent book is Peter Boyce, 2003, *Human Factors in Lighting,* 2nd ed., London: Taylor and Francis.

5 Human Information Processing

5.1 INTRODUCTION

Human information processing has become a very important area in HFE. It may seem surprising that there was little interest in cognition and human information processing until 1967, when Ulrich Neisser published his book on cognitive psychology. Today, cognitive psychology and cognitive science are fundamental to the design of information systems, including the design of information displays, pervasive and ubiquitous computing (computers everywhere), handheld computing, and mobile phones.

To illustrate the significance of human information processing, we will first introduce Hick's law and a model of human information processing. Some models for decision making will be presented next. These models either rely on principles of cognitive psychology, or on observations of naturalistic decision making of people in real work settings.

5.2 EXAMPLE: THE TROUBLE WITH INFORMATION

A major problem with the design of computing systems as well as consumer products is that there is often too much information in the interface design, in particular the controls for manipulating the machine. Recently, my microwave oven stopped working and I had to replace it. The old one had two controls, one for power and one for time (see Figure 5.1). Turn the time knob and the microwave is already humming.

I talked to sales people in shopping centers, trying to find a similar microwave oven. They just shook their heads. In the end I gave up and bought a microwave oven with 28 controls. Some controls were specialized to perform certain functions. For a person who had not made popcorn for 5 years, the first control—in the most prioritized location—was useless and irritating. My old microwave took 2 seconds to turn on; the new one took 20 seconds. Several problems with the interface design may be identified:

1. Too many controls—not all are required.
2. Rather than turning a dial, you have to key in the cooking time.
3. Too many contingent actions—even the "start" button has to be pressed. It is no longer automatic after step 2 above.
4. Functions that seem deceptive. For example, if you want to heat up your cup of coffee—press "beverage". It will then cook for 1 min 20 sec. But the microwave oven does not know your cup is half full, and that you want the temperature to be less than burning hot—about 65°C.

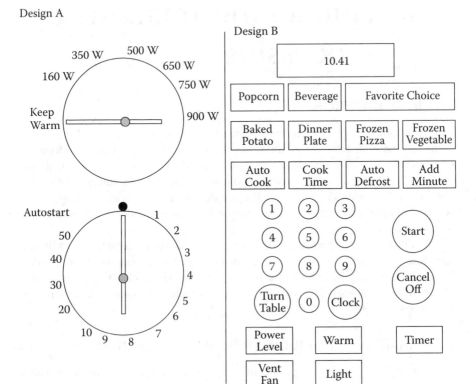

FIGURE 5.1 Interface design of two microwave ovens. Design A combines simplicity and functionality. Design B leads to confusion and many user errors.

It puzzles me that these microwave designs have become so popular. However, consider that the extra controls cost very little to manufacture. To the customer it looks like a better buy—more for the same investment. But the investment in time due to difficulty of use is completely ignored. The effect of too much information is easily explained by Hick's law.

5.3 HUMAN REACTION TIME AND HICK'S LAW

A common performance measure in HFE is human reaction time (RT). The reaction time varies depending upon what people are reacting to. The most common RT scenario in real life is probably the start of a running competition: ready, steady, go. In HFE reaction time is used to measure how complex decisions are. Complex decisions take a long time and easy decisions take a short time. RT is used to reformulate or redesign decisions so that they become easier, quicker, and more reliable.

Hick's law stipulates that the reaction time is a function of the number of choices in a decision.

$$RT = a + b \log_2 N$$

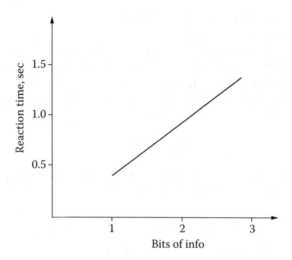

FIGURE 5.2 Hick's law. Reaction time is plotted as function of information uncertainty (or entropy) in bits.

where *RT* is reaction time, *N* is the number of alternatives, and *a* and *b* are constants. The equation is plotted in Figure 5.2.

Hick's law can be verified in a reaction time experiment in a laboratory environment. The experimenter can use a box with eight lights, where any of the lights can come on at any time. Your task when a light comes on is to press a contact switch just under the light. Eight lights correspond to 3 bits of information uncertainty, or entropy. Let's say that your reaction time was 1.4 sec. The experiment is then repeated with 4 lights (2 bits of information) and the RT is 0.9 sec, and finally with 2 lights (1 bit of information) with RT = 0.4 sec.

The interesting finding is the straight linear relationship in Figure 5.2. From the figure it seems that the brain is wired to respond linearly to the number of bits of information. This is a very useful finding, since it makes it possible to predict the information processing time. For example, let's say that we are designing a computer interface with pull-down menus and we want to understand how the presentation of menu items with different information content can be optimized so that the search time is minimized. This question brings us to the next heading: information theory.

5.4 INFORMATION THEORY

The amount of information in a stimulus depends on the probability that the stimulus carries relevant information. In this context one also speaks of information uncertainty or entropy. Shannon and Weaver (1949) pioneered information theory. They defined information as uncertainty or entropy. Some information has very little information content. For example, a statement such as "The sun went up this morning" carries no information, since the sun goes up every morning. The probability is $p = 1.0$, and therefore there is no information uncertainty. A statement such as "There was an earthquake in Paris" carries much information, since it is very unlikely to happen—Paris is not on a fault line! The probability of an event therefore affects the amount of information.

Shannon and Weaver (1949) presented a model for calculation of information in terms of bits. If 2 stimuli are equally likely to occur with p = 0.5, there is an information uncertainty of 1 bit. If there are 4 possible events, each with a probability of p = 0.25, there are 2 bits. The general formula for calculating the number of bits of information is hence tied to the probability:

$$H_s = \log_2 N$$

where H_s is the amount of information and N is the number of equal probability outcomes. For N events the probability that each may occur is therefore $p = 1/N$.

$$H_s = \log_2 (1/p)$$

To summarize the information for all N events we obtain:

$$H_s = \log_2 (1/p)$$

EXAMPLE 1

Start with a deck of 64 cards—16 cards in 4 suits. Ask a person to think of card. Your task is to identify the card the person is thinking of. You can ask the person for hints. First, let us note that the amount of information uncertainty in the single card that you will try to identify is as follows:

$$H_s = \log_2 N = \log_2 64 = 6 \text{ bits}$$

We can also do the calculations using probabilities. The probability to find a single card is p = 1/64:

$$H_s = \log_2 (1/p) = \log_2 (1/1/64) = \log_2 64 = 6 \text{ bits}$$

Given the hint that the card is red, the uncertainty is reduced from 64 to 32: 1 bit. Given that the card is a heart: 1 bit; lower 8 hearts: 1bit; lower 4 cards: 1 bit, lower 2 cards: 1 bit; ace of hearts: 1 bit; 6 bits altogether. This is in agreement with the calculations above.

Unequal Probabilities

For events that have unequal probabilities:

$$h_i = \log_2 (1/p_i) = -\log_2 (p_i)$$

where p_i is the probability of the ith event and h_i is the information of the ith event. The variable h_i is also called surprisal—obviously, the lower the probability for a single event to occur, the greater the surprisal.

EXAMPLE 2

A bent coin is tossed and it comes up heads 90% of the time and tails 10% of the time. Calculate the amount of information in this coin, considering both outcomes. Consider:

$$H_s = \sum p_i \log_2 (1/p_i)$$

To perform the calculation we use the conversion from $\log_2$ to $\log_{10}$;

$$\log_2 x = 3.332 \log_{10} x$$

$$H_{head} = p \log_2(1/0.9) = 0.9 \times 0.0453 \times 3.332 = 0.135 \text{ bits}$$

$$H_{tails} = p \log_2 (1/0.1) = 0.1 \times 1 \times 3.332 = 0.332 \text{ bits}$$

The total information uncertainty of tossing the coin is $0.135 + 0.332 = 0.467$ bits.
   Consider a regular coin:

$$H_s = 2 \times 0.5 \log_2 (1/0.5) = 1.000$$

Because the bent coin comes up with heads almost all the time, there is less surprisal than in a regular coin.
   We can now understand that the amount of information can be calculated, and we can apply Hick's law to compare different designs, realizing that the more information, the longer the reaction time will be for making decisions. A calculation exercise is presented at the end of this chapter.

5.5 HUMAN INFORMATION PROCESSING

A traditional approach to human information processing is presented in Figure 5.3. According to this model, people sense the environment through seeing and hearing, then make decisions, and finally act on the decisions. Here we refine the model by breaking down the components further.

Figure 5.3 illustrates the human information processing cycle, as conceived by Card, Moran, and Newell (1983). There are three processors in this model:

1. A perceptual processor (to see and hear)
2. A cognitive processor (to think)
3. A motor processor (to act)

The average time to process information in the 3 processors is as follows: 100 ms for perception; 70 ms for cognition; and 70 ms for action. These numbers depend upon the task; a complex target with many details takes a longer time to process than a simple target. For motor processors, however, the variability in time is not

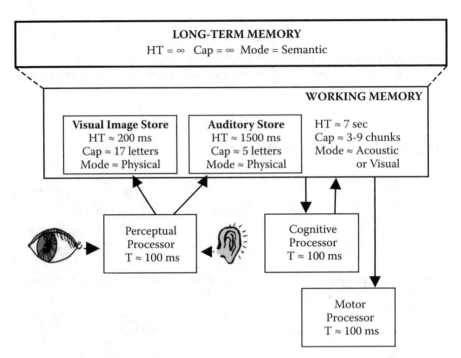

FIGURE 5.3 Human information processing. Adapted from a model by Card, Moran, and Newell (1982). HT = half-time of memory; Cap = capacity; T = average processing time.

so large; once a decision has been made, it takes a relatively short time to act upon it. Taking these issues into consideration, the range of variability in perceptual and cognitive processors is about 25–200 ms, and in motor processors about 30–100 ms (Card et al., 1983).

At the top of the figure is the long-term memory (LTM). The LTM supposedly has unlimited storage capability. This is probably not true—people do forget things that are of less relevance, although under hypnosis and in dreams, information which has not been retrieved for years can surface.

The mode of information in LTM is semantic; that is, it consists of concepts. Most people store and retrieve information from LTM in terms of concepts. For example, you can teach a child how to read a watch, no matter whether the face of watch has numbers or markers.

There are exceptions among people. My father had the rare gift of photographic memory. When he sat for exams as a student he could visualize any page in a book, somewhat like reading a "visual copy" of the book. He had an image store, but he could also understand the writing. Unfortunately I did not inherit his capability; instead I rely heavily on semantics—the meaning of things.

The LTM supports the working memory (WM), also called short-term memory (STM). The WM is what we use when we perform a task. We note to the right of Figure 5.3 that the half-time (HT) of the working memory is 7 sec. This means that the working memory fades away very quickly, and after 7 sec, half of the information has been forgotten. This is in a sense rather practical, since we do not want to burden our memory with information about routine tasks. For example, when driving your car from home to work you rely on your working memory to make routine decisions. After you arrive at your destination, chances are that you do not remember anything of the journey, unless something unusual happened such as an accident. Baddeley (1992) referred to the working memory as a mental scratch pad that temporarily stores information while actively operating on it. It is like a pull-down menu in a computer system. Once we get into a situation we can select and execute the appropriate strategies. While driving downtown, you know where the busiest streets are, you know where to look out for children; and for each of these two scenarios a "pull-down menu" with driving strategies can be activated. Thereby the actions and responses are preprogrammed.

The driving information in WM gets updated from LTM as it becomes useful during driving. For example, as I drive into the campus of the Nanyang Technological University, there is a sharp curve to the left. Automatically (without thinking about it) I retrieve the appropriate motor control schema—I slow down—so that I can steer through the curve safely. This particular type of working memory is also referred to as "running memory"—it is constantly updated, but there is no need for retention (Moray, 1986). This is typical for many human–machine interaction tasks; for a skilled operator they become routine tasks, with nothing much to remember afterwards (Wickens and Hollands, 2000).

A classic paper by George A. Miller (1956) sparked the cognitive revolution. Miller claimed that the working memory capacity is 7 ± 2 chunks. Chunks are units of information, such as letters, words or situations. For example, the letter string MBITTAWRT has 9 chunks, and formatted in this manner may just barely exceed the STM of a smart reader. As people learn tasks, they form large chunks—a chunk can be a group of letters or a situation. Therefore experienced operators are more efficient than inexperienced operators. A grand master of chess can see strategies and makes projections that a novice chess player has difficulty perceiving.

Chase and Simon (1973) published a seminal study on how chess players chunk information. They used two different types of test subjects in their experiment: masters of chess and beginners. The players were given 10 sec to memorize a chess board with 20 to 25 pieces. They were then asked to set up the game on another board. The masters were 93% accurate, while the beginners were only 33% accurate. In the second part of the experiment they were shown chess boards with the pieces in random positions. Surprisingly, the score of all players dropped to 20%, illustrating that the superior players were subject to the same STM constraint as the weaker players. Only when faced with a meaningful chessboard could they outperform the less experienced subjects.

Bratko, Tancig, and Tancig (1986) classified chess games by using positional patterns that are commonly used in chess. They came to the conclusion that for a group of masters of chess, there was an average of 7.54 (large) chunks per board. Novices of chess have not learned to characterize the different positions. They don't understand the semantics of the game, and would therefore use many more chunks than the masters, thereby overburdening their WM.

These studies inspired other researchers to investigate how efficient people can become at chunking information. Ericsson, Chase, and Faloon (1980) performed an experiment with one undergraduate student at Carnegie Mellon University. The purpose was to investigate how chunking can help someone to memorize a string of numbers. This college student, of average memory abilities and intelligence, performed a memory span task for about 1 hour a day, 3 to 5 days a week, for 20 months. The task was to memorize as many digits as he could. The digits were read to him at a rate of 1 digit per second; he then recalled the sequence. If the sequence was reported correctly, the next sequence was increased by 1 digit; otherwise it was decreased by 1 digit. During the course of the 20 months of practice, his digit span steadily improved from 7 digits on the first day to about 80 digits at the end. Furthermore, his ability to remember digits after the sessions also improved. In the beginning he could recall virtually nothing after an hour's practice; after 20 months of practice he could recall more than 80% of the digits presented to him. As time passed by, the test person became very skilled at chunking the numbers, and early on he started to use mnemonics (schemes to aid the memory). For example, the numbers 3492 was recorded as 3 minutes and 49.2 seconds (the new world record for the mile); 893 was 89.3 (a very old man). Running times and ages accounted for almost 90% of his mnemonic associations. Over time he started to organize his retrieval structure by segmenting the numbers into subgroups. He used two 4-digit groups followed by two 3-digit groups.

At one time, after three months of practice, his experimental session was switched from digits to letters of the alphabet, and his memory span dropped back to about six letters. The authors concluded that it is not possible to increase the capacity of the STM with extended practice. Clearly, increases in memory span are due to the use of mnemonic associations in the LTM.

Coming back again to the letters presented above, MBITTAWRT: formatting the letters in groups of three—MBI TTA WRT—makes the sequence easier to recall. A telephone number such as 5282772 is easier to memorize as 528 2772 or as 52 82 772. Much research has gone into the design of postal codes as well as telephone numbers. The number 8 is slightly easier to recall than other numbers; therefore it is popularly used to denote toll-free numbers around the world. Other numbers that are particularly memorable have been published by Chapanis and Moulden (1990).

Even better than the chunking IBM, ATT, and TRW. Symbols, icons, and labels that are used frequently can simplify chunking, since we may refer to well-understood scenarios and concepts, thereby making it easy to form a mental model for organizing information.

Returning again to Figure 5.3, we note that the mode of information storage in the working memory is acoustic or visual. This can be exemplified by a game of bridge. During the game the four players must keep track of the discarded cards. They do so by silently repeating to themselves: two of clubs, five of spades, jack of diamonds, and so forth. By repeating many times, they refresh the memory trace in the WM; or perhaps they even hope to put the information into LTM. Otherwise they will be fighting a losing battle against the decay of the WM. The more chunks of information there are, the quicker the information dissipates. The halftime (HT) for 1 chunk of information is about 100 sec, and for 3 chunks of information it is about 7 sec.

As we pointed out above, these numbers are broad figures; they are not absolute. The capacities and the HTs of various people's memories varies depending on what type of information is stored and on how much information is stored.

PERCEPTUAL PROCESSORS

When we make decisions, we first perceive information, then extract features of the information. In Figure 5.3, the perceptual processors send data from the eyes and the ears to two data banks, the visual image store and the auditory image store. The data in these banks decay very quickly. The HT for the visual image store is 200 ms; for the auditory store the HT is 1500 ms. With visual impressions, the image is first available from the retina of the eye and is stored in the brain for a very short time. The image is veridical, which means that it is true to the real world. However, due our limited information processing capability, we can only attend to a few parts of the image. As I look at the 30-odd students in my classroom, I can pay attention to only 1 at a time. The attention is like a search light. By putting a light beam on an object, we focus on that particular object but forget about the rest, for the time being.

Sometimes there are features in the environment which make us attend to those particular features more than we attend to other details. From advertising we know that large elements, complex figures, and dynamic images draw our attention more than the opposite; small, simple, and static images are not attended to as much. The peripheral vision plays a large role in this context. Take the case of moving images. When there is an object moving in our peripheral vision, we cannot help but turn our heads to look at it. This is an innate pre-attentive reflex—we look automatically even before we pay attention to what is going on (Neisser, 1967). Such reflexes can warn us about many dangers, as our forefathers were warned about wild animals.

The visual image store is also called iconic memory. It has a capacity of about 17 letters. The memory capacity can be investigated in a laboratory experiment; such experiments are commonly performed in psychology departments. Using a tachistoscope, one can flash an image with several letters on a screen for about 500 ms. As the flash duration is so short, the test person will not have a chance to memorize the letters, and can only reproduce them by "looking" at what is left of the quickly decaying image. The task is to mention as many letters as possible; 17 letters is the approximate limit.

The auditory image store, also referred to as echoic memory, works in a similar fashion. The echoic memory stores sound impressions, and its HT is about 1500 ms. During this 1.5 s we can still "hear" what was said, which is why this type of memory is called echoic memory. The HT of the echoic memory is much longer than for the iconic memory. This makes it possible to listen to the words together in a sentence and comprehend them as an entire sentence, rather than as isolated words.

We have now completed the entire information processing chain in Figure 5.3, from perceptual input to cognition (or decision making) to action (or motor output).

For a routine task it is possible to predict how long a time these elements will take. A simple task, such as moving the cursor back and forth between two targets on a screen (without missing the targets) will take approximately 240 ms; 100 ms for the perceptual processor, 70 ms to make a decision about where to go next, and

70 ms for the motor processor to accomplish the movement. For more complex tasks, the perceptual and cognitive processing will take more time, although the motor response may still be completed fairly quickly. Once a decision has been made, action is usually quick.

5.6 HEURISTICS ARE USED FOR COPING WITH THE LIMITATIONS OF THE WORKING MEMORY

As mentioned above, the limitation of the WM (or STM) is 7 ± 2 chunks. This is our greatest vulnerability as processors of information. In real life we often run into situations where the processing demand is greater than 9 chunks. If so, we quickly find a way of coping with the information overload by using a rule of thumb, or a heuristic. This makes it possible to simplify a decision and thereby put less of a burden on the working memory. Let us assume that you are trying to multiply 147×52 in your head. You will have to be really quick because the half time of the information is about 7 sec. This means that you may forget the intermediate multiplication products more quickly than you can generate them. So we simplify the problem by using a heuristic. Let's see, 147×52 is almost the same as 150×50 which equals 7,500. This answer is often good enough, unless there is a need for an accurate response.

Or let's assume you are reading a sentence in a book and there are some strange words you do not understand. Skip the words and you may understand the meaning of the sentence anyway.

Recently there has been much research on heuristics and biases. This deals with how heuristics save us from overloading the brain (Gilovich, Griffin, and Kahneman, 2002). Slovic and Lichtenstein (1988) noted that cognitive limitations force decision makers to construct simplified models of their problems. Gigerenzer and Selten (1999) highlighted that heuristics enable fast, frugal, and accurate decisions, and that, contrary to conventional wisdom, the use of heuristics need not be disadvantageous. Simple heuristics can exploit regularities in the environment. They are usually task specific; unlike operations research models, they are not general-purpose tools. Below we give a brief overview of some well-known heuristics.

SALIENCE BIAS (PAYNE, 1980)

The operator pays the most attention to salient information, such as bright signs, loud noise, large lettering, top of the page, and so forth. People in advertising understand this very well: large signs work much better than small signs. In the yellow pages we find a surprising number of companies whose names start with A, such as AA Automotive Repairs.

AS IF HEURISTIC (JOHNSON, 1973)

The operator acts as if all information were equally valuable. Some information is obviously more important for diagnosis and decision making, and other information is less important. However, we treat all information as if it were of equal value. Medical doctors, for example, base their diagnosis of diseases on several different symptoms. One doctor may be looking for six symptoms, and if he finds all six, there is little doubt. But let's say that the doctor found only four symptoms out of six. What should the doctor do? It depends on contextual information, such as the type of disease that is being diagnosed. A tropical disease would be reasonable in Singapore but not in Sweden. Often, however, a Swedish doctor (as well as anybody else), will ignore probability information—and may decide for the tropical disease—as if all information were equally important.

IGNORING ARITHMETIC CALCULATIONS

We have already made reference to this above. In multiplying 147×52 we take a short cut and multiply 150×50. This is almost the same, and maybe good enough for the problem that we are trying to solve. Thereby the working memory is not exhausted.

AVAILABILITY HEURISTIC (TVERSKY, 1974)

High probability events are favored over low probability events. For example, you may take your car to the garage for repair because it refuses to start. The mechanic has just repaired two cars with similar symptoms and they both involved a replacement of the timing belt. Since these were recent repairs, he will start off with the hypothesis that the timing belt may be the cause of the car failure.

CONFIRMATION BIAS (EINHORN, 1978)

This is somewhat similar to the availability heuristic. Let's assume that you are hiring new employees for your company. The first impression is very important. One person comes in and immediately makes a bad impression. You will then be biased to keep searching for confirming negative information throughout the interview and deemphasize the positive information. This strategy makes the task much easier. One clear answer is generated without any ifs or buts.

REVERSE CAUSAL REASONING (EDDY, 1982)

A implies B also means that B implies A. Increased temperature leads to increased pressure in a pressure cooker —true. Increased pressure leads to increased temperature—not true; but this is the type of mistake we may make in reasoning about the process.

OVERCONFIDENCE IN DIAGNOSIS (KLEINMUNTZ, 1990)

This applies to most judgments that we make in daily life. Ask a football fan of Manchester United what he thinks is the probability that Manchester will beat

Arsenal. Although statistics of recent football matches show a probability of 0.6, he may say 0.9.

OVERESTIMATION OF SMALL NUMBERS AND UNDERESTIMATION OF LARGE NUMBERS

It is difficult to judge small probabilities. A driver may expect that the probability of a police speed control is 0.02 (2%). In reality, speed controls are much rarer, but since we think of them as being significant events, the number is exaggerated.

SUMMARY

To summarize, we use heuristics all the time. In most cases, we don't even reflect on how a decision was made, and we are not aware of the particular heuristics we adopt to deal with events in real life. The use of heuristics increases our ability to deal with decisions in real time and to arrive at reasonable solutions to everyday problems. Sometimes there may be unwanted effects and very poor decisions, as illustrated in the case study below of the Three Mile Island accident.

EXAMPLE: CONFIRMATION BIAS AT THREE MILE ISLAND NUCLEAR POWER PLANT

At the time of the nuclear power plant accident at Three Mile Island in the U.S., operators focused on a display that indicated that a relief valve had closed. This turned out to be wrong information: the relief valve was in fact still open. They then searched for confirming evidence that the water level was too high, although it was actually too low. In so doing, their attention was diverted from many contradictory factors, and the hazardous situation kept building up until it became a disaster (Wickens and Hollands, 2001). There are three reasons for the confirmation bias:

1. People have cognitive difficulties in dealing with negative information.
2. To change a hypothesis requires effort.
3. The final formulation of the decision becomes a self-fulfilling prophecy. We start off with a cautious attitude and keep building up the confirming evidence.

The main challenge is to make decision makers consider disconfirming evidence. This is a very difficult problem to solve in a nuclear power plant. Perhaps it would be possible to emphasize disconfirming information by using different displays. Such displays could express the semantics in the situation, summarize various decision making alternatives, and keep the alternatives alive until the final decision is made.

5.7 FROM FORMAL DECISION MAKING TO NATURALISTIC DECISION MAKING

In the discussion above, we have seen that it is difficult to formulate a proper or complete model for information processing. In this section we will describe a couple of extreme cases: the classical school of decision making and the naturalistic approach to decision making. In classical decision making, much emphasis is placed on finding an optimal solution based on stable goals, values, and environmental factors. However, the classical view does not explain the cognitive processes underlying decision making. The naturalistic decision-making paradigm attempts to study how decisions are made by people under real-world conditions, including aspects that are often missed under laboratory conditions.

CLASSICAL SCHOOL OF DECISION MAKING

A reasonable decision maker is supposed to choose an alternative that maximizes the expected value $\Sigma \rho_i v_i$, where ρ_i and v_i are the probability and value of different decision alternatives, summarized over i consequences of the given alternative (Gigerenzer and Selten, 1999). The expected value of a decision is offered by the normative school as the gold standard for good decision making (Wickens and Hollands, 2000). This means that the optimum decision would consistently produce the maximum value if repeated many times.

In recent years, there has been a departure from formal decision making to a more opportunistic approach. There is a realization that alternatives are difficult to formulate, and an optimal solution may not exist. It is difficult to fully diagnose the entire situation, assign probabilities and values, and consider all possible outcomes. Simon (1962) addressed the importance of sub-optimizing, or making a "satisficing" choice. Decisions can never be perfect; rather we must search for a decision that is good enough.

CHARACTERISTICS OF NATURALISTIC DECISION SETTINGS

In everyday situations, decisions are embedded in larger tasks. Decision research in the laboratory tends to lose the greater perspective of a meaningful context. In natural settings, making a decision is usually not an end in itself, but a means to achieving a broader goal. Orasanu and Connolly (1993) identified eight factors that characterize decision making in naturalistic settings:

Ill-Structured Problems

When a task is ill-structured, the decision maker may not fully understand the problem, yet she needs to generate an appropriate response. There may be no definite procedure to follow, and there can be several equally good ways of solving the problem. Engineering design is an example of an ill-structured problem.

Uncertain Dynamic Environments

The environment of the task may change rapidly. This results in an unstable situation where the conditions for decision making also change. Assume that you are sailing five miles from shore. Dark clouds are emerging; maybe you are sailing into a storm. Should you lower the sails and go engine or raise larger sails, so that you can quickly get to the shore? Well, let's see how the situation unfolds.

Shifting, Ill-Defined, or Competing Goals

The decision maker may be guided by many different goals, some of which may be unclear or in conflict with other goals. Thus the decision maker has to trade off different goals. In buying a car one may consider conflicting criteria such as sportiness, second-hand value, need for repair, inexpensive maintenance, and price. These are conflicting criteria, and trade-off decisions must be made.

Exploring Alternative Actions

In a naturalistic setting, a decision maker may take a long time assessing the situation, and the final decision is composed of several smaller decisions that take place over time. In confirming decisions, constraints are introduced now and then: "OK, now that we decided to get a sport utility vehicle, can we agree that we don't like black?" The decision maker will engage in a series of events to handle the problem or simply to find out more about the situation. A car buyer usually takes several months to come up with the final decision.

Stress

Decision makers who are under high levels of personal stress are often tired and unmotivated. Under these conditions, they may use incomplete strategies to arrive at a decision.

High Stakes

In real situations, the outcomes of a decision often involve real stakes that matter to the decision maker, who will therefore take an active role in ensuring a good outcome.

Multiple Players

In a team, individuals assume different roles. Problems arise when they do not share the same goals.

Organizational Goals and Norms

The organization may impose general goals and standard operating procedures which individuals have to comply with. These may be in conflict with the personal goals of the employee.

The phenomena observed in complex natural environments differ substantially from the research that is produced in the university laboratory setting. Below we will give examples of models developed in naturalistic settings.

5.8 RASMUSSEN'S MODEL

Jens Rasmussen (1983; 1986) was one of the first to formulate a model of naturalistic decision making. He distinguished between Skill-Based, Rule-Based, and Knowledge-Based decision making and task performance (see Figure 5.4).

A person may enter a situation that is very familiar. In manufacturing, for example, an operator on the assembly line picks different parts from bins, assembles them, and puts the finished product on a conveyer belt. He could perform the task in his sleep, as it were. It is automatic and there are usually not many decisions. This is referred to as Skill-based decisions behavior. There is no need for formal decisions and the execution of the task is automatic.

The next level is called Rule-based decisions. This is no longer automatic. In this case there are several well-understood rules for decision making.

- If situation A, then I do X.
- If situation B, then I do Y.
- If situation C, then I do Z.

For example, let's assume that you are driving on a winding and icy road, and you are steering the car along a curve to the right.

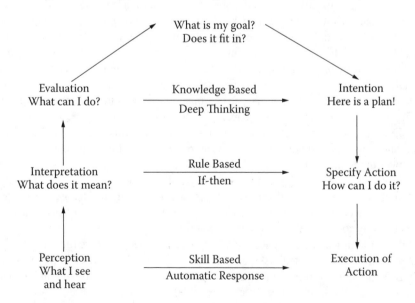

FIGURE 5.4 Rasmussen's model of skill-based, rule-based, and knowledge-based behavior.

- Situation A: The car is stable. Turn the steering wheel to the right.
- Situation B: The car is skidding on the road. First steer to the left to counteract the skidding. Then steer to the right.

Rule-based decisions are usually very effective, since they are quick and they can deal with a variety of conditions.

Knowledge-based decisions are typical for complex environments and for unfamiliar tasks. In this case the operator may first have to think about the purpose and the goal of the task (the box at the top of Figure 5.4). This may require deep thinking and analysis, and the outcome is not obvious. Consider, for example, an oil refinery. This is a complex environment, where there are typically 2000 sensors for measurement of temperature, pressure, and flow. These sensors are located throughout the plant, and they are often coupled, in the sense that if there is a problem at one location, it will also affect sensors upstream and downstream. The operator in the control room is sitting in front of a large computer screen, where he can monitor the situation. In an alarm situation there may be 200 simultaneous alarms and it is very difficult to identify the root cause.

The operator will first need to understand what is happening. He will observe information and data on the screen and try to locate the fault. He may be able to solve the problems immediately through rule-based decision making: if the alarms for Sensors A and B are activated, then do X; if the alarms for Sensors B, C, and D are activated then do Y; and so forth.

Sometimes the operator may understand the consequences of the fault, but may not be able to decide what he should do about it. There may be a problem of understanding company goals and priorities. For example, management may have decided that operators can only make decisions on minor problems, but must call supervisors if there are major problems. But if there is a catastrophic event, the operator has to shut down the plan immediately. These rules may seem clear, but in real life it is difficult to understand what is a major problem and what is a catastrophic problem. There are also the conflicting goals of productivity and safety, and a major dilemma is to understand how far one can manipulate the settings of the oil refinery to improve productivity without jeopardizing the safety of the operation.

Even if the operator eventually resolves the goals and the priorities of the operation, he may still not understand what to do about it. Let's say that he is trying to lower the temperature. What are the criteria for lower temperature and the associated values for flow and pressure? Are there policy implications for the company?

If these issues are finally resolved, the operator needs to change the control settings. In a complex plant there are usually several ways. It may be necessary for the company to implement a policy for control settings.

With training and experience, many knowledge-based tasks are turned into rule-based tasks and rule-based tasks into skill-based tasks. Take for example a fighter pilot. With increased training the knowledge-based routines are so well understood, that the pilot can develop a set of effective rules. In fact, the main purpose of pilot training is to develop automatic responses and reflexes to difficult scenarios. Time is of the essence, and any hesitation to consider policies and goals will introduce time delays, which may be disastrous for the pilot.

One main distinction between Rasmussen's model and earlier decision making models is that Rasmussen *was interested in classifying tasks*; he was not so interested in investigating human cognition. If a task is knowledge-based, it may be problematic; the question then is how the task can be redesigned. This is an engineering approach. To classify tasks we need to understand about the task's requirements and the operator's skill level, but detailed understanding of the cognitive processes is not necessary.

EXERCISE

A Swedish colleague claimed that jobs should be designed so that they have a large knowledge-based content; thereby the operator would have greater job satisfaction. To propose knowledge-based decisions the operator must think really hard. This is a creative process, and people who have an opportunity to be creative have greater job satisfaction. Important decisions must be left to operators, while repetitive actions should be automated and performed by machines.

A counter argument is that an operator may just be stressed out. There is probably more satisfaction from a fluid action, or a good work flow.

Discuss the pros and cons. Can knowledge-based tasks create more satisfaction than rule-based tasks? Give examples of different occupations and different tasks. Compare the tasks of (1) a writer writing a novel, (2) a pianist performing a concert, (3) a fighter pilot in a dog fight, (4) a politician giving a speech, and so forth. Divide these into skill-based and rule-based tasks, and explain which tasks you think may be satisfying to the operator.

5.9 NORMAN'S GULFS OF EXECUTION AND EVALUATION

Rasmussen's concepts had a great influence on research in cognition and decision making. Norman (1988) published a model called the Gulfs of Execution and Evaluation (see Figure 5.5). This model has been used extensively in human–computer interaction.

According to this model the user of an interface must be able to do several different things: formulate her goal; formulate her intention; specify her action; execute the action; perceive the system state; interpret the system state; and evaluate the outcome. In order to make this happen, a designer of an information system needs to consider carefully how she can design the interface so that:

- The system state and the action alternatives are visible
- There is a good conceptual model with a consistent system image
- The interface has useful mappings that reveal relationships between stages
- The user receives continuous feedback on his actions

The model is actually fairly similar to Rasmussen's. One difference is that Norman's model starts off in the lower right corner, whereas Rasmussen's starts in the lower

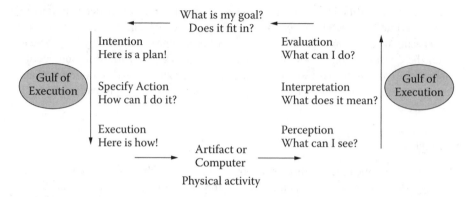

FIGURE 5.5 The gulfs of evaluation and execution (Norman, 1988).

left. More importantly, the implications of the model are interpreted quite differently. Norman was clearly more focused on systems design, whereas Rasmussen only presented a conceptual framework, which was later interpreted in many different contexts, leading to several design proposals, including Norman's.

5.10 RECOGNITION-PRIMED DECISION MAKING

The model of Recognition-Primed Decision making (RPD) was developed by Klein (1989) and Klein et al. (1993). It was originally based on observations of fire fighters in their natural environment. Klein found that in 80% of firefighters' decisions, they recognized the situation and could use standard actions in fighting the fire. The more experienced a firefighter was, the greater the repertoire of actions he could employ. RPD also has similarities to Rasmussen's rule-based behavior: if situation A, then do X; if situation B, then do Y; and so forth.

In RPD there are three different decision situations.

1. The first situation is as described above, where operators recognize a situation and act like they have acted before.
2. Sometimes the situation may be a little different from past situations, and the operator will then go though a mental simulation of what could happen if he decided to act according to the familiar pattern. Only then may he decide to accept the common routine action.
3. An operator simulates the action again, but decides that a routine action is no longer appropriate.

One important distinction of RPD is that decision makers do not go through all possible alternatives before they make a decision. They use reliable heuristics (in this case for fire fighting) and in most cases they are correct. The model has therefore features of both Simon's satisficing criteria (1969) and Rasmussen's decision ladder in Figure 5.4.

MICRO-COGNITION AND MACRO-COGNITION

In a development of RPD, Klein (1994) made a distinction between micro-cognition and macro-cognition. Micro-cognition is typical for cognition in an experimental situation in a psychology lab: puzzle solving, searching a problem space, selective attention, choosing between options, and estimating uncertainty values.

Many studies have been performed on puzzle solving (also called crypt-arithmetic problems) to understand what kinds of strategies people use to arrive quickly at a result. One example of such a problem is this:

$$\text{DONALD} + \text{GERALD} = \text{ROBERT}; \text{ solve for } D = 2$$

From these studies we can learn quite a bit about strategies for problem solving and cognition, and we can learn how to set up quantitative predictive models of cognition (such as goals, operators, methods, and selection [GOMS]; see Chapter 7), but the results are difficult to apply to real-world problem solving. This is not to say that quantitative models are useless. GOMS has been used for design of certain consumer products, such as keyboards and calculators (Card, Moran, and Newell, 1980).

Macro-cognition is typical of real life situations, such as planning and replanning, problem detection, building courses of action, attention management, recognizing situations, and managing uncertainty. These types of problems are difficult to model quantitatively, and the analysis is qualitative.

5.11 SITUATION AWARENESS

In order to function well, people must have an accurate picture of the situation they are in and how it is evolving. For example, a car driver must be aware of the surrounding traffic. A driver with situation awareness will foresee that some children playing next to the street are about to run out into the street to catch a ball.

A model of situation awareness was developed by Endsley (1995). It has been used extensively for analysis and design of systems. She provided the following definition: "Situation awareness is the perception of the elements in an environment within a volume of time and space, the comprehension of their meaning, and the projection of their status in the near future."

The main concept is a human information model, with three steps as follows:

1. Perception of elements in the current situation
2. Comprehension of the current situation
3. Projection of future status

Based on these the person makes a decision and performs an action (see Figure 5.6). Prediction is essential in all human–machine interaction and HCI tasks; if a user understands what is coming up, then it is easier to plan future action. Fewer errors will be committed and performance times will decrease.

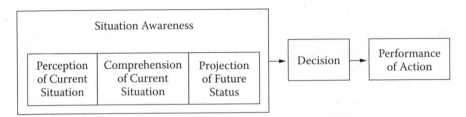

FIGURE 5.6 The main features of situation awareness. Adapted from Endsley (1995).

The information related to situation awareness is different from a regular information processing model; it is much more specific and related to task performance. While most of the information is accumulated in STM, an expert may also draw from information in LTM and quickly download patterns for information processing and action. This will apply to chess players, to car drivers, to weather forecasters, and to air traffic controllers. Some people have good situation awareness because they are experts and they have learned to master a situation.

Situation awareness is difficult to measure. One cannot just ask people if they have situation awareness; they wouldn't know if there are aspects that they are unaware of.

A fighter pilot who is aware of and can prioritize among the various threats at an early stage may have an advantage over enemy aircraft. One main challenge is to design displays that can help pilots in predicting future actions. Assume that we have designed two alternative displays. The displays can be tested in a flight simulator. Typically a pilot will fly a scenario and at a particular time the simulation will be stopped and the pilot will be asked questions about the display content, and more importantly, what his next actions would have been. The display that can best support predictions of future actions is preferred.

5.12   THE SIGNAL DETECTION THEORY PARADIGM

Sometimes it is difficult to hear and see because the signal is weak and there is uncertainty. Imagine that you are walking down Main Street and at 50 m distance you recognize your friend and start waving your arm. At 20 m distance you realize that you made a mistake. These types of events can be modeled using signal detection theory (SDT). According to STD, there are two possible states of the world: either there is a signal or there is no signal. In both cases, the operator may respond "Yes" (there is a signal) or "No" (there is no signal) (see Figure 5.7). The classic example in the military involves operators watching radar displays or listening to sonar beeps. The radar signal is disguised by noise, and the operator cannot always be certain if the blip on the screen is an airplane, some other object, or merely noise. Much research has gone into this area with the purpose of designing displays, training operators, and ensuring that the operator is alert and not missing signals.

OPERATOR RESPONSE	Signal	Noise
Yes - There is a signal	Hit p (Yes/S)	False Alarm p (Yes/N)
No - There is no signal	Miss p (No/S)	Correct Rejection p (No/N)

FIGURE 5.7 The signal detection theory paradigm.

Watching the screen can be a very boring task. Nothing may happen for hours, causing the operator to become drowsy. Maintaining vigilance—that is, the ability to sustain attention over a long period of time—can be quite challenging.

According to SDT, there are two situations: signal or noise. Typically the noise is superimposed on the signal like the noise blips on a noisy radar screen. The operator may respond in two ways: "Yes" (there is a signal) or "No" (there is no signal). If there is a signal and the operator says yes, the situation is referred to as a hit. This may be expressed as a probability of saying yes, given signal plus noise, or p(Yes/S). False alarm is a situation where there is no target but the operator says "Yes." If there is no signal and the operator says "No," we have a correct rejection. Finally the operator may miss the signal, thereby resulting in a miss.

There are many applications of signal detection theory in real life.

- A radiologist examines an x-ray to determine if a tumor is malignant (signal) or benign (noise).
- A nuclear power plant supervisor decides whether the present alarms mean that there is a malfunction in the power plant (signal) or that the situation is still normal (noise).
- A polygraph expert determines whether the data from a polygraph indicates that a person has lied (signal) or told the truth (noise).
- Aircraft maintenance personnel trying to find cracks in the aircraft frame. The cracks are difficult to see. In order to detect faults, they may use an eddy current device, and they have to decide on the presence (signal) or absence (noise).
- Airport security inspectors looking at an image of the contents of a carry-on bag, and deciding whether there is a concealed weapon (signal) or other objects (noise).

The signal detection theory may be explained using probability density functions. In Figure 5.7, there are two probability distributions, one for noise and one for signal. The horizontal axis measures the evidence variable X, and the vertical axis the frequency. The signal and noise distributions are overlapping, and this makes it difficult to judge what is signal and what is noise. One may decide that above a certain value of X everything will be called a signal and below the value of X everything is noise. The limit value of X is called criterion (beta). As we can see in Figure 5.8, the noise level is sometimes prominent and will be interpreted as signal, while sometimes the signal level is weak and will be interpreted as noise.

The location of β depends on the scenario for decision making. How should a judge assess a suspected criminal? If there are uncertainties he will most likely move

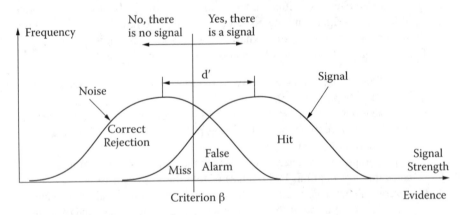

FIGURE 5.8 Signal–noise detection theory. There are two distribution curves of noise intensity and of signal intensity. The horizontal axis is the signal strength, and the operator will decide on a cut-off point at a location β. Values to the right represent a signal, and values to the left represent noise. The sensitivity d' measures the distance between the two curves.

the criterion β to the right in Figure 5.8, so that the number of false alarms is minimized. We don't want to put innocent people behind bars. The number of misses will correspondingly increase, meaning that there will be more crooks on the streets. A high value of β is referred to as a conservative (the judge will be conservative in his judgment).

A similar application of signal detection theory is eyewitness testimony. Here the observer decides whether a person who has been caught by the police is the same person the observer saw at the crime scene.

Assume another situation of a surgeon deciding to operate on a patient. In this case we would like to move the criterion to the left—a so-called risky β. Thereby we have minimized the number of misses. Unfortunately the doctor will also end up performing procedures on healthy people.

In the quality control section of a factory there are similar decisions. There may be very stringent criteria on quality, such as six sigma (three quality errors per million), which requires applying a very conservative β value. Unfortunately, we may then also end up rejecting many good products. We should therefore be guided by cost criteria—that is, the total cost of the faulty items and the cost of discarding good items. Achieving six sigma may not be a reasonable goal!

In the best of worlds we would like to avoid trade-off decisions. If the sensitivity— also called d' (d prime)—can be increased, we will have solved the problems. Sensitivity or d' measures the distance between the mean values of the two distributions. Ideally, if we could separate the distributions totally there would be no false alarms and no misses.

In some cases a medical doctor will know that the diagnosis is 100% accurate, and therefore there will be no false alarms or misses. Sometimes, however, medical evidence is very uncertain. Rhea et al. (1979) reported that radiologists missed about 20–40% of tissue abnormalities, which would lead to a tumor diagnosis. Parasuraman (1985) reported that radiologists did not change the data when screening patients as compared to when investigating referrals. Since referral patients have already been screened, one would think that the criterion β would be set lower resulting in fewer misses.

The sensitivity (d') or the separation between the two curves is affected by operator experience. An inexperienced radiologist will have a lower value for d' than an experienced radiologist, which means that the two curves will come closer together. An experienced person has learned what to look for, and as a result d' is greater, and the rates of both misses and false alarms are lowered.

Sensitivity can also be used to evaluate equipment used to diagnose patients. Swets et al. (1979) found that computerized tomography (CT) has greater sensitivity than a radio nuclide scan apparatus.

5.13 VIGILANCE AND SUSTAINED ATTENTION

As mentioned above, many research studies have been performed with people looking at displays or listening to rare signals. The job of military radar operator is

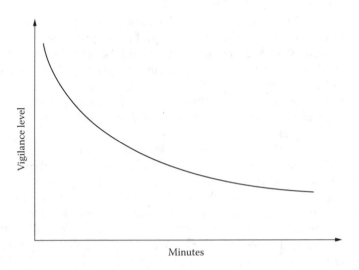

FIGURE 5.9

typical of tasks that involve watching over long periods of time to detect intermittent and unpredictable signals, which may happen once every half hour or so. It is a very monotonous task and operators have a tendency to fall asleep on the job when it gets too boring (reference). In addition, the job involves total isolation from cowork-ers, which makes it even less stimulating. Many studies have been conducted to analyze vigilance problems (see Figure 5.9).

Figure 5.9 shows that at the beginning of a task the vigilance level remains adequate for some time, but starts dropping after 20 to 30 minutes. The loss in performance over time is called the vigilance decrement, whereas the steady state level of vigilance performance is called vigilance level. The problem with vigilance increases if the signal has low strength; if there is spatial or temporal uncertainty, meaning that one cannot predict where or when the signal will occur; and if only a few events happen in the background.

EXAMPLE

At one time I worked for Human Factors Research, Inc., in Santa Barbara, U.S. This company was famous for vigilance research, and the project described here was one of many. It had been observed that security guards at some military installations in the U.S. had a tendency to fall asleep at work. They were sitting all day long looking at video images from security cameras that overlooked the fence and the entrance to the building. Nothing ever happened, so it was a boring job. The objective of the study was twofold:

1. To demonstrate that people could actually break in unseen by the guards
2. To suggest solutions to solve the vigilance problems

As part of the study, my colleague Selz was given the task of breaking into this secure environment without being observed by the security guards. He crawled along the fence, found an opening, and eventually managed to get inside the building. At some time the security guards looked at the display and noticed something moving—Ah! A groundhog! (This became a new nickname for Selz.) The break-in was a dangerous undertaking; he could have been caught and shot at by the security guards. By successfully breaking in, Selz proved a very important point: vigilance decrements constitute a very serious problem. The military organization in charge of these highly classified installations had to come up with new ways of solving the vigilance problems.

To reduce vigilance decrements, one can do many things:

- Show examples of targets on the screen and thereby increase mental availability.
- Increase target salience—for example, the size of the target.
- Remove social isolation.
- Add irrelevant tasks to increase the physiological activation level, such as playing games.
- Provide trial testing with feedback on hits, correct rejections, false alarms, and misses.
- Schedule work and rest periods so as to reduce fatigue. For example, there can be a rotating job assignment so that operators spend at the most one hour at a time in front of the screen.
- Present the signal in two modes or channels at the same time, such as visual and auditory. This is common in sonar operation, where operators listen to a "ping" and also see the target on a screen.
- Drink coffee and take other stimulants.

EXERCISE

Redesign the pull-down menu Format in MS Word (see Table 5.1).

One common design principle for pull-down menus is to put the most commonly used information on top of the menu, thereby reducing the probability of having to look further down. In Table 5.1 I estimated my own probabilities p of using all the functions. Replace these numbers with your own estimated probabilities. Calculate the surprisal and the total information, $p \log_2 1/p$, in the menu. With these calculations as a basis, discuss how you could redesign the pull-down menu.

TABLE 5.1
Estimated Probabilities of Using Certain Commands in the Pull-Down Menu

Function	p	$\log_2 1/p$	$p \log_2 1/p$
Font	0.20	—	—
Bullets/numbers	0.02	—	—
Alignment	0.01	—	—
Line spacing	0.10	—	—
Change case	0.35	—	—
Replace font	0.20	—	—
Slide design	0.05	—	—
Slide layout	0.05	—	—
Background	0.01	—	—
Object	0.01	—	—

RECOMMENDED READING

Wickens, C.D. and Hollands, J.G., 2000, *Engineering Psychology and Human Performance*, 3rd ed., New York: Prentice Hall.

6 Design of Controls, Displays, and Symbols

6.1 INTRODUCTION

Much research on control and display design has been sponsored by the U.S. Department of Defense. One purpose was to develop design principles that could be used in standardizing the design of military equipment. Military Standard 1472F is a standard that prescribes HFE design (U.S. Department of Defense, 2002). Today, much research is undertaken by companies to support the design of consumer products and computer systems. The aim is to improve the design of controls, displays, and symbols. In this chapter we present the design principles that may be applied to the design of appliances, cars, equipments, and tools.

We first present principles for the selection of controls, and then examine those that apply to computer input devices. Several design principles are presented, including coding of controls, control movement stereotypes, and the control–display relationship. We then present principles for the design of symbols and labels that are used in controls and displays.

In a manufacturing plant, operators handle a variety of objects, including controls, handtools, and parts to be assembled. The design principles derived for controls may be applied to most things that an operator uses; in fact, they apply to anything that an operator touches with his hands, such as parts used for assembly work. The coding of controls principle can also be extended to the coding of parts in manufacturing.

EXAMPLE

While traveling on the X2000 high-speed train between Linköping and Stockholm in Sweden, I wanted to wash my hands. I went to the toilet and discovered that there were several control problems.

First, it was difficult to lock the door (see Figure 6.1). I tried to turn the crank but it did not move. I bent down to look; it turned out that there was a second crank. A small sign on the door said, "Don't use the top crank—this was for the conductor." Imagine the number of complaints before the management put up the sign! But it was so small that it was difficult to read.

I later looked up X2000 on the Web and found comments from a passenger in Australia who traveled from Sydney to Melbourne. He claimed, "The toilet has a few problems: There are two locks on the door, one is labeled in Swedish and one in English. The English one works."

The second problem was that I could not find the water tap. There was a blank piece of metal next to my foot—a foot control! I stepped on it several times, but no water! From where I was standing I could not see any other controls, so I bent down and found the tap hidden on top of the water basin. Interestingly enough the control was electronic and touch sensitive, thereby violating all my expectations; poor affordances in design! It is surprising that the design engineers of X2000 could not get the toilet–user interface right. Compared to the engineering innovations in designing the high-speed train, this would seem a very trivial issue. But then we forget: few people have the skills to think of human factors design.

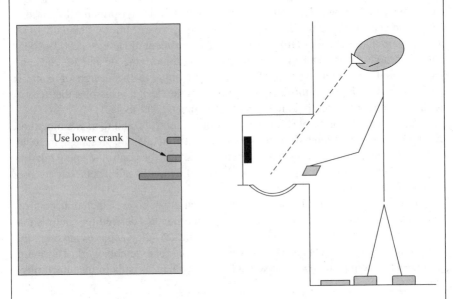

Use lower crank

FIGURE 6.1 Illustration of the control problem in the X2000 train.

6.2 APPROPRIATENESS OF MANUAL CONTROLS

Manual controls should be selected so that they are appropriate to the task and intuitive to use. Some controls can make a task easy to perform, whereas others make a task difficult. One way of analyzing control requirements is shown in Table 6.1, where controls are classified by the number of settings and by the force required to manipulate the control. For example, if a control does not require much force and there are only two discrete settings, the recommended types are toggle switch, pushbutton, or key lock. If there are several control settings, a rotary selector would be a good choice.

If a large actuation force is necessary, one should select a control where it is easy to apply force. Finger-actuated controls will not do. Hand pushbuttons, foot pedals, levers, or cranks could be used.

Note that the controls in Table 6.1 are all mechanical and the rules were published over 30 years ago. Presently many controls are programmed and they have become

---

TABLE 6.1
The Choice of Control Depends on the Force and the Number of Control Settings

Forces and Settings	Type of Control
Small Actuation Force	
2 discrete settings	Key lock, pushbutton, toggle switch
3 discrete settings	Rotary selector, toggle switch
4–24 discrete settings	Rotary selector switch
Small range of continuous settings	Knob, joystick lever
Large range of continuous settings	Crank, rotary knob
Large Actuation Force	
2 discrete settings	Hand pushbutton, foot pedal
3–24 discrete settings	Detent lever, rotary selector switch
Small range of continuous settings	Handwheel, joystick lever
Large range of continuous settings	Crank, hand wheel

Adapted from Chapanis and Kinkade (1972).

---

inexpensive. The microwave oven in Figure 5.1 is another example. Unfortunately this gives the designer endless options. And it is easy for the designer to go wrong; the toilet on the X2000 train is an example.

6.3 STANDARDIZATION OF CONTROLS

Over the years many controls has become standardized. In driving a car we are so accustomed to steering wheels and foot pedals, that it would be difficult to imagine any other arrangements. Examples of standardized controls include:

- Steering wheel for steering
- Joystick for airplanes
- Foot pedals for braking and acceleration
- Manual lever for aircraft throttle
- Lever control for gear shift

A user would be confused and annoyed to find other types of controls. But many of these industry standards were created at the beginning of the century, and were put in place without any human factors investigation of what type of controls would be best.

Today we understand how to evaluate controls and can conduct extensive research. One example was the design of pushbutton telephones, which replaced

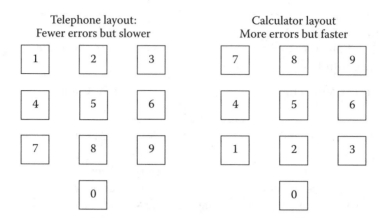

FIGURE 6.2 Two alternative layouts of a telephone keyboard.

rotary dial telephones. Bell Laboratories of AT&T investigated nine different alternatives for the layout of pushbuttons. Figure 6.2 shows two of the alternatives (Conrad and Hull, 1968). The telephone layout was chosen over the well-established calculator layout, although the latter was already a de facto standard in offices. The main reason was that experiments showed that users make fewer errors in dialing with the telephone layout, and since the dialing of wrong telephone numbers is costly, the telephone layout was selected, despite the fact it was a little slower than the calculator layout. As a result we now have two types of numeric keyboard in an office. It may be confusing, but considering the trade-off, it may still be the best overall design.

With the emergence of computers, many control functions have become less visible (like the water tap in the train), and as a result they are more abstract. What is difficult to see is also difficult to think of. Systems that used to be controlled manually are controlled remotely by a computer. For example, processing plants used to be controlled manually by opening and closing valves. Today computers are used. The control action as well as the system response can now be represented graphically on a computer screen. In developing such a system there are two problems: the selection of an input device and the design of the graphical representation of the control process. The first problem is the easier of the two, and is described below. The second problem takes detailed understanding of the process and the operator's task.

6.4 SELECTION OF COMPUTER INPUT DEVICES

There has been much research on the design of input devices, such as the mouse, the track-ball, the joystick, the touch screen, the light pen, and the graphics tablet. Some of the advantages and disadvantages of these devices are summarized in Table 6.2 (Greenstein, 1997).

From Table 6.2 it is obvious that different input devices have different advantages and disadvantages. For touch screens and light pens, one has to point with a finger

TABLE 6.2
Advantages and Disadvantages of Standard Pointing Devices

	Touch Screen	Light Pen	Touch Tablet with Stylus or Puck	Graphics Tablet	Mouse	Track-Ball	Joystick
Hand–eye coordination	+	+	0	0	0	0	0
Training requirements	+	0	0	0	0	0	0
Ability to attend to display	+	0	+	0	0	+	+
Unobstructed view of display	–	–	+	+	+	+	+
Freedom from parallax problems	–	–	+	+	+	+	+
Flexibility of placement in workplace	–	–	+	0	0	+	+
Comfort in extended use	–	–	0	0	0	0	0
Capability to emulate other devices	0	0	+	+	0	0	0
Suitability for							
Rapid pointing	+	+	0	0	+	0	–
Accurate pointing	–	–	0	+	+	+	0
Pointing with confirmation	–	0	0	+	+	0	0
Drawing	–	0	–	+	0	–	–
Tracing	–	–	–	+	–	–	–
Continuous tracking, slow targets	0	0	+	+	+	+	0
Continuous tracking, fast targets	–	–	0	0	0	0	–
Alphanumeric data entry	0	–	–	0	–	0	–

+ = Advantage; 0 = Neutral; – = Disadvantage.

or a stylus; this provides excellent hand–eye coordination. Pointing is a very direct way of expressing preference. A child will point at something and say, "I want that." This is such basic behavior that training is not necessary. Touch screens and light pens are therefore the most direct devices (Whitefield, 1986). There are, however, disadvantages. The pointing finger will partly obscure the view of the display, and the input resolution is poor. For the touch screen the resolution is the width of the finger, and for the light pen it is the width of the pen.

Touch screens are particularly appropriate for use in public environments, such as information displays at train stations and ticket vending machines at airports. This type of device has no moving parts, and it is sturdy and robust.

The mouse, the track-ball, and the joystick have the best input resolution; therefore they are the best for accurate pointing. This is because the input resolution can be programmed by changing the gear ratio between the device movement and the cursor movement.

For some devices there is flexibility with respect to placement on the work table. Track-balls and joysticks are excellent in this respect, because they are small and easy to move. One disadvantage with the mouse and the graphics tablet is that they occupy prime table space to the right of the keyboard, in the space where users like to write. As shown in Figure 6.3, it is also possible to place the mouse on top of the keyboard, thereby minimizing the use of prime work space.

Operators of touch screens and light pens sit with an extended arm, which could induce muscle fatigue.

Graphics tablets are primarily used for drawing. But they can also be programmed with special functions or subroutines. For example, a manufacturing process can be depicted on the screen as well as on the tablet with iconic representations and flow lines. A graphics tablet can thus serve as a command input device, and the display can provide feedback on the input. The integrated control– display arrangements can emulate the entire manufacturing process. This would be a functional and naturalistic setup, since there is a direct correspondence between the tablet input, the display feedback, and actual events in the work environment.

There are several input tasks, including pointing, pointing with confirmation (double-click), drawing, and tracking (see Table 6.2). The main advantage of a mouse is in pointing with confirmation; track-balls are less appropriate for this. Joysticks

FIGURE 6.3 Mouse over keyboard.

are superior for military tracking tasks (Parrish et al., 1982). Touch screens and light pens are primarily good for pointing, because they are very intuitive.

The input devices described above are the most common. Many new inventions are continually advertised in computer magazines. Before one can make any judgment about these devices, it is necessary to test them. This would imply experiments with human test subjects to perform tasks such as pointing, drawing, and tracking. The best device would be the one that requires the least time and produces the least number of errors in performing the task.

6.5 CONTROL MOVEMENT STEREOTYPES

People have expectations about what to do with controls. In the U.S., a light switch is moved upwards to turn on the light. For a person raised in Europe, it is the opposite expectation—the switch is turned down. The point is that control movement stereotypes are trained expectations, and many have been learned since childhood. Some of the most common stereotypes are shown in Table 6.3. For example, to turn something on there is the expectation of a control movement up (in the U.S.), to the right, forward, or clockwise (Van Cott and Kinkade, 1972).

To raise an element vertically, such as an overhead crane in a manufacturing plant, we would expect to move a control vertically upwards, as can be done with a control that extends horizontally. This is a clear stereotype, since there is a one-to-one correspondence between the movements of the control and the controlled element. For a vertical control the best stereotype would be to pull the lever back,

TABLE 6.3
Control Movement Stereotypes: Common Expectations for Control Activation

Controlled Element	Human Control Action
On	Up, right, forward, clockwise
Off	Down, left, backward
Right	Clockwise, right
Raise	Up, back
Lower	Down, forward
Retract	Up, backward, pull
Extend	Down, forward, push
Increase	Forward, up, right, clockwise
Decrease	Backward, down, left, counterclockwise
Open valve	Counterclockwise
Close valve	Clockwise

just as a joystick in an airplane is pulled back to raise the airplane. But this is a less clear stereotype for an overhead crane. The control movement is horizontal, but the controlled element moves vertically. Many individuals would make a mistake by pushing the control forwards (unless they have been thoroughly trained). Thus our first option for a crane control would be a horizontal control lever.

The stereotype for opening a valve is always to turn the control counterclockwise (unscrew), and to turn the control clockwise to close the valve. It would seem that manufacturers of bathroom taps could learn from this principle. There is now a proliferation of different designs, and unaccustomed users (e.g., hotel guests) cannot understand how to operate the controls. Many accidents happen in bathrooms, especially with old people, who slip and fall in the shower because they are startled by the water temperature. Standardization of tap design would be very helpful.

EXAMPLE: CONTROLS FOR AN OVERHEAD CRANE IN MANUFACTURING

Controls for overhead cranes in manufacturing often violate control movement stereotypes. I was recently involved as an ergonomics expert in a legal case. In this case a worker had been injured while trying to catch the hook of an overhead crane when the hook was lowered. He was not successful in catching the hook; he lost his balance and fell about 3 m to the floor, severely injuring his hip. The hoist was being lowered by a fellow worker using a control box which was strapped to his stomach (Figure 6.4). On the control box there were four identical lever controls with a neutral detent position. The main hoist and the auxiliary hoist conformed to common stereotypes (push to lower). However, movement of the hoist along the bridge and the trolley was controlled by moving the levers in the same direction, which is confusing. Ideally, these controls should have been operated by a joystick which could move the hoist in the x and y directions simultaneously.

This was not the only problem. There was a second set of controls in a crane cab which was located under the ceiling of the manufacturing facility. The four controls in the cab were obviously confusing, since someone had pasted labels next to the control levers to indicate how they were supposed to be moved. For three of the controls (the auxiliary hoist, the main hoist, and the trolley) there was actually compatibility with the direction of movement. But the bridge control was incompatible; a forward movement of the lever made the bridge move to the right and a backward movement made the bridge move to the left (Figure 6.4).

A third problem concerned the fact that the layout and the ordering of the controls in the crane cab differed from the hand-held control box. An operator who was familiar with the control arrangement in the cab would have problems using the control box, and vice versa; there would be an increased likelihood of errors in activating the control. In human factors terms, this outcome is referred to as negative transfer of training (Wickens and Hollands, 2002; Patrick, 1992). Operators are likely to revert to the type of behavior they learned first, especially if they are under stress. One can therefore expect many more errors in an emergency situation, and one error is likely to lead to another. It is therefore very important to use control movement stereotypes, and to analyze the compatibility between control movements and the controlled element.

There are also advantages for productivity. A good control arrangement will always save time in performing a task. A poor control arrangement will always take longer time even after several years of practice (Fitts and Seeger, 1953).

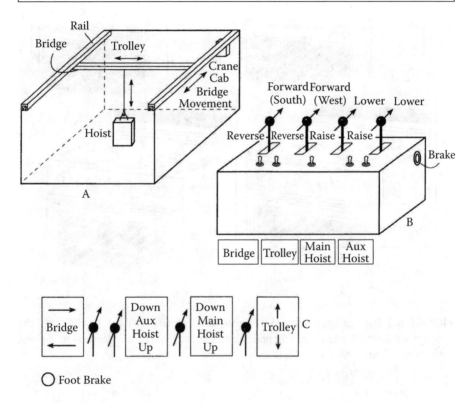

FIGURE 6.4 (A) The layout of the manufacturing area; note the crane cab in the top right-hand corner. (B) The control box, carried with a strap on the stomach. (C) Layout of the control arrangement in the cab; note the labels that operators taped next to the controls.

6.6 CONTROL–RESPONSE COMPATIBILITY

Chapanis and Lindenbaum (1959) performed a classic study on control–response compatibility. They studied the preferred locations of controls for burners on stove tops. This study was later complemented by Ray and Ray (1979), and their results are represented in Figure 6.5. In the figure the controls are numbered 1, 2, 3, and 4 from left to right. There are, however, four alternative layouts of the burners. The question is: Which one is the best? To answer this question, Ray and Ray (1979) used 28 female test subjects. There were 560 trials on each stove, and the preferences (P) and errors (E) were calculated as percentages for each stove.

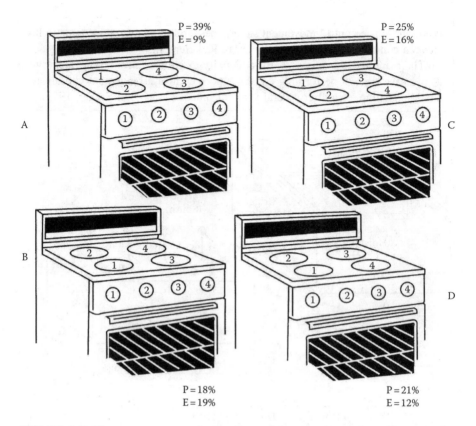

FIGURE 6.5 The arrangement of controls and burners is used to illustrate the concept of control–response compatibility. Preference (P) and errors in control activation (E) are given as percentages.

The problem with the design of the stove top is that there is no clear control–response compatibility. Ideally, there should be a one-to-one relationship between the controls and the responses (burners). It would be easy to redesign the stovetop so that there is control–response compatibility. For example, the rear burners can be offset slightly to the side. The controls can then be lined up one-to-one with the burners, and the association is immediate. Chapanis and Lindenbaum (1959) proved that there was not a single error in control actuation with this arrangement. For a less compatible arrangement such as in Figure 6.5, the user must look a few times to decide. Instead of a simple reaction time there is a double or triple reaction time, which takes 3–4 s rather than 1 s.

The stovetop represents a familiar and common problem in control–response compatibility. In manufacturing, it would be expected that workstations should be designed with similar considerations. In designing a workstation for manual assembly one must line up part bins so that they are compatible with the assembly process.

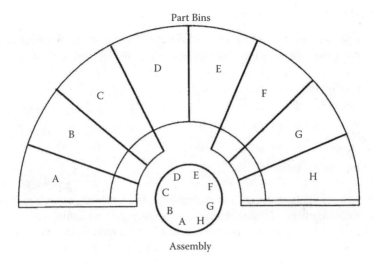

FIGURE 6.6 Bin–assembly compatibility for the assembly of car brakes, as used at General Motors, Saginaw Division, Buffalo, NY, U.S.

For example, in the assembly of car brakes (Figure 6.6) there is a one-to-one correspondence between the location of parts in the bins and the corresponding location of parts in the assembly. Such bin–assembly compatibility reduces assembly time (Helander and Waris, 1993). Similar principles apply to the design of controls and displays for process control—the controls and displays must be lined up to be compatible.

EXAMPLE: POOR CONTROL–RESPONSE COMPATIBILITY

A few years ago I inspected the control design of a fighter aircraft. It had two major controls: the left hand was on the throttle, which had 21 control functions, and the right hand on the stick, which had 18 control functions. This is a design concept called HOTAS—hands on throttle and stick. The pilot keeps his hands on the two controls all the time, and all functions are accessible from the two controls. This is not an easy task to learn, and there are probably better design options.

In one aircraft I inspected, the direction of movement of the cursor control for selection of objects on the display screen was incompatible with the movement of the cursor on the screen. If you moved the control to the left, the screen cursor moved up. Move the control down and the screen cursor moves to the left, and so forth. Despite pointing it out to the military personnel, the attitude was that training will take care of the problem. It may be so, but at what cost?

6.7 CODING OF CONTROLS

Controls can be coded by adding features to them. This makes them easier to distinguish. There are six common types of control coding (Sanders and McCormick, 1993):

1. Location
2. Color
3. Size
4. Shape
5. Labeling
6. Mode of operation

These principles apply to controls in automobiles and airplanes, as well as industrial and office environments. In manufacturing assembly, control coding can be applied to hand tools, parts bins, and parts—practically to anything that is purposefully touched or handled by the operator. We first explain the different types of coding and then give some examples.

CODING BY LOCATION

Coding by location is the most powerful principle. For example, in automobiles the locations of many controls have been standardized and drivers have clear expectations of where to find certain functions. These expectations build up with increasing experience in driving. Most car drivers can immediately find the location of the ignition even when blindfolded; the location has been well coded (McGrath, 1976). Likewise, in the operating theatre, the location of surgical tools is standardized, so that the surgeon can minimize pick-up time as well as human error.

CODING BY COLOR

In color coding, items are colored differently depending upon the function and the task. One potential problem with color coding is that it only works in a well-illuminated environment. Color coding of controls on underground mining machines, for example, does not make sense as it is very dark in mines.

Color coding requires a longer reaction time than location coding, since it is first necessary to reflect on the meaning of the color before the task can be performed. This typically involves a double reaction time.

Some control colors have stereotypical meaning. It has become common to make emergency controls red. For example, in industrial standards an emergency stop control for a machine or a conveyor line must be red. However, different countries may have different color stereotypes. Courtney (1986) surveyed a large sample of Hong Kong Chinese to determine the strength of associations between nine concepts and eight colors and then compared these data with a similar study of Americans (Bergum and Bergum, 1981). Results of the two studies are compared in Table 6.4.

There were some substantial differences between the two populations. While cold is associated with blue among Americans, the preferred color among the Chinese is white. For the concepts of hot, danger, and stop, red was the dominant color

TABLE 6.4
Concepts and Most Frequently Associated
Color for Two Populations: Hong Kong
Chinese and Americans

Concept	Chinese Color	%	Americans Color	%
Safe	Green	62.2	Green	61.4
Cold	White	71.5	Blue	96.1
Caution	Yellow	44.8	Yellow	81.1
Go	Green	44.7	Green	99.2
On	Green	22.3	Red	50.4
Off	Black	53.5	Blue	31.5
Hot	Red	31.1	Red	94.5
Danger	Red	64.7	Red	89.8
Stop	Red	48.5	Red	100.0

among both populations. However, the percentage values for Americans were much greater than for the Chinese. Courtney pointed out that the reason for the lower percentage values among Chinese is that for them red is the symbol for happiness. The strength of the happiness association detracts from the safety association.

A series of cross-cultural comparisons showed that the color preference for different process control symbols differs among the Asian population (Liang et al., 2004). Color coding on machines must then be designed to fit the target user population.

CODING BY SIZE

To distinguish easily between different controls, size can be one coding option. A fighter pilot in combat is highly stressed, but can distinguish between three different sizes of control knobs: small, medium, and large (Chapanis and Kinkade, 1972). Size coding can of course be used in other environments as well.

CODING BY SHAPE

Controls can be coded by shape (Figure 6.7). In this case an operator can distinguish up to 12 different shape-coded control knobs under conditions of stress (Woodson and Conover, 1964). These controls have been standardized for aircraft design. The best control design is when the control shape resembles the control function. In Figure 6.7 the flap control resembles the flap and the landing gear resembles a wheel so that the association is immediate. But for more abstract functions, such as the shape codes for carburetor air and revolutions per minute (rpm), the mental association between control function and control will take a much longer time to establish.

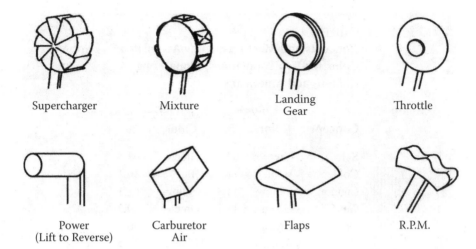

| Supercharger | Mixture | Landing Gear | Throttle |

| Power (Lift to Reverse) | Carburetor Air | Flaps | R.P.M. |

FIGURE 6.7 Shape-coded controls for airplanes.

EXAMPLE

Some years back I worked on a project sponsored by U.S. Bureau of Mines to standardize controls for the roof bolters that are used in underground coal mines. In one mine we found the roof bolter in Figure 6.8. The figure shows Brian to the left and the roof bolter with 15 hydraulic controls. The controls used to be identical. However, the operator had welded on extensions and added on shape coding to the controls. This was obviously a very difficult machine to handle and the shape coding simplified the operation.

Shape coding of controls can also be used in industrial situations. In fact, sometimes operators add their own shape coding. During investigations following the Three Mile Island nuclear accident, many intriguing principles of shape coding used by power plant operators were discovered. At one plant, operators had coded identical control levers with beer bottles.

CODING BY LABELING

A label may be used to describe a control. The label can be put above, underneath, or on top of the control. The location of the label does not really matter as long as it is clearly visible and the wording reads from left to right (Chapanis and Kinkade, 1972). Vertical labels take longer to read and should not be used. One problem with labels is that they might not survive in a harsh industrial environment. In particular, printed characters may be soiled or destroyed. Embossed labels are therefore often used (Loewenthal and Riley, 1980). As with color coding and shape coding, the use of label coding implies a double reaction time; the label has to be read and understood before action can be taken.

FIGURE 6.8 The shape-coding of the controls of this roof bolter had been welded on by the mine workers. This simplified the operation of the roof bolter.

CODING BY MODE OF OPERATION

Controls can also be coded by the mode of operation. This implies that each controls has different feel or that each control has a unique method of operation. A car driver can distinguish between the accelerators and brake because they have different control resistance, dampening, and viscosity. The same principles may be used for controls in industrial settings. The operator can verify that the correct control has been activated and also interrupt control activation if there is an obvious error.

EXAMPLE: There are often usability problems with mode control. Many consumer products have mode control. I have at least two mode controls on my fax machine: one is labeled "Mode" and the other "Function." These mysteriously combine with number sequences dialed on one of the two keyboards, and a myriad of combinations of control input. To enter the year, month, day, and time, there are 22 control inputs. The problem with a mode control is that not only do I forget what mode I am in, but if I press the wrong button I end up in the wrong mode. How to reset the functions and start again? I found out that if I call the fax machine, it resets itself!

There is nothing intuitive about this machine, so I am forced to use the 110 page manual which lists about 200 different settings. All I want to do is to fax a letter! Why has this become so incredibly complex? This machine and many others with mode controls suffer from *functionitis*, a common disease that affects

---

systems designers. Here is how the designers think: "Let's program every function we can think of. It does not cost anything extra for the buyer." So here I am with 199 functions I never asked for.

When I installed the machine I must have pressed the wrong button, because Russian command words showed up in the ten-character display. There were about ten different command words, none of which could be understood. So I wrote them down and identified their meaning in a Russian–English dictionary that I found on the Web. I could then translate the commands and switch the language to English. This little exercise took me about 2 hours—wasted time indeed.

CODING OF OTHER ITEMS TOUCHED BY THE HAND; HAND TOOLS AND PARTS IN MANUFACTURING

Coding principles can be applied to any items that are touched or held; it could be parts to be assembled as well as hand tools. Hand tools and parts can be coded by location, color, and labeling. For example, color coding can be used as a scheme for organizing a workstation, by applying the same color to parts bins and hand tools that belong together. Color coding of parts is nothing new. It has long been used in electronics for marking resistors, transistors, and capacitors, and this simplifies electronic assembly. In fact, these parts are also shape-coded, so that it would be difficult to confuse a transistor with a capacitor. Handtools are often colored with different colors. This makes it easier to find them.

6.8 EMERGENCY CONTROLS

The design and location of emergency controls requires particular attention, since it is crucial to be able to find them quickly (Atherton, 1986). Emergency situations are stressful, and operators are likely to make mistakes. Emergency controls must therefore be particularly well designed to allow fast action without any errors. Some design recommendations are summarized in Table 6.5.

TABLE 6.5
Recommendations for the Design of Emergency Controls

Position emergency controls away from other frequently-used controls, thereby lessening the risk of inadvertent activation

Make emergency controls easy to reach; put them in a location that is natural for the worker to reach

Make emergency controls large and easy to activate; e.g., use a large rather than a small pushbutton

Color emergency controls red

Many types of emergency control are used. Some industrial machinery have "dead man's" switches. As long as this type of switch is actively pressed, the machinery keeps going. If the pressure is released, the machinery stops. Some types of industrial machinery have an automatic switch or function in case the worker inadvertently comes into the danger area. For example, rotating tire-building machines have emergency trip cords located above the operator's feet. If the operator is pulled into the machine the feet will catch the trip cord and the machine will stop.

EXAMPLE: ACCIDENTAL ACTIVATION OF SEAT EJECTION CONTROL IN AIRPLANES

Emergency controls must be placed away from other frequently-used controls in a location that is easy to distinguish. This principle has been of great concern in the design of airplane cockpits. It used to be the case that fighter pilots were killed by accidentally ejecting themselves when the airplane was still on the ground. The pilot would be catapulted 100 m and fall flat to the ground before the parachute had time to open. It also happened that pilots would eject themselves into the ceilings of hangars. The eject control button has now been relocated to a safe place (under the seat between the legs). This location has an additional benefit, since the pilot's arms are kept out of the way during the ejection.

6.9 ORGANIZATION OF ITEMS AT A WORKSTATION

All items in a workstation that require handling need to be organized efficiently. This includes controls, hand tools, parts to be assembled, and part bins. Workers sometimes take the initiative to organize a workstation. But one cannot rely on this since workers do not understand HFE—and neither do their supervisors. It is better if there is a deliberate design effort involving both the engineers and workers collaborating together in arranging a workstation. This would result in an optimal solution for all users.

Predetermined time-and-motion studies (PTMS), such as methods time measurement (MTM), MOST, and WORK FACTOR, have primarily been used for predicting and quantifying the time it will take to assemble a product (Konz and Johnson, 2004). PTMS measurements can be used to divide a large task into several parts, thus balancing the work between different workers. However, PTMS could have a much wider usage. It could be used to evaluate the design of a product and alternatives for organizing a workstation. But this is rarely done, perhaps because there are so many different options for workstation design; and it is difficult to understand where to start. Some guiding principles are clearly needed, especially ergonomics-related principles, as described below.

KEEP THE NUMBER OF ITEMS TOUCHED BY THE HAND TO A MINIMUM

Minimize the number of hand tools, the number of different parts, and the number of controls. The number of parts and the number of necessary tools depend on the product to be manufactured. It is important for product designers to understand the implications of their designs in terms of manual labor. Why use five varieties of screw when two are enough? Why not combine parts such as incorporating washers with the screws?

ARRANGE THE ITEMS (CONTROLS, HAND TOOLS AND PARTS) SO THAT THE OPERATOR CAN ADJUST HIS OR HER POSTURE FREQUENTLY

Sometimes the location of items ties up workers in impossible work postures. There are many examples of industrial machinery which must be operated using a foot control. For example, in using an industrial punch press the operator must hold the work item with both hands and press the foot control to initiate the pressing action. Using just one foot causes one-sided strain that is likely to lead to back problems. It must be possible to move the foot control so that it can be operated with either foot at the worker's convenience.

CONSIDER PREFERENCES IN HAND MOVEMENTS AND HANDEDNESS

People can move their hands both faster and with much better precision in an arc than horizontally or vertically. Imagine that you are drawing a straight line on a piece of paper. It is difficult to get the line straight if it is drawn horizontally or vertically. It is easier to draw if the paper is turned at an angle so that the hand can move outwards from the body, such as in the movement envelopes shown in Figure 6.9. This is because there are only a few active joints in the arm; typically only the elbow joint moves. But for drawing a horizontal or a vertical line there are many more active joints and many muscles that have to interact, which makes the movement more complex.

Handedness is important in the design of hand tools, particularly those intended for tasks which require skill and dexterity. Assembly tasks do require skill and dexterity, and thus hand tools for left-handed individuals are needed.

ORGANIZING ITEMS IN THE WORKPLACE

Distinguish between Primary and Secondary Items

Primary items are those that are used most frequently, and secondary items are those that are not used as frequently. List all the parts and classify them as primary or secondary items.

Divide the Tasks into Subtasks, Each Forming a Logical Unit

For very short tasks this may not be important, but for more comprehensive tasks it is desirable.

Divide the Worktable into Several Areas, One for Each Subtask

This may be practical only for comprehensive tasks where there are many items to keep track of. Organizing the items for each subtask is practical and makes it easier for the operator to think of the task.

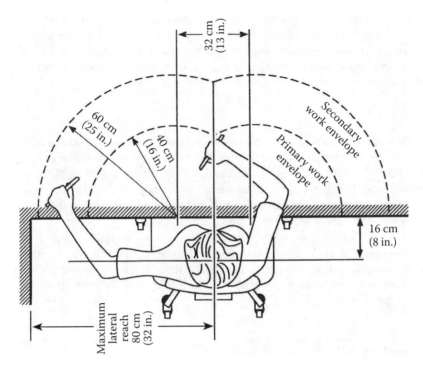

FIGURE 6.9 Arrangement of a workstation, showing primary and secondary movement envelopes.

Identify Primary and Secondary Movement Envelopes on the Worktable

The functional reach for a 5th percentile female worker is about 40 cm (16 in), and this determines the limit of the primary work envelope (Figure 6.9). Put primary items in the primary envelope. Secondary items should be put in the secondary envelope, but within a reaching distance of about 60 cm (24 in).

Place Items such as Bins and Tools in a Convenient Location So That They Can Be Used Sequentially for Each Subtask

A sequential order helps in organizing the task and facilitates task learning and productivity. A well-organized workstation will save time, resulting in better productivity.

6.10 DESIGN OF SYMBOLS AND LABELS

In this section we explain how symbols and labels can be designed so that are easy to understand, especially when employed in visual displays. The term visual displays refers to a wide variety of displays, from posters and signs to computer displays. These displays have one thing in common: they carry visual information which must be given a semantic interpretation, so that the reader understands what to do. We also discuss the design of safety warnings, and explain why many workers ignore safety signs.

SYMBOLS

Symbols are often used in industry to identify controls, machine functions, and states of processes. Symbols are also widely used as traffic signs and for public information at airports and train stations. The idea is that a picture can convey 1000 words, so a symbol can be more succinct than a label with many words. The other assumed advantage is that symbols do not have to be translated and can be understood by individuals throughout the world.

As exemplified below, many international machine manufacturers prefer to use symbols, since labels would have to be translated to the local language. But some symbols are difficult to understand, particularly those that relate to abstract machine functions that may be hard to visualize or recall. In such instances, it is better to use a label (Collins and Lerner, 1983).

Example: Standardization of Symbols

At one time I participated in a meeting organized by the Society of Automotive Engineers (SAE). The purpose was to standardize symbols for off-road vehicles such as construction vehicles and agricultural machines. In the first meeting, a proposal was made to standardize 130 different symbols, all of which had different meanings. As the only ergonomist in the group, I asked if there was any information about whether the symbols would be understood by the users. This was obviously the wrong question. The sole purpose of this group was to standardize symbols so that labels did not have to be translated to 100 different languages. My involvement seemed futile and I resigned immediately from the group. However, I brought with me several proposed symbols, some of which are depicted in Figure 6.10. These symbols were intended for use in construction vehicles. To evaluate them we asked 40 construction workers in the U.S. to translate the symbols into words and actions (Helander and Schurick, 1982).

Some of the symbols were easy to understand. The two arrows representing the up and down motion of a controlled element was an obvious symbol. The oil level, which was understood by 75%, is an example of composite symbols: a level and a drop of oil. The electric starter was understood by only 40%. But the worst was a caution warning which was understood by only 20%. My point was proven: symbols must be evaluated, and the SAE committee should have taken on this effort.

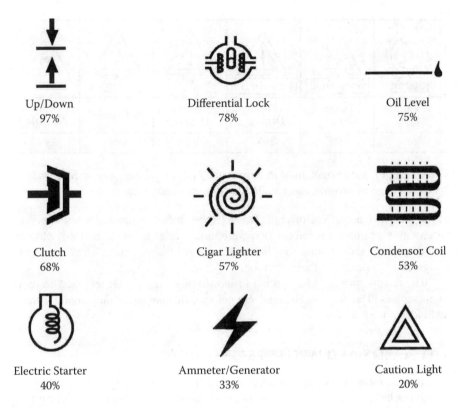

FIGURE 6.10 Symbols and the percentage of construction workers who understood their meaning.

Symbols may be difficult to understand for individuals in industrially developing countries, possibly due to lack of education or previous exposure to other symbols. Liang et al. (2004) found various symbols were given different interpretations by process control operators in Malaysia, Singapore, and China. The International Standards Organization (ISO) has suggested that international symbols must be tested in a minimum of five different countries and they must be understood by an average of 66% of users (Zwaga and Easterby, 1982).

A very large number of research studies have been conducted and they are nicely summarized by Lehto (1992) and Lehto and Miller (1986). These address traffic signs, public symbols, military symbols, and computer icons.

One early study of symbolic traffic signs in Sweden investigated the percentage of drivers who could recall a road sign 1 minute after passing it (Johansson and Backlund, 1970). Some of the results are displayed in Figure 6.11. The danger sign could be recalled by 26% of the drivers only, probably because it is too general and can refer to a variety of situations. The sign for frost damage is very common on Swedish rural roads. Most drivers have the experience that the damage is usually

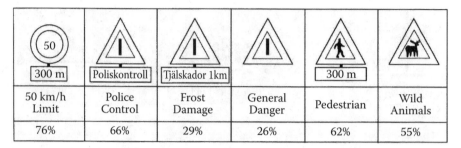

50 km/h Limit	Police Control	Frost Damage	General Danger	Pedestrian	Wild Animals
76%	66%	29%	26%	62%	55%

FIGURE 6.11 Road signs used in a field study on a rural highway in Sweden, and the percentage of drivers who recalled having seen each sign.

slight and will not affect driving. On the other hand, traffic signs with specific information or important information were much easier to recall; and it is obvious that they were read. In summary, this study showed that individuals will remember a sign that is relevant to their situation.

The results may also be applied to industrial situations. Generalized warning signs such as "Danger" or "Be Safe" are not specific enough, as they do not instruct individuals what to do.

6.11 LABELS AND WRITTEN SIGNS

The main constraint in designing labels is that the message must be short; otherwise it will not be read (Vora et al., 1994; Loewenthal and Riley, 1980). The difficulty is to find a short message that expresses the situation and employs good semantics.

Across the university campus of SUNY Buffalo in the U.S., there is a sign next to light switches carrying the following message:

Please turn out lights

when not required.

Most people probably read only the first line, because they understand the message. But some individuals will ignore the label because it is too long. A problem is that the message is confusing. Why would anybody turn off a light if it is not required? A better design would be the following:

Please turn off lights

Broadbent (1977) noted that statements can be expressed in an affirmative, a passive, or a negative manner, as illustrated below.

*   Affirmative: The large lever controls the depth of the cut.
*   Passive: The depth of the cut is controlled by the large lever.
*   Negative: The small lever does not control the depth of the cut.

Broadbent pointed out that an affirmative, active statement is easier to understand than a passive statement. Negative statements require a double reaction time, since the user must first understand what not to do, and then, only by inference, is the appropriate action clear. The same observation is also valid for traffic signs. Positive signs stating the preferred action are easier and quicker to understand than prohibitive signs.

6.12 WARNING SIGNS

There are several distinct stages in processing information on warning signs (Figure 6.12). An individual is first exposed to a warning sign. As a result there is an image of the sign on his or her retina, but the person has not yet paid attention. There are several factors that make people look at a sign (Hale and Glendon, 1987). The size of the sign is clearly important: the bigger the better. Location is also important. One should position a sign where people tend to look. Drivers, for example, usually try to look as far down the road as possible to maximize the sight distance. This means that when the road turns to the right they will look along the right-hand side of the road. This is therefore the optimal location for a road sign. Similarly, one can determine where employees in an industrial plant look and find optimum locations for warning signs.

Having processed the sign, we would expect that the semantics of the sign will make it possible for the worker to draw a clear conclusion. The individual must also agree with the conclusion. If there is no agreement he will not take any action. Following an agreement, the individual must select and execute one of many alternative responses.. This selection of response depends largely on experience; an experienced individual has a greater response repertoire (Hale and Glendon, 1987). We discuss some of these issues in more detail below.

INFORMATION PROCESSING OF WARNING SIGNS

Based on the model in Figure 6.12, we can now introduce several additional factors that are important in assessing the effectiveness of signs (Lehto, 1992).

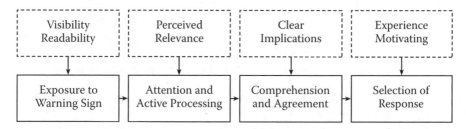

FIGURE 6.12 The four stages involved in the information processing of warning signs.

Information Overload

Information overload is particularly obvious during driving. It is a common experience when driving in an unfamiliar city that there may be too many traffic signs competing for one's attention (Lehto and Miller, 1986). Many drivers make mistakes since they do not have the time to attend to and read every sign.

Attention and Active Processing

Individuals will attend to a sign if they perceive that the sign is relevant. Unfortunately, many individuals regard warning signs as irrelevant. Because the hazard is not perceived, the warning sign is not read. This is also one of the basic problems in motivating workers to work safely: the hazard is simply not perceived, so there is no need to work differently. Zimolong (1985) provided some interesting data on construction workers. Painters who stand on ladders for much of their working time do not perceive ladders as being unsafe. However, according to accident statistics, painters have more injuries in falling from ladders than any other occupational group. Similarly, scaffold assemblers do not think of scaffolding as being unsafe, although 50% of all fatalities in construction work involve scaffolding.

It is possible that workers develop strong coping mechanisms that make them underestimate hazards, particularly if these hazards are inherent in their own job. It is then difficult to motivate, say, painters to consider a warning sign for ladders. The warning contradicts the worker's mental model of what is safe and unsafe. The same basic problem prevails in safety training. Participants feel the training is directed at others but not at themselves.

Another basic problem is that accidents happen fairly infrequently and, therefore, there are not enough current warning examples. This is more of a problem for young workers who may have never seen an accident. Older workers are more perceptive and motivated to comply with safety instructions (Dedobbeleer and Beland, 1989). Perhaps there are ways of enhancing the perceived relevance by exposing young workers to a "benign experience" (Purswell et al., 1987). One could make something happen that could reinforce the sense of hazard. Perhaps citations written by a company safety officer would do the trick. More research is required to investigate these matters.

Comprehension and Agreement

In comprehending the meanings of words, there is a trade-off between detailed description and the use of simple words. Simple words may not be illustrative enough, whereas detailed descriptions will not be read. Wogalter et al. (1985) suggested that there should be four fundamental elements in a warning sign:

1. Signal word: to convey the gravity of the risk; for example, "Danger", "Warning" or "Caution."
2. Hazard: the nature of the hazard.
3. Consequences: what is likely to happen if the warning is not followed?
4. Instructions: the appropriate behavior required to reduce the hazard.

An example of an effective minimalist warning would be:

Danger
High Voltage Wires
Can Kill
Stay Away

The main reason for a short warning is the limited capacity of the human short-term memory. Typically, the short-term memory can store about seven "chunks" or concepts (Miller, 1956; Simon, 1974). But short-term memory is constantly upgraded to include current items, and half of the memory is therefore replaced in 3–4 sec. Warning signs, as they are read by a driver in a car or by workers in industry, are considered only for a very short period of time. There is a quick decision to do or not to do something. The scenario is processed quickly in short-term memory, after which the situation is forgotten. Humans take short-cuts in information processing. These are necessary to sort out the important issues expediently and deal with scenarios at hand on a continuous basis. There is really no need to store such information permanently.

Selecting and Performing a Response

An individual may have fully comprehended a warning sign and may also be in full agreement, but may select to do something different because there is a cost of compliance. For example, the decision to press the stop button on an industrial robot is offset by the realization that it may take an hour to start up the process again (Helander, 1990). Employees often select not to be safe. Safety glasses, steel-toed boots, respirators, and other personal protective equipment are perceived of as being inconvenient and uncomfortable (Krohn et al., 1984; Hickling, 1985). The cost of compliance is too high, unless of course the company decides to change the cost equation by enforcing safety rules. Some manufacturing companies have acquired a reputation for doing this very efficiently.

Another issue is whether the action implied by a warning sign can be incorporated in the regular work task. For most drivers, safety behavior has become an integral part of driving. For example, drivers trace the movements of children and bicycles instinctively. Stopping has become an act that is totally integrated into the driving task. Would it be possible to incorporate safe behavior in a regular work task? Maybe not, unless the novice worker is coached extensively by a qualified trainer (supervisor or peer) on the shop floor, similar to driver training.

EXERCISE

International airports have a need to develop symbols for air travelers that every-body can understand. Develop symbols for the following functions: airport train, car rental, customs office, information, money exchange, restaurant, room for smokers, shower cabin, sleeping cabin, storage of luggage, and toilet for the disabled.

Describe the difficulties in developing symbols. What types of symbols are easy to propose and what types are difficult? Draw the symbols and ask a sample of potential air travelers what they mean. Calculate the percentage of correct responses. Which symbols would you select? Do you need to redesign any of the symbols?

RECOMMENDED READINGS

Wogalter, M.S., Laughery, K.R., and Young, S.L., 2002, *Human Factors Perspectives on Warnings*. Santa Monica, CA: Human Factors and Ergonomics Society.
Woodson, W.E., 1981, *Human Factors Design Handbook*, New York: McGraw-Hill. (This is a comprehensive collection of design guidelines with many illustrations.)

7 Design of Human–Computer Interaction

7.1 INTRODUCTION

Research in human–computer interaction (HCI) started in the 1960s. One of the pioneering studies was presented by Douglas Engelbart, who introduced the computer mouse at the Fall Joint Computer Conference in San Francisco in 1968 (Engelbart, English and Berman, 1967). The mouse was developed at Xerox Palo Alto Research Center (PARC). Several other important design innovations originated from PARC, including the graphical user interface and direct manipulation. These design principles were applied to the interface of Xerox Star computer, which was introduced in April 1981. The design effort in developing this particular computer was without parallel. Several hundred experiments were performed with human test subjects to validate design details (Card and Moran, 1986; Verplank, 1988). Computer users are still fortunate to rely on this inspirational design.

The Xerox Star was a commercial failure, but the design concepts were copied by Steve Jobs and Steve Wozniak, who launched the innovative Lisa in 1983 (U.S. $10,000) followed by the Apple Macintosh in 1984 (U.S. $2,500). The Microsoft Windows software package, which built on similar principles, was released in 1985.

Since then, computers have revolutionized the way we work, communicate, and participate in all forms of activities, including leisure. We have found new ways of collecting data, solving problems, and making decisions. In doing so, we have been forced to adopt new work methods, which fit the requirements of computers, but not necessarily the requirements of users. The adoption of computers has been a gradual process, and it may seem that we have accepted the evolution of personal computer without much reflection or criticism. This may not have happened were it not for the efforts at Xerox PARC to design usable interfaces.

Software usability has become an important part of software development. Today, Microsoft, Oracle and IBM and many other companies employ several hundred usability experts, and they produce some of the best software on the market. Sometimes, however, computer routines are actually not productive, and we would do better if we could retain the old work routines. Landauer (1995) pointed out that many tasks that we now use computers for take less time without computers.

Human–computer interaction is today a large scientific field supported by research in many areas including anthropology, cognitive psychology, cognitive science, ergonomics, experimental psychology, human factors, learning, linguistics, philosophy, and sociology (Helander, Landauer, and Prabhu, 1997).

Much research has also been published in various application areas, such as in manufacturing, aviation, military command and control, e-commerce, and computer games. Lately there have been new areas, such as HCI for PDAs and mobile phones. In these cases the screen is so small that the interaction is limited to the few options that can be displayed, which makes it more difficult to handle.

7.2 SOME USEFUL DESIGN PRINCIPLES IN HUMAN–COMPUTER INTERACTION

In this chapter we give a brief overview of some important design issues in HCI. Six design issues are listed below, but in mentioning these, we are only touching on the surface.

1. Provide an interface that does not violate the user's expectations or mental model.
2. Design a consistent interface. This will improve expectations and reduce user errors.
3. Reduce the memory requirements and memory load of the user.
4. Give feedback to the user, to inform users about what is going on.
5. Design the interface to cater to all types of users. Consider that there are great differences among individuals in memory and cognitive capabilities.
6. Use direct manipulation to simplify the way information may be accessed and manipulated by the user. This is one of many great ideas that were implemented at Xerox Star.

We will now describe these principles in greater detail.

DO NOT VIOLATE THE USER'S MENTAL MODEL

The human–computer dialogue must not violate the user's mental model, the way users think about the problem. It should not be necessary for the user to translate the information on the screen, so that it fits with his mental model. As an example, consider the situation where a company would like to write software for process control of a mineral flotation process that is used to enrich minerals. This software could then partially replace the operators who worked with this task.

The programmer who was assigned the book of writing the software realized that she was not an expert on mineral enrichment and flotation. She wanted to understand whether the operators have a mental model of the process, and if she should write the software so that it would agree with the mental model—and the procedures that were commonly used, or maybe there are many mental models, one for each operator? If so, she would like to find out if there are common features between the different models.

To find out, we can observe how operators make decisions about process control parameters—about temperature, when to add chemicals, and when to finish. A cognitive task analysis would be a good way to document the operators' expertise and what their mental models are.

What happens if all mental models are different? In this case the HCI designer can either select the best model, or try to formulate a new model that will not violate the existing mental models. This is a difficult task in HCI; it will require much

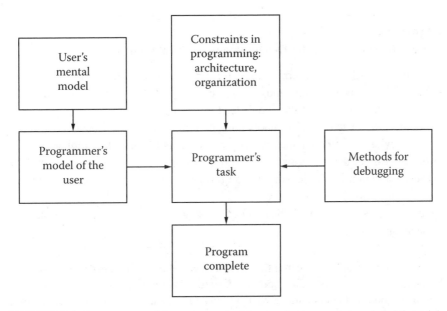

FIGURE 7.1 Programmers have many priorities, such as organizing and debugging the program. The diffuse picture of the user's needs may come second hand.

analysis and evaluation before one can propose a good design solution. But once this is accomplished, the software will have commercial value.

The programmer and the user have different goals, and therefore different mental models (see Figure 7.1). Often the programmer will impose her model in developing software. The programmer has a different point of departure, which is based on the structure of the program and the code, and the user needs may sometimes come second. The top priority is often to just debug the program. As a result users' needs are deemphasized and usability flaws may sneak into the program without being noticed. Table 7.1 lists some of the issues that should be taken into consideration in software design. In developing a user interface, the programmer and the usability expert must think of many issues (see Table 7.1).

CONSISTENCY IN USER INTERFACE DESIGN

Lack of consistency is a major problem in interface design. This forces the user to adopt several different methods, rather than one consistent method, to solve similar problems. For example, the information may be presented at different locations on adjacent screens, thereby forcing the operator to search for the information. The requirement for consistency is applicable for all levels of design: screen layout, command words, and organization of dialogue.

Much research has been devoted to the choice and abbreviation of command words (summarized in Barnard and Grudin, 1988; Paap and Cooke, 1997). The problem is how to generate good descriptors that users will find useful. There is a similar problem in the yellow pages: should one look for attorney, lawyer, legal, solicitor, barrister, or what?

TABLE 7.1

Issues to Consider in the Development of Usable Software

Use the user's language, terminology, format, and system presentations.

Words used in menus, commands, help information, and error messages must be chosen carefully, so that they don't lead the user astray.

The meaning of icons and symbols must be intuitive.

Input devices must be compatible with the task. As illustrated in Table 6.2, different input devices are good for different tasks.

The information on the screen must be formatted so that it is easy to find the relevant information on the screen.

Objects on the screen should be grouped in categories that support the operator's decision making.

Some researchers advocated the idea of user-defined command words—that is, let the users themselves define command words. This is not a good idea. Users are not very good at understanding what is best for them. Although they understand their task, they do not understand HCI. Several studies investigated the likelihood that any two people would generate the same command name; the probability of doing so ranged from 0.07 to 0.18 (Carroll, 1985).

One study showed that a single user employed 15 different principles for abbreviation of command words (Carroll, 1985). In other words, user-defined abbreviations would not work. Instead, the interface designer has to select a strategy for abbreviations. There are four main strategies:

1. *Phonetic strategy:* Use an abbreviation that sounds the same. For example, APND for APPEND.
2. *Contraction:* Keep the first and last character of the word. For example, EXTE instead of EXECUTE.
3. *Delete Vowels:* Remove vowels in the word. For example, RMV for REMOVE.
4. *Abbreviation:* Use a minimum number of characters from the beginning of the word, such as OP for OPERATE.

Although many studies have indicated that abbreviations are best, there is an even more important principle: consistency. Use only one way to abbreviate!

Consistency is also important for other design details, such as consistency in location of information on the screen. A study at IBM (Czerwinski, 2000) showed that if menu words had different locations from one screen to the next, the menu search time increased by 73%. Consistency in schematics and organization of dialogue can be investigated using a state transition diagram (see Figure 7.2).

The state transition diagram starts off in the upper left corner with a display of the current month, week, and day. From the month one can select another month,

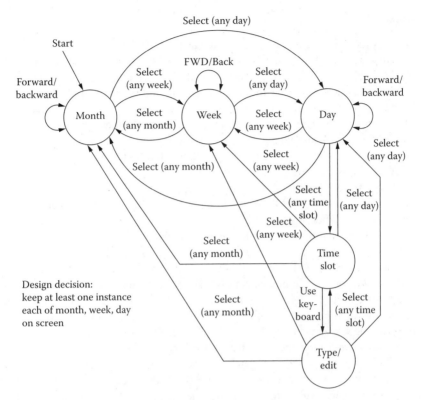

FIGURE 7.2 State transition diagram for an electronic calendar.

week, and day. From the day, one can select a time slot, and from the time slot one can do text entry. Using the state transition diagram, one can evaluate the loops and get an overall view of the organization of the program. The functionality of the calendar can easily be investigated, thereby identifying important access points and interactions that we may have forgotten.

Memory Requirements

The computer dialogue should be designed so as to minimize the memory require-ments, or the number of chunks in the user's short-term memory (STM). Miller's magic number 7 ± 2 applies here (see Chapter 5). For example, if there are more than eight different choices in a menu, some users will forget the first few alternatives and will have to read the menu alternatives all over again.

It is possible to format the information on the display in such a way that the operator will find it easy to group and chunk information. Symbols, icons, and labels which are used frequently can help to simplify chunking of information, thereby making it easy to form a mental model. The pioneering study by Tullis (1997), which was explained in Figure 2.3, proved that chunking information can be very cost effective; it reduced the performance time by 40%.

FEEDBACK

Give users feedback on everything, as often as it is practical. Users must understand where they are in the program, what has been done, and what is correct and incorrect. With feedback users can understand how well they are doing, and it is possible to correct errors. It is also easier to learn the system and develop a mental model.

One type of feedback is "intelligent help." Here the computer keeps track of the type of errors committed. The program will draw conclusions with respect to how advanced the user is and then adapt help messages and other feedback to the level of the expertise. Typically the expert will prefer short symbolic help messages, but the novice will require more complete information.

Intelligent help is difficult to design. An example is Clippy, the paperclip that travels on a motorcycle in MS Word. I have yet to find a computer user who likes Clippy's advice. Often, users are irritated by Clippy's uninvited appearance on the screen, because the offered help is not relevant to the user's task. It is indeed difficult to make inferences from a user's action in order to provide intelligent advice.

Sometimes in the ambition to help users the wrong feedback is given. A recent usability study on e-government portals showed that there were various feedback messages that were inappropriate or not meaningful to users (Khalid, 2004). Users could apply for a business license online. While waiting for their password to be confirmed, the screen went blank for about 30 s. Some users thought their passwords were incorrect and used the back button to retry. Only then a message appeared: "Processing in progress. Please wait." However, they were not informed how long they should wait. Another example is that, when users clicked on the "Proceed" button in the license form, they expected the cursor to change its shape from an arrow to a hand, but there was no such feedback.

INDIVIDUAL DIFFERENCES

People differ in their abilities to handle computers and software. A classic paper by Egan (1988) summarized research on differences between high-performing and low-performing individuals. In his studies he took great care to compensate for differences in training and experience. The groups he studied were homogenous, and consisted of student test subjects, professionals and experienced programmers. Egan selected three important applications:

1. Text editing—routine task
2. Information search—non-routine task
3. Programming—abstract reasoning

There were three dependent variables: time required to perform the task, number of attempts, and percent correct at first attempt.

For text editing the performance time for the slowest user compared to the fastest was 5:1; for the information search task, 9:1; and for the computer programming task, 22:1. Egan realized that the fastest and slowest users were not always representative of the population, so he removed the top and bottom 25 % of the individuals from the comparison. Then he compared the difference between the 75th percentile

TABLE 7.2
Results from Egan's Studies

Task	Max/Min	Q_3/Q_1
Text editing	5:1	2:1
Information search	9:1	2:1
Programming	22:1	2:1

Three tasks were compared: text editing, information search, and programming. The ratio of maximum to minimum performance and the ratio of the 75th to the 25th quartile performance are given.

and 25th percentile. The performance ratio for all three tasks was then about 2:1, or 100% increase in performance from 25th to 75th percentile user (see Table 7.2).

Egan went further to investigate exactly what cognitive factors could be responsible for the differences in HCI performance. Several cognitive skills were tested using standard testing procedures (Ekstrom et al., 1976). Typically, persons who performed well in text editing scored higher on spatial memory than those who performed poorly. The differences in information search and programming ability were explained by differences in spatial memory as well as logical reasoning ability.

Spatial memory can be tested by asking participants to memorize a map of a city. Participants first study the map for 5 minutes. Then the map is taken away, and they are asked questions such as: Where is the church located? Where is McDonald's? and so forth.

The significance of spatial memory is that it helps people to navigate. Users who can remember how to navigate to a function on the computer will definitely save time. But many of us are poor at memorizing the location of different functions. In MS Word, do you know how to sort words according to the alphabet? Look under the pull-down menu for Table.

Poor spatial memory can be a debilitating handicap. The question then emerges: What can we do as designers of HCI to help people with low spatial memory perform better on computers? One way would be to make all files and functions visible on the screen. Thereby there is no need for navigation. This was indeed the idea behind direct manipulation, the concept for the Xerox Star and the Apple Lisa. To avoid navigation problems it is possible to design a direct manipulation interface, so that users do not have to depend on their spatial memories.

Differences in cognitive performance are much greater than differences in manual performance. Typically among workers on a factory assembly line there is a maximal difference in manual assembly performance of about 25% (Konz and Johnson, 2004). The cognitive differences are greater than 100% and are much more consequential. People with low cognitive performance can, as we explained, be helped by a different style of interface design.

DIRECT MANIPULATION

Principles for direct manipulation were established by the time the Xerox Star was developed. Ziegler and Fähnrich (1988) documented three aspects of direct manipulation.

1. *Directness through visibility.* All information must be visible on the screen, and it should be possible to refer to information by pointing and clicking.
2. *Semantic directness.* Symbols such as icons (e.g., the shopping cart icon) should be easy to understand and intuitive to use. Designing a meaningful icon requires much work and validation tests. Some functions are so abstract that it is difficult to find symbols. It is then better to use words.
3. *Directness through fluid action.* It should be possible to execute a sequence of commands with the mouse without interrupting the movement. This means that there is a fluid action through click and drag.

Direct manipulation worked well on the Apple Lisa, which was released in 1983. Since that time the software for common applications has expanded tremendously; and there is no longer space to display all functionalities on the screen. Pull-down menus and similar interfaces are necessary. Between 1992 and 1995 the number of commands in Microsoft Word increased from 400 to 1100. Dr. Nathan Myhrvold, Microsoft chief technology officer, explained the situation as follows: "Software is a gas—it expands to fill the container it is in" (Gibbs, 1997). Myhrvold also explained the customer's dilemma: "Users are tremendously non-self-aware. Corporate customers often demand that Microsoft simultaneously add new features and stop adding new features" (Mann, 2002). They realize that fewer features will increase usability, but they cannot resist the added functionality—they are like kids with a cookie jar.

7.3 DEVELOPMENT OF USABLE DESIGN

Good design requires analysis and design iteration. "If Ernest Hemingway, Neil Simon, Pablo Picasso and Frank Lloyd Wright could not get it right the first time — what makes you think you will" (Heckel, 1984). "The most important tools the architect has are the eraser in the drawing room and the sledge hammer at the construction site" (Frank Lloyd Wright).

To avoid problems in system design, the interface must be planned at an early stage in development (Gould, 1988). All aspects of software design must be developed in parallel. Just like concurrent engineering is essential in product design, concurrent design must be used for software. This brings together many specialists including systems designers, programmers, usability experts, and marketing experts. Thereby common goals can be established at an early stage for the development work.

To understand users' mental models, cognitive task analyses are helpful. These can be complemented with information obtained from user interviews, observations, and participatory design. With this information one can understand users' needs as well as their knowledge. Consumers have finally become quite aware of usability issues and they are likely to raise concerns about complex designs. The reason why Apple iPod and Google have been so successful is because of their simplicity in design (*The Economist*, 2004). Google has a 27-word interface and iPod has one rotary control. *Consumer Reports* now devotes about 40% of its reviews to the usability and usefulness of a product.

Just as in product design, prototypes are very helpful in software development. The purpose of these is to gather information on usability, which can then be used as a basis for improvements and redesign.

During the initial phase paper prototypes can be used. Although a paper prototype may not be appropriate for user testing, usability experts can use it for a cognitive walkthrough, as explained below. During the later development phases software prototypes can be evaluated in user testing. The results can be expected to improve functionality as well as usability.

According to Karat (1997) there are significant economic benefits in improving usability. Usability is critical to product design. Products with poor usability will not sell; it may already be too late to improve usability, since the costs for post-release changes are not affordable for most companies. An early introduction is critical, so there is rarely a second chance. The loss of momentum and customer goodwill may even make it difficult to continue in the same market area.

Usability Engineering contributes to various stages of the product design cycle and benefits accrue throughout the product life cycle. But it is less expensive to implement usability in the beginning of the design effort. Gould (1997) proposed a checklist to ensure early usability testing of software (see Table 7.3).

TABLE 7.3
Checklist for Achieving Early User Testing

We have made informal, preliminary sketches of a few user scenarios—specifying exactly what the user and system messages will be—and showed them to a few prospective users.
We have begun writing the user manual, and it is guiding the development process.
We have used simulations to try out the functions and organization of the user interface.
We have used mockups.
We have invited people to comment on usability components.
We had prospective users think aloud as they used simulations and prototypes.
We used hallway and storefront methods.
We did formal prototype user testing.
We met our behavioral benchmarking.
We did field studies.
We did follow-up studies on people who use the system.

Adapted from Gould (1998).

7.4 USABILITY

It is essential to emphasize usability in software development. Indeed, usability has become so important that a new profession has been established: usability experts and usability engineers. Many belong to the Usability Professionals Association (UPA), a body of professionals who take a pragmatic view of their work by focusing on design rather than cognitive theories. UPA organizes regular conferences to keep abreast with developments in the field.

The International Standards Organization has published a famous standard, ISO 9241-11. This standard defines usability as follows (ISO, 1995):

Usability is the extent to which a product can be used by specified users to achieve specified goals with effectiveness, efficiency and satisfaction in a specified context of use.

Effectiveness refers to the accuracy and completeness with which users achieve specified goals. Efficiency pertains to the resources expended in relation to the accuracy and completeness with which users achieve goals. Satisfaction refers to comfort and acceptability of use.

Over time the definition has seen many modifications. Nielsen (1994) specified five characteristics of usability: ease of learning, efficiency of use, memorability (ease of remembering), error frequency, and subjective satisfaction. Jordan (1998) was more interested in the user's own experience of usability: guessability, learnability, experienced user performance (EUP), system potential (optimum performance), and re-usability.

Lately there has been a debate focused on "fun" in HCI. Things are fun, claims Carroll (2004), when they attract, capture, and hold our attention by provoking new or unusual perceptions, arousing emotions in context. They are fun when they surprise us, and present challenges or puzzles as we try to make sense of them.

Carroll (2004) proposed a redefinition of usability to incorporate fun and other significant aspects of user experience. According to Carroll this new concept should rely on an integrated analysis of the *user's experience*. This is likely to lead to greater technological progress, than merely itemizing a variety of complementary aspects of the user's experience. The user experience is unified; it is a Gestalt impression. Users do not decompose their impressions in terms of usability, aesthetics, and fun. Norman (2004), who previously emphasized usability, is now quick to add that usable products that are unattractive do not sell well. Emotions, aesthetics, and usability are indeed intertwined (Helander and Khalid, 2005).

7.5 METHODS FOR USABILITY EVALUATION

Usability evaluations are conducted to identify the problems and difficulties experienced in using software. There are two main categories of usability methods: analytical evaluations and user-based evaluations.

Analytical evaluations are also called usability inspection methods. In this case the product is evaluated by a professional analyst, such as a usability expert. There are three common methods:

- Heuristics evaluation
- Cognitive walkthrough
- Model-based analysis such as goals, operators, methods, and selection (GOMS)

User-based evaluations involve testing by users. There are several methods:

- Usability testing
- User participatory design
- Thinking aloud
- Questionnaires and interviews

These methods are discussed below.

Heuristics Evaluation

Heuristics evaluation refers to a set of rules or design goals that are used to evaluate an interface. These are usually formulated at a high level with abstract goals, and it is up to the designer to determine how each goal can be implemented in terms of interface design. For example, one high level goal is "Speak the user's language." The analyst will then interpret what this implies in terms of interface design. To implement this heuristic we must understand the following:

- What exactly is the users' language. How do users refer to various items?
- How do users think about their task?

Based on this knowledge the analyst will understand better what words to select for dialog design and for menus.

Shneiderman (1998) proposed eight "golden rules" for interface design, but he remarked that there is a need to validate and interpret these rules for different applications. The golden rules are shown in Table 7.5. Nielsen and Molich (1990) and Nielsen (1994) proposed ten usability heuristics. These have become widely used and are the best known set of usability heuristics (see Table 7.5). Many other researchers have also developed recommendations. Bruce "Tog" Tognazzini's list of first principles of interaction design is interesting reading (Tognazzini, 2004).

Cognitive Walkthrough

Cognitive walkthrough (CW) is a method for evaluating user interfaces by analyzing the mental processes that are required of users (Lewis and Wharton, 1997). It is performed by an analyst, and real users are therefore not used in this type of analysis. CW can be performed at any stage of the design cycle. At the early stages of design,

TABLE 7.4
Shneiderman's Eight Golden Rules

1. Strive for consistency
2. Enable frequent users to use shortcuts
3. Offer informative feedback
4. Design dialogues to yield closure
5. Offer error prevention and simple error handling
6. Permit easy reversal of actions
7. Support internal locus of control (user in control of the system)
8. Reduce short-term memory load

sketches, storyboards, and paper mock-ups can be used to visualize the design and the expected actions. At a later stage a rapid prototype generated by computer can be the basis for CW.

Helander and Skinnars (2000) performed a CW of the interface design for a Saab Aerospace JAS 39 fighter cockpit (see Figure 7.3). At the time of the analysis the JAS 39 had been flying for a couple of years, but there were continuous improvements of the interface.

To perform CW the analyst (Helander) worked jointly with the designer (Skinnars). We decided to analyze navigation routines. One of the problems in modern warfare is that the timing must be very accurate. To attack a target several aircrafts take different routes and meet at the target. They must then arrive at the target at exactly the same time. In order to delay or speed up an aircraft the pilot will program an alternative route to the target.

Depending upon his experience a pilot may select different ways to navigate. First he must calibrate the geographical location. An experienced pilot may check landmarks and click on them (on the map) when he passes. A less experienced pilot does not have the confidence to look out and identify landmarks; instead he will rely on the information on the electronic map. The question is whether the interface design can support both ways of navigating, so that in both cases the action required to coordinate the aircraft's location with the map is intuitive and quick to perform.

In general, after the task sequence has been established, the analyst will determine whether the interface can support the appropriate action. If a difficulty is found, the reason for the difficulty is identified in terms of interface design parameters.

There are five key features of cognitive walkthrough, which distinguish cognitive walkthrough from other methods.

1. CW is performed by an analyst and reflects the analyst's judgment. It is not based on data from test users (in our case, pilots).
2. CW examines specific user tasks, rather than analyzing the global aspects of the user interface.
3. The analyst investigates correct sequences of user actions, asking if the user can perform these sequences using the intended interface design.

TABLE 7.5
Nielsen's Ten Usability Heuristics (Nielsen, 1994a)

1. Visibility of system status
The system should always keep users informed about what is going on, through appropriate feedback within reasonable time.

2. Match between system and the real world
The system should speak the users' language, with words, phrases and concepts familiar to the user, rather than system-oriented terms. Follow real-world conventions, making information appear in a natural and logical order.

3. User control and freedom
Users often choose system functions by mistake and will need a clearly marked "emergency exit" to leave the unwanted state without having to go through an extended dialogue. Support undo and redo.

4. Consistency and standards
Users should not have to wonder whether different words, situations, or actions mean the same thing. Follow platform conventions.

5. Error prevention
Even better than good error messages is a careful design which prevents a problem from occurring in the first place.

6. Recognition rather than recall
Make objects, actions, and options visible. The user should not have to remember information from one part of the dialogue to another. Instructions for use of the system should be visible or easily retrievable whenever appropriate.

7. Flexibility and efficiency of use
Accelerators—unseen by the novice user—may often speed up the interaction for the expert user such that the system can cater to both inexperienced and experienced users. Allow users to tailor frequent actions.

8. Aesthetic and minimalist design
Dialogues should not contain information which is irrelevant or rarely needed. Every extra unit of information in a dialogue competes with the relevant units of information and diminishes their relative visibility.

9. Help users recognize, diagnose, and recover from errors
Error messages should be expressed in plain language (no codes), precisely indicate the problem, and constructively suggest a solution.

10. Help and documentation
Even though it is better if the system can be used without documentation, it may be necessary to provide help and documentation. Any such information should be easy to search, should focus on the user's task, should list concrete steps to be carried out, and should not be too large.

FIGURE 7.3 The interface of a Saab Aerospace Gripen 39 cockpit.

4. CW will also identify likely trouble spots in an interface and suggest a reason for the trouble. In our case we will ask what will make the pilot commit an error.
5. The analyst identifies problems by tracing the likely mental processes of the hypothetical user, not by focusing on the interface itself. Thus the analyst must understand the background knowledge of users and try to infer how to influence or change mental processes.

Lewis and Wharton (1997) discussed the necessary steps in conducting a CW as follows:

1. Preparation
    *   Define the user's knowledge while using the interface.
    *   Choose a sample task. The task should be a critical aspect of the system, and it should be a realistic representation of what users will do. Specify the correct action sequence(s) for the task.
2. Analysis
    *   For each action, construct a success story that explains why a user would choose that action. The analyst should again note the assumptions made about the user. In particular it is important to understand the user's knowledge and experience, and how this affects his decisions

and actions. In analyzing a success story, four questions need to be considered:

a. Will the user be trying to achieve the right effect?
b. Will the user notice that the correct action is available?
c. Will the user associate the correct action with the desired effect?
d. If the correct action is performed, will the user see that progress is being made?

- Record problems. Identify failure stories. The analyst should identify and document all the problems in a failure story, the reason for failure, and the assumption about the user's behavior. Consider and record design alternatives. Modify the interface design to eliminate the problem.

GOALS, OPERATORS, METHODS, AND SELECTION (GOMS)

GOMS is a task analysis method. It uses a notation and syntax that is similar to traditional programming languages (Card, Moran, and Newell, 1980). A GOMS model consists of the following elements:

1. *Goal*—a specification of what a user wants to achieve, such as finding a word in a manuscript that is being edited.
2. *Operators*—the actions that the analyzed software allows a user to take, such as keyboard input, mouse input, or voice recognition input.
3. *Method*—methods that can be used to accomplish a goal. For example: to find a specific word in a manuscript, such as NASA-TLX, one can either employ the method of using the Find routine, which is listed under Edit in MS Word, or one can use the method of simply scrolling down the text and searching for the word.
4. *Selection*—rules that a user follows in deciding which method to use in a particular instance. For example, a user may utilize the following decision rules:
   - If it is a long document, use the Find routine.
   - If it is a short document, use the Scroll function, since it will be quicker.

GOMS is actually the wrong acronym. It should be GMSO. The user will first select a goal, then consider what methods are available to accomplish the goal, then select a suitable method, and finally use an operator (e.g., a keyboard) to expedite the selected method.

To find the word in the document, we can program the two methods (see Table 7.6). Method A takes five steps. Method B takes three steps if the word is on the first page. If the word is on the second page it takes five steps. If it is on the third page it takes seven steps, and so forth.

Method B works well for a one page document if the word is easy to find. Some words stand out and are easy to find. A word in caps, such as "GOMS," looks different and is very easy to find. We can quickly scan one entire page. But to search for the word "method" one may have to scan a paragraph at a time. This will add

---

TABLE 7.6
Two Alternative Methods for Finding a Document

Method A. Find a word using the find routine
   Step 1. Accomplish goal: click on Edit
   Step 2. Select Find
   Step 3. Input word
   Step 4. Click on Find Next
   Step 5. Return with goal accomplished
Method B. Find a word by visually scanning pages
   Step 1. Accomplish goal: search current page
   Step 2. Decide if word is on page
   Step 3. If word is on page, return with goal accomplished
   Step 4. If word is not on page, scroll to next page
   Step 5. Go to 2

---

to the number of steps, and the method of scrolling and visual scanning will no longer be a good option.

A typical GOMS model describes user tasks and user knowledge in the form of production rules. These rules can be tedious and time-consuming to formulate, particularly when the tasks are complex.

GOMS itself does not identify the user goals. The user must identify the goals. There are several traditional methods available for doing this: task analysis, interviews with users, observation of users using similar systems, or intuition on the part of the analyst (Kieras, 1998). There are always several ways to decompose a task and write a GOMS program. The analyst must judge how users view their task, what their goals are, how they decompose the task into subtask, and what steps the user will select to accomplish his methods (Kieras, 1998).

Predicting Human Performance through GOMS

The results of a GOMS analysis may be used to predict human learning. The logic behind it is that the greater the number of statements (steps), the longer the learning time. Kieras (1998) noted that for a program with about 100 statements the learning time (LT) is as follows:

$$LT = 30\text{--}60 \text{ minutes} + N \times 0.5 \text{ minutes}$$

where N is number of statement.

As an example, Gray (1998) calculated the learning time for a computer game. He showed that even ten-year-old boys take as long as their parents would take. The main difference is that some boys have endless motivation. Parents are usually not aware of how long a time their children spend to perfect a game, and therefore they elevate them to the status of computer genius. Unfortunately this is rarely the case.

Usability Testing

Usability testing has been established for more than 25 years. It is focused on identifying usability problems in a product. The test is performed by a participant in a usability laboratory. There are several characteristics of a usability test.

1. The focus is on usability
2. Participants are end users or potential users
3. Participants may be asked to think aloud as they do the task (see below)
4. Data are recorded and analyzed

The sophistication of the analysis depends on the purpose of the study. For a simple pilot study it is not necessary to use a laboratory with a one-way mirror and data logging software. There are inexpensive and quick methods, such as the use of a single video camera. There is also software available that will store all screens, cursor movements, and voice records on a PC; Camtasia Studio is one example (TechSmith, 2004).

Collecting Data for Usability Testing

One can use performance measures as well as subjective measures and video recordings. Figure 7.4 shows the usability lab at Nangang Technological University. Performance measures typically involve the time to perform the task, the number of errors, the types of errors, and the number of times the experimenter had to step in to help the test person. Software can be used to record mouse movements, keyboard input, and screen content. Some of these data can also be difficult to analyze.

Keyboard input, for example, is difficult to analyze. Many years ago Alan Neal at IBM recorded operator keystrokes. He soon realized that he was recording a tremendous amount of data, and there was seemingly little use for the results (Neal, 1977).

Subjective measures involve ratings (such as questionnaires—see below) or verbal comments of the participants regarding usability. Video cameras can be used to record the user's facial expressions and other types of reactions. However, this type of data can be fairly difficult to analyze.

Selection of Test Persons

Van Welie et al. (1999) commented that user knowledge of an application will affect usability assessments. To evaluate systems it is therefore important to carefully select test persons. Experienced users will often find different usability problems than will inexperienced users. It may be important to test both types of users.

Analyzing Data from a Usability Test

In analyzing the data from a usability test one should consider that the different measures (dependent variables) are often related to one another. For example, a poorly designed icon would create user errors because the wrong icon was selected. It would also increase the task performance time and decrease user satisfaction (Dumas, 2003). The dependencies between the measures can be investigated using

FIGURE 7.4 The usability lab at Nanyang Technological University. There are several video cameras hidden in the ceiling and behind the one-way mirror. User actions and voice comments are recorded on video recorders and analyzed. The user performance is evaluated and related to the interface design. Design improvements can then be suggested.

correlation analysis. If there are significant high correlations, it would make sense that the three measures show the same trends.

One way of analyzing the data is to divide it into three components that follow the ISO definition; namely, efficiency, effectiveness, and satisfaction measures. To measure the efficiency, one can use the performance time—the time it takes to perform a certain task. Effectiveness measures refer to the task success. To what extent did a user manage to accomplish what she was supposed to accomplish? Satisfaction (or dissatisfaction) measures pertain to the user's ratings and verbal comments.

Dumas and Redish (1999) proposed a method to organise usability problems by scope and severity. Scope answers the question, How widespread are the problems? Are the problems local and restricted or are they global, so that they affect most of the interaction? An example of a global problem is when it is difficult to find menu items in a pull-down menu. A local problem, on the other hand, is restricted to one screen.

To classify the severity of usability problems, Dumas and Redish (1999) proposed a four-point severity scale:

1. Problems that prevent completion of a task, such as consistently selecting a wrong menu option and not knowing what to do.
2. Problems that create significant delay and frustration. For an example: no feedback to confirm the action a user has taken.
3. Problems that have a minor effect on usability, such as using the same word for two different actions.
4. Enhancements that can be added in the future, such as a user request for a tutorial to comment on a few issues.

Severity rating scales have three properties in common (Dumas, 2003):

1. Rating scales are derived from software bug reporting scales. The most severe category would result in data loss or task failure, while the least severe category would not need an immediate fix because it is considered not so important.
2. Scales alone are not sufficient to assess severity. It is also important to assess whether additional measures need to be taken in order to determine the severity of the problem. Rubin (1994) suggested four factors: frequency, impact, persistence, and market impact. Dumas and Reddish (1999) introduced a second dimension: the scope of the problem (local or global).
3. None of the scales are able to indicate how to treat the differences between participants.

User Participatory Design

Many researchers are in favor of user participatory design, since it gives an opportunity for end users to specify what is important. This is an area where much research has been done, and Muller et al. (1997) made reference to no less than 61 different methods for participatory design. There are, for example, future workshops, storytelling workshops, forum theatre, ethnographic methods, specification games, co-development, and many more. Most of the methods are used for software design, and they apply to the different design cycle phases: requirements definition, conceptual design, and detailed design.

Some tests have the purpose of investigating usability, but they are quite different from regular usability testing. There are, for example, the following methods: interface theater, cooperative evaluation, pluralistic walkthrough, storyboard prototyping, work mapping, and participatory heuristic evaluation.

Researchers are divided in their opinions. Even though end users understand the application, they don't understand usability and HFE. Hence there are argument pros and cons. Arguments in favor of participatory design:

- Accurate information about tasks
- Democratic decision making
- A sense of participation
- Potential for better user acceptance

Arguments against participatory design:

- Takes a long time
- Forces designer to comprise design goals
- Users don't understand HCI
- User's don't understand the negative effects of poor design

With the right method, the right leadership, and the right users, participatory design can be a great success (Muller et al., 1997). I believe that participatory design can be very relevant, interesting, and fun. Secondary goals can also achieved, namely a sense of friendship and collaborative spirit among the participants.

THINK-ALOUD PROTOCOLS

Think-aloud is a method where a participant verbalizes what she is trying to accomplish, as she is looking for information and trying to solve problems (Ericsson and Simon, 1993). Think-aloud is often done as part of usability testing. The participants are asked to think out loud as they perform the task. Ericsson and Simon (1993) cited several authors who were in favor of this method.

- "Try to think aloud. I guess you often do so when you are alone and working on a problem" (Duncker, 1926).
- "Think, reason in a loud voice, tell me everything that passes through your head during your work searching for the solution to the problem" (Claparede, 1934).
- "Don't plan what to say or speak after the thought, but rather let your thoughts speak, as though you were really thinking out loud" (Silveria, 1972).
- "In order to follow your thoughts we ask you to think aloud, explaining each step as thoroughly as you can" (Smith, 1971).

The purpose of think-aloud is to identify the user's intentions and actions for the purpose of improving the interface design:

- Were the intentions correct? If they were not correct, how can the interface be improved so that the user will have the correct intentions?
- Were the actions correct? Were they in agreement with the user's intentions? If the actions were wrong, how can the interface be improved to avoid wrong actions?

One problem with the think-aloud method is that a secondary task is imposed, in addition to the primary task of solving the problem presented on the screen. The secondary task—speaking—will hence add to the overall workload, and users may therefore perform worse than they would if they did not speak. Rosson and Carroll (2002) commented that although it is not a natural behavior for most computer users to think aloud, it reveals many unobservable cognitive sources of usability problems.

Questionnaires

Questionnaires can be used to measure users' subjective evaluations of a product or a software program. A questionnaire often includes a five-point or a seven-point scale, which can be used to measure how good or bad the user finds the situation. The number is useful because one can sort the good and poor parts of the interface. For example, one may ask the user the question in Figure 7.5. The surprising result is that many users from North American have difficulties with this particular icon, while people in Europe and Asia recognize it as a warning sign, since it is commonly used on the roads.

A questionnaire must also be tested to see if it is usable. There are often many opportunities to misunderstand questions. Note that for the question in Figure 7.5, the wording is not correct: "Very Effective" is not the opposite of "Very Poor." One should either use poor–good or ineffective–effective. In addition, a seven-point scale provides better resolution than a five-point scale. Respondents have a tendency of avoiding the extreme ratings 1 and 5, which leaves only a 3-point scale.

Try to word the questions so that the same scale can be used throughout the questionnaire. For example a question about aesthetics can be formulated as follows: "Do you think that the screen image is beautiful?" This will make it necessary to use an adjective pair such as ugly–beautiful, rather than ineffective–effective. Therefore the scale cannot be easily compared to the one in Figure 7.5. However, the same scale can be used if the question is phrased like this; "Please rate the effectiveness of the screen image in portraying a beautiful setting."

Always test a questionnaire with a few users. Some questions are bound to be misunderstood, and must be replaced or rephrased.

Many software usability scales have been published. These can be used to inspire the development of new scales; see, for example, Brooke (1996) and Kirakowski (1996).

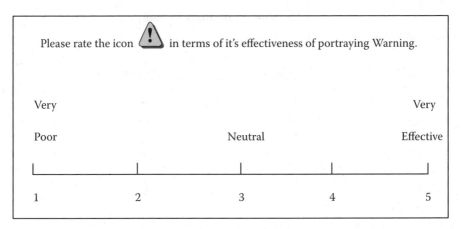

FIGURE 7.5 Rating of icon effectiveness using a seven-point Likert scale.

COMPARISON OF USABILITY METHODOLOGIES

Many studies have been undertaken to compare the effectiveness of different usability evaluation methods (UEMs). Fu et al. (1999) proposed that different UEMs have different kinds of problems, such as usability testing and expert reviews.

Fu et al. (1999) studied the types of problems found in usability evaluation by human factor experts and users. The usability problems were broken down into three categories using Rasmussen's (1996) concepts of skill-based, rule-based, and knowledge-based problems.

- *Skill-based performance.* The errors committed are due to problems in perception and/or motor response.
- *Rule-based performance.* The errors committed relate to the selection of the wrong procedure, and often lead to omission of steps in a procedure.
- *Knowledge-based performance.* These errors are related to the user's lack of knowledge. Users perform incorrect actions, because they do not understand the problem.

In analyzing an interface, HCI experts could use cognitive walkthrough and users could participate in usability testing. Human factor experts would then be expected to find rule-based and skill-based problems (icons or layout of interfaces) while users could identify knowledge-based problems (due to lack of knowledge). This certainly makes sense—an outsider would have difficulties understanding what the users know.

Lewis and Wharton (1997) compared the different usability evaluation methods (see Table 7.7).

TABLE 7.7
Comparison of Different Usability Evaluation Methods (Lewis and Wharton, 1997)

Characteristic	Cognitive Walkthrough	Heuristics Evaluation	GOMS	User Testing	Thinking Aloud
Test users	No	No	No	Yes	Yes
Task specific	Yes	No	Yes	Yes	Yes
Traces correct paths	Yes	No	Yes	Maybe	Maybe
Assigns reasons for errors	Yes	Maybe	No	Maybe	Yes
Analyses user mental processes	Yes	No	Yes	No	Maybe
Estimates learning time	Maybe	Maybe	Yes	Maybe	Maybe
Estimates performance time	No	No	Yes	Yes	Maybe

There are obviously advantages and disadvantages with the different methods:

- Cognitive walkthrough has an advantage in that it can analyze user's mental processes and can be used to understand how lack of user knowledge and experience can make a certain interface difficult to understand.
- In heuristics evaluation the evaluators do not make any assumptions about user knowledge, and therefore cannot analyze reasons behind user errors.
- One disadvantage with GOMS is that it cannot analyze errors. A GOMS model lists production rules that are necessary to accomplish a task. However, errors may happen in a thousand different ways and these can be modeled with GOMS.
- User testing has the benefit of face validity. Since the method tests users one can easily remove the causes for errors. However, other interface flaws that do not lead to errors but may slow down the use of the interface are not identified.
- The think-aloud method can give some insights into how users reason, make conclusions, and commit errors. In particular, it can be efficient in removing causes for error.

There have been many comparison studies of the effectiveness of the different methods. Many of them were published in the beginning of the 1990s, for example, Virzi et al. (1993). However, since 1990 several of the methods have been improved. For example, the procedure for cognitive walkthrough has been simplified, and is now much easier to perform (Lewis and Wharton, 1997).

COST-EFFECTIVENESS OF USABILITY

Usability contributes directly and indirectly to several product goals:

- Improves the look and feel of the user interface
- Leads to less documentation
- Reduces the cost for help desk
- Reduces the need for training
- Incorporates user requirements

Karat (1993) documented a case study of a system that required 1 hour of training as compared to one week for another similar system. The investment in usability saved the organization millions of dollars for the first year alone. There are several business areas where usability demonstrates significant benefits. Great usability does the following:

- Improves product definition and product performance
- Increases user satisfaction and user productivity
- Reduces development time and development costs
- Increases sales and revenue
- Reduces training and help desk costs
- Reduces maintenance costs and personnel costs
- Improves work process control and audit trails

According to Karat (1997), usability is part of the critical path for product development. Companies cannot afford to put products on the market that do not have the required usability—there may not be a second chance. The costs for post-release changes are unaffordable for most companies. Even the losses of momentum and customer goodwill make it difficult to continue in the same market area. Most companies realize a ten-fold return on user-centered design (Maya Group, 2002). Usability Engineering hence contributes to various stages of the product life cycle. Benefits accrue throughout the product life cycle. But it is less expensive to make corrections in the beginning.

EXERCISE: DESIGNING A CLOCK RADIO WITH A MAC OR WINDOWS INTERFACE

The purpose of this project is to design the human interface for a clock radio with a Mac- or Windows-type interface (see Table 7.7). All information and system output except the radio signal must be displayed on the computer screen. Users must be able to access all software functions using a pointing device (or keyboard and mouse if you prefer).

For a hypothetical design review you should do the following:

1. Prepare examples of screens that users will see, as well as command sequences to perform the various functions.
2. Justify your choices on the basis of human factors and HCI design considerations.
3. Analyze the command sequences in terms of principles of direct manipulation. Use the concepts in this chapter that describe direct manipulation.
4. Would you like to propose additional concepts of directness that are more appropriate for the task or in general for this type of interface?

Designing a Clock Radio with a Mac or Windows Interface

Feature	Required User-Controllable Functions
Clock	Set current time Display current time
Alarm	Set alarm time Display alarm time Set alarm time
Snooze	Activate snooze function Set snooze interval
Sleep	Activate sleep function Set sleep interval (this is the interval between pressing the sleep button and automatic radio shut-off)
Radio	Turn on and off Adjust volume Set to AM or FM tuning Tune to station of choice Display current station Store up to six preset stations Retrieve up to six preset stations

Part II

Human-Body-Centered Ergonomics

8 Anthropometry in Workstation Design

8.1 INTRODUCTION

A basic philosophy in ergonomics is to design workstations that are comfortable, convenient, and productive to work at. Ideally, workstations should be designed to fit both the body and the mind of the operator. This chapter focuses on the body, which certainly is the easier of the two problems. We will demonstrate how adjustability of chairs, stools, benches, and task arrangements can help to accommodate people of different body sizes. Using anthropometric design principles it is possible for a variety of people to find physical comfort at a workstation. On the other hand, not taking these physical requirements into consideration may create bad work postures leading to fatigue, loss of productivity, and sometimes injury.

Anthropometry is not only a concern about appropriate working height, but also about how the operator can easily access controls and input devices. In an automobile it should be possible for a small driver to reach the controls on the dashboard while being held back by the seatbelt. Similarly, the controls of machine tools must be easy to reach. The lathe shown in Figure 8.1 was originally described by Singleton (1962). It is a classic design and makes a clear argument. To control this particular piece of equipment, the ideal operator should be 137 cm (4.5 ft) tall, 62 cm (2 ft) across the shoulders, and have a 235 cm (8 ft) arm span, which is close to the shape of a gorilla!

8.2 MEASURING HUMAN DIMENSIONS

There are large differences in body size due to gender and genetics. Men are, on average, 13 cm (5 in) taller than women and are larger in most other body measures as well. Genetic differences are evident from a comparison of individuals living in different countries. For example, the average male stature in the U.S. is 167 cm (66 in), whereas while that in Vietnam is 152 cm (60 in). A car designed for the U.S. population would fit only about 10% of Vietnamese, unless of course the differences can be compensated for by using an adjustable seat (Chapanis, 1974). However, some of the differences between countries are decreasing, suggesting that there are factors beyond genetics. For example, during the last 20 years the average Japanese teenager has become 12 cm taller (Pheasant, 1998). This is largely attributed to changes in eating habits; in particular, animal proteins have become much more common in the Japanese diet.

147

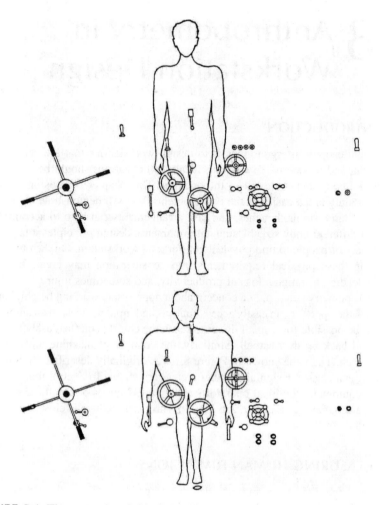

FIGURE 8.1 The controls of this lathe are not within easy reach of the average man. The bottom figure shows the ideal size operator (Singleton, 1962).

According to a study done in the U.K., the average male manager is 3–4 cm taller than the average male blue-collar worker (Pheasant, 1998). There could be many reasons for this. It may be that taller people are more often promoted to managers, or that taller people are a little more intelligent, or that managers come from a higher social class and thus had better education and also eat more animal protein. It is difficult to attribute causes, but probably all of these reasons contribute. Of particular interest to ergonomics is that a male manager may have a different physical frame of reference than the individuals who work for him. For example, a managerial chair is oversized and uncomfortable for a female secretary, and vice versa. A manager may have difficulties in understanding problems related to physical accommodation, simply because they do not apply to him.

Anthropometric measures are usually expressed as percentiles. The most common are the 5th, 50th, and 95th percentile measures (Table 8.1). Anthropometric data are usually normally distributed (Figure 8.2) (Roebuck, 1995). A normal distribution is characterized by its mean value (M) and its standard deviation (SD). As long as we know these two values of distribution, it is possible to calculate any percentile value. For example, the 95th percentile equals the mean value plus 1.65 SD and the 5th percentile equals the mean minus 1.65 SD (8.2).

The common procedure is to design for a range of population from the 5th percentile (small operator) to the 95th percentile (large operator). The choice of 5th and 95th percentiles is traditional, although one could argue that a greater percentile range should be used. But many ergonomists consider it impossible to include extremes of the population, such as dwarfs and giants, in the common design range. For example, the seat of a height-adjustable chair in the U.S. must adjust between 16 and 20.5 in, which roughly corresponds to the range established by 5th percentile females and 95th percentile males, although some users may be smaller

TABLE 8.1
Explanation of Percentile Measures

Percentile	Description
5th	5% of the population is smaller
50th	Average value
95th	95% of the population is smaller

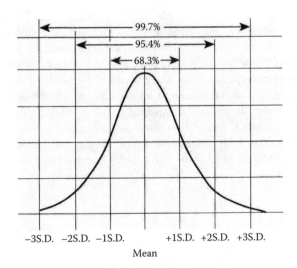

FIGURE 8.2 Anthropometric data are usually normally distributed.

or larger (see Human Factors Society, 1988). Similarly, it would not be practical to make door openings 8 ft tall, although this may be required by a giant.

The greater the design range, the greater the cost. It is more expensive to design for the 5th to 95th percentile range than for the 10th to 90th percentile range. The percentile value selected is largely a political decision, and companies may adopt different policies. One potentially controversial question is whether one should design for the worker population at hand, e.g., the 5th to 95th percentile male, or if one should extend the range to 5th percentile female workers in order to provide equal physical access to females.

Workers in a specific manufacturing plant may have different body size that are not typical of the population at large. These were the concerns in a study we performed for IBM Corporation in San Jose (Helander and Palanivel, 1990). At this location there were about 1000 female microscope operators, many of whom had recently arrived to the U.S. from Asia and were shorter than the 5th percentile U.S. female. Many operators had to stretch to be able to get to the eyepiece and they could not put their feet on the floor. We measured 17 different anthropometric measures of 500 operators and calculated the means, standard deviations, and 5th and 95th percentile measures. These measures were used to specify the appropriate measures for the microscope workstation.

We will explain below how anthropometric measures can be translated into workstation design measures by using the anthropometrics design motto:

- Let the small person reach.
- Let the large person fit.

EXERCISE: SELECT BODY DIMENSIONS FOR DESIGN OF A CAR INTERIOR

The anthropometric motto implies that reach distances should be designed for the small, 5th percentile individual, whereas clearance dimensions should be designed for the large, 95th percentile individual. For example, assume that you are designing a car interior, such as in Figure 8.3. Several of the measures are reach dimensions (5th percentile) and several are fit dimensions (95th percentile). Figure 8.3 shows a side view of the interior of a truck cab and a front view of a driver's seat. The numbered dimensions correspond to those listed in Table 8.2. Select body dimensions which would be critical for determining the cab dimension distances. Record the body dimension number in the space provided. For each cab dimension, indicate whether the 5th or 95th percentile body dimension should be the design value.

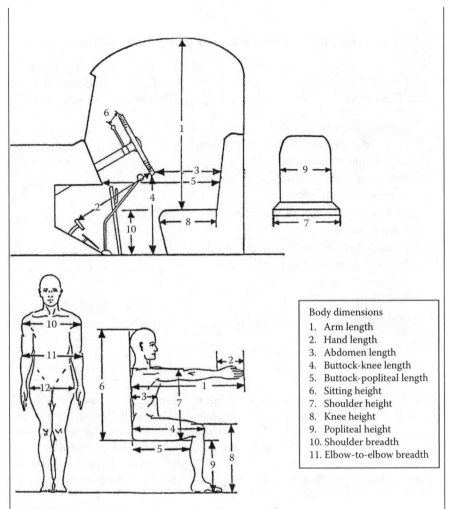

FIGURE 8.3 Vehicle measure and anthropometric measures.

TABLE 8.2
Cab Dimensions, Body Dimensions, and Design Value (5th or 95th)

Cab Dimension	Body Dimension	Design Value (5th/95th) Reach or Fit
1. Distance from seat to roof	_____	_____
2. Distance from top of foot pedals to lower edge of the steering wheel	_____	_____
3. Horizontal distance from lower edge of steering wheel to seat back	_____	_____
4. Vertical distance from lower edge of the steering wheel to floor	_____	_____
5. Distance between dashboard and seat back	_____	_____
6. Distance from steering wheel rim to directional signal	_____	_____
7. Width of cab seat	_____	_____
8. Seat depth	_____	_____
9. Width of seat back	_____	_____
10. Height of seat front to floor	_____	_____

A simple case of anthropometric design is illustrated in Figure 8.4. The 5th percentile female and 95th percentile male measures are illustrated for a sitting workplace. Note that the popliteal height (from the sole of the foot to the crease under the knee) is 36 cm (14.0 in) for 5th percentile females and 49 cm (19.2 in) for 95th percentile males. These values may actually differ slightly in different anthropometric tables. Note also that the popliteal height (and other measures) are taken without shoes, so that for design purposes one must add the height of the heel

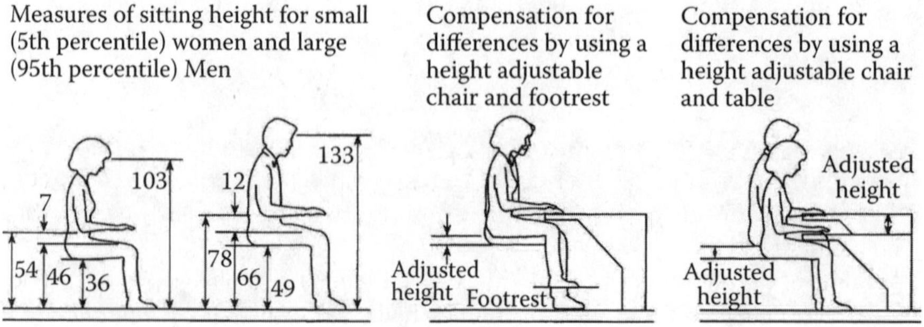

Measures of sitting height for small (5th percentile) women and large (95th percentile) Men

Compensation for differences by using a height adjustable chair and footrest

Compensation for differences by using a height adjustable chair and table

FIGURE 8.4 Comparison of anthropometric measures (cm) for a sitting 5th percentile female and a sitting 95th percentile male—height-adjustable chairs and tables can be used to compensate for these differences.

of the shoe (about 3 cm). The appropriate range of adjustability for a chair-seat height is then 39–52 cm (15–20.2 in). The distance from the floor to the elbow is obtained by adding the popliteal height, sitting elbow height, and shoe height (3 cm). This measure is 57–81 cm (22–32 in) and it can be used to select appropriate table height.

As illustrated in the right-hand part of Figure 8.4, there are two different ways to compensate for anthropometric differences. One can use either a height-adjustable chair plus a foot rest, or a height adjustable chair plus a height-adjustable table. Both arrangements will make it possible to support the feet and have the table at elbow height. The height-adjustable table is more expensive than the foot rest, but it is more comfortable to rest the feet on the floor than to use a foot rest.

In many offices (including the author's) the table height has been set once and for all. Although it is possible to raise and lower the table height, this usually requires some effort, and a change may not be worth it. A height-adjustable chair is rarely changed more than once per day. For individuals who have their own workstation, ease of adjustability is therefore not so crucial. But for people who share a work-station, for example shift workers, adjustability becomes essential (Shute and Starr, 1984).

Most of the microscope workers at IBM worked three shifts; therefore, adjust-ability of the workstation was deemed important. Microscope work is an exacting task. It is necessary to adjust (1) the eye pieces to the exact level of the eyes, (2) the table so that it is convenient to reach the microscope controls, and (3) the chair to be able to put the feet on the floor. This is a complex case of adjustability. The three adjustability elements interact. For example, raise the chair height, and the elbow-height and eye height will also increase.

8.3 DEFINITION OF ANTHROPOMETRIC MEASURES

The most complete and up-to-date source of anthropometric measures has been published in Taiwan by Wang et al. (2001). The U.S. National Aeronautics and Space Administration (NASA, 1978) published a reference publication with measures of 306 different body dimensions from 91 different populations around the world. About half of the populations are airplane pilots, which illustrates the great importance attributed to the anthropometric design of cockpits. Anthropometric investigations have been supported by the Air Force in the U.S. and many other countries, but surprisingly there is a lack of civilian anthropometric measures. In the U.S., there has actually never been a comprehensive civilian anthropometric investigation. The measures listed in Table 8.3 are adapted from the data reported by McConville et al. (1981), who extrapolated civilian body measures by using data from the military. The measures are also illustrated in Figure 8.5.

Some of the anthropometric measures have Latin names. This is practical, since they refer to bone protrusions on the human body. For example, the *tibial height* is the height of the proximal medial margin of the tibia, a bone protrusion on the tibia right under the knee cap. The *acromion height* refers to the highest point of the shoulder blade, and the *popliteal height* (mentioned above) is the height from the sole of the foot to the crease under the knee between the upper and the lower leg.

TABLE 8.3
U.S. Civilian Body Dimensions in cm of Relevance for Workplace Design

	Female			Male		
	5th	50th	95th	5th	50th	95th
Standing						
1. Tibial height	38.1	42.0	46.0	41.0	45.6	50.2
2. Knuckle height	64.3	70.2	75.9	69.8	75.4	80.4
3. Elbow height	93.6	101.9	108.8	100.0	109.9	119.0
4. Shoulder (acromion) height	121.1	131.1	141.9	132.3	142.8	152.4
5. Stature	149.5	160.5	171.3	161.8	173.6	184.4
6. Functional overhead reach	185.0	199.2	213.4	195.6	209.6	223.6
Sitting						
7. Functional forward reach	64.0	71.0	79.0	76.3	82.5	88.3
8. Buttock–knee depth	51.8	56.9	62.5	54.0	59.4	64.2
9. Buttock–popliteal depth	43.0	48.1	53.5	44.2	49.5	54.8
10. Popliteal height	35.5	39.8	44.3	39.2	44.2	48.8
11. Thigh clearance	10.6	13.7	17.5	11.4	14.4	17.7
12. Sitting elbow height	18.1	23.3	28.1	19.0	24.3	29.4
13. Sitting eye height	67.5	73.7	78.5	72.6	78.6	84.4
14. Sitting height	78.2	85.0	90.7	84.2	90.6	96.7
15. Hip breadth	31.2	36.4	43.7	30.8	35.4	40.6
16. Elbow-to-elbow breadth	31.5	38.4	49.1	35.0	41.7	50.6
Other dimensions						
17. Grip breadth, inside diameter	4.0	4.3	4.6	4.2	4.8	5.2
18. Interpupillary distance	5.1	5.8	6.5	5.5	6.2	6.8

Adapted from McConville et al. (1981). 1 in. = 2.54 cm. Measurements are in cm with bare feet; add 3 cm to correct for shoes.

The anthropometric measures illustrate that there are large differences between genders. For many measures, the 5th percentile (small) male is about the same size as the 50th percentile (average) female. For example, the inside diameter of the hand grip (measure 17) is 4.3 cm for a 50th percentile female and 4.2 cm for a 5th percentile male. This measure is important for the design of hand tools that will fit the size of the tool to the size of the hand. Women often complain that they have to use handtools designed for men, resulting in muscle fatigue of the hand and the arm, lower productivity, and possibly injuries as well (Greenburg and Chaffin, 1977). As a result, the U.S. Department of Defense and industry (e.g., General Motors) have taken measures to supply hand tools of different sizes for males and females.

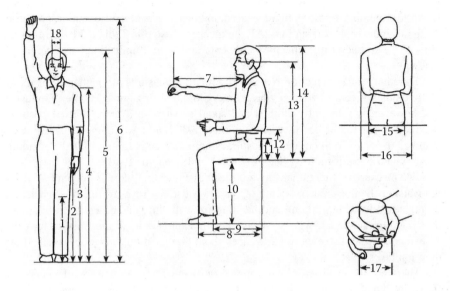

1. Tibial height. This measure is important for manual materials handling. Items located between the tibial height and the knuckle height must usually be picked up from a stooped position.
2. Knuckle height. This height represents the lowest level at which an operator can handle an object without having to bend the knees or the back. The range between the knuckle height and the shoulder height is ideal for manual materials handling and should be used in industry.
3. Elbow height. This is an important marker for determining work height and table height.
4. Shoulder (acromion) height. Objects located above shoulder height are difficult to lift, since relatively weaker muscles are employed. There is also an increased risk of dropping items.
5. Stature. This is used to determine the minimum overhead clearance required to avoid head collision.
6. Functional overhead reach. This is used to determine the maximum height of overhead controls.
7. Functional forward reach. Items that are often used within the workstation should be located within the functional reach.
8. Buttock–knee depth. This defines the seat depth for chairs and clearance under the work table.
9. Buttock–popliteal depth. This is used to determine the length of the seat pad.
10. Popliteal height. This is used to determine the range of adjustability for adjustable chairs.
11. Thigh clearance. Sitting elbow height and thigh clearance help to define how thick the table top and the top drawer can be.
12. Sitting elbow height. Sitting elbow height and popliteal height help to define table height.
13. Sitting eye height. Visual displays should be located below the horizontal plane defined by the eye height.
14. Sitting height. This is used to determine the vertical clearance required for a seated work posture.
15. Hip breadth. This is used to determine the breadth of chairs and whole body access for clearance.
16. Elbow-to-elbow breadth. This is used to determine the width of seat backs and the distance between arm rests.
17. Grip breadth, inside diameter. This is used to determine the circumference of hand tools and the separation of handles.
18. Interpupillary distance. This is an important measure in determining the adjustability of eyepieces on microscopes.

FIGURE 8.5 Illustration of the anthropometric measures given in Table 8.3.

A Guide to Human Factors and Ergonomics, 2nd Edition

Das and Sengupta (1996) provided a comprehensive overview of anthropometric design of industrial workstations. All the measures listed in Table 8.3 have implications for manufacturing. The measurements and their implications are explained below.

To reduce measurement error, anthropometric measures are gathered for minimally clothed men and women who are standing or sitting erect. People in industry are, however, usually fully clothed and stand or sit with a more relaxed posture. With shoes on, the height measures in Table 8.3 should be increased by approximately 3.0 cm. To compensate for postural slump, 2.0 cm is subtracted from standing height and 4.5 cm for sitting height (Brown and Schaum, 1980).

The measure of functional forward reach assumes that there is no bending from the waist or the hips. When bending from the waist, the forward function reach can be increased by about 20 cm, and bending from the hips increases reach by about 36 cm (Eastman Kodak, 1983). Since a person cannot bend at the waist or hips for an extended time, these extra allowances should be used only for occasional, short duration tasks.

There are many different anthropometric databases in use, some of which are fairly dated and may not reflect the fact that the population keeps getting taller. But it may also be the case that some anthropometric data are inaccurate and researchers have not used enough precautions in obtaining accurate measurements. Anthropometric measures are well defined, and there are standard procedures for taking them. There are also special tools and equipment available for taking the measures (see Roebuck, 1995).

In the past, much research and many anthropometric surveys have been initiated by the U.S. Air Force, which presently is developing tools for three-dimensional modeling using computer-aided design (Landau, 2000). There are already several computer programs for analysis of posture in workplaces based on anthropometric data. These programs include: JACK by Unigraphics, RAMSIS by Human Solutions, SAFEWORK by Safework Inc., and Manne Quin PRO by NexGen Ergonomics. The Manne Quin PRO is based on an algorithm described in Chaffin, Andersson and Machin (1999).

Depending on the application, anthropometry is used differently (see Figure 8.6). For example, in designing cars, it is common to start with the hip joint or hip reference point (HRP), and then to take the measures up to the head and hands and down to the feet.

Some automobile manufacturers (of racing cars) start off with the accelerator reference point (ARP) and then lay out the rest of the body going upwards. In the design of fighter planes, it is important to put the eye at the right height. This is because there are many displays which must be visible, including head-up displays (HUD) that are projected in the windshield. Since the pilot is tied back to the seat, one can make a very accurate estimation of where his or her eye will be. In this case the design will start with the eye or eye reference point (ERP), and the rest of the body can be modeled going downwards.

For assembly work in manufacturing, we advocate the use of a hand reference point (HARP). The ideal location of the hands depends on the task. For heavy manual jobs, the hands should preferably be about 20 cm below elbow height, but for

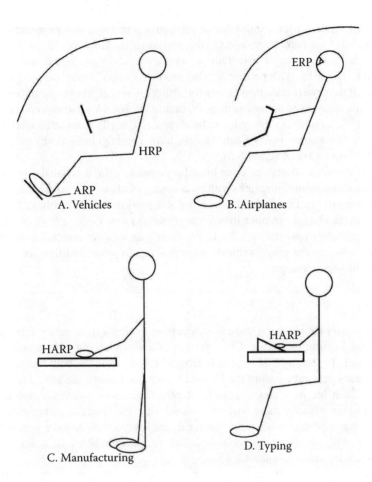

FIGURE 8.6 Anthropometric design can use different reference points.

precision tasks with supported underarms, the hands should be about 5 cm above elbow height. Therefore, to design a workstation, one needs first to determine the most convenient hand height for the task in question. The rest of the body can then be determined by finding measures down to the feet and up to the head. Typists have a similar work situation. It may be preferable to start off with a HARP and then lay out the rest of the body.

It is common to design for the range from the 5th to the 95th percentile. In doing so, one may have to add different anthropometric measures. For example, for a sitting workstation with the table top at the elbow height it is necessary to add two measures: popliteal height and sitting elbow height. The addition of anthropometric measures actually produces an inaccurate estimate, since very few individuals are 5th percentile throughout. Typically, a person with a short back may have long legs, or vice versa. Kroemer (1989) showed that the correlation coefficient between stature and sitting eye height is $r = 0.73$, between stature and popliteal height $r = 0.82$, and between stature and hip breadth $r = 0.37$ ($r = 1.00$ is a perfect positive correlation between two measures; r = 0 implies no relationship between two measures). If two 5th percentile measures are added, the resulting measure could be about the 3rd percentile. However, if two 95th percentile measures are added, the resulting measure might be the 97th percentile. This problem (with the table top height) could be solved if there were a single measure for sitting elbow height in the anthropometric tables. This measure is not one of the 306 defined in NASA's anthropometric tables (NASA, 1978), but in the future it can be measured directly from three-dimensional models of the human body. Such models have recently been developed by the CAESAR project (see Section 8.5).

Despite several sources of error in anthropometric data, it is usually possible to estimate anthropometric measures with an accuracy of about 1–2 cm. This is satisfactory for industry. In fact, individuals sitting at a workstation do not have the sensitivity to judge changes smaller than 2 cm (Helander and Little, 1993). If the chair height is raised or lowered by 1.5 cm, the chair user will not notice the difference. The reason is that the proprioceptive nerve endings in joints and ligaments are not sensitive to small changes.

EXERCISE

You can verify the inaccuracies in proprioceptive feedback quite easily. Sit straight in a chair in front of a table. Close your eyes and stretch out your arm in front of yourself. Try to make the arm as horizontal (at 90° from the body) as possible. Open your eyes and measure the height of your hand above the table top. Make a note. Then let your arm rest on the table. Then close your eyes and repeat several times. Usually there will some variability; the hand varies up and down about 3 cm. The inaccuracies are greater if you ask yourself to raise your arm to 105°—that is, 15° above the horizontal. This is a more difficult task, since there is no clear mental reference for 15°.

8.4 PROCEDURE FOR ANTHROPOMETRIC DESIGN

A procedure for anthropometric design is presented below.

1. Characterize the user population. What anthropometric data are available? Can existing anthropometric data be used with the present population? If there are no valid data, consider creating a database by obtaining measures of the existing workforce.
2. Determine the percentile range to be accommodated in the workstation design. If the workforce is dominated by either men or women it would make sense to design for the predominant gender, for example by using 5th to 95th percentile male or 5th to 95th percentile female measures. On the other hand, it may be an issue of equality to provide accessibility for the other gender group. If so, one would design from the 5th percentile female to the 95th percentile male population.
3. Let the small person reach and let the large person fit. Determine reach dimensions (5th percentile) and clearance dimensions (95th percentile)

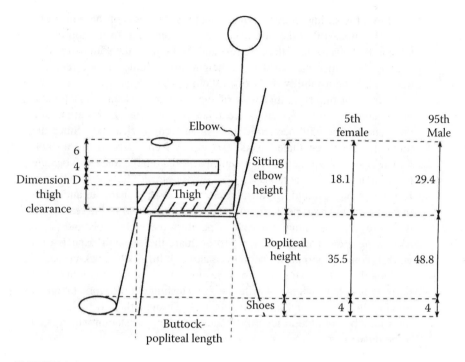

FIGURE 8.7 Anthropometric measures used to calculate the adjustability of seat height and table height.

for the work situation that is analyzed. An example is given in Figure 8.7. In this manufacturing task, the operator is sitting on a chair with his or her hands at elbow level and manipulating objects 6 cm above the table height. Two important reach measures are the popliteal height from the chair seat to the floor and the buttock–popliteal depth (see Table 8.3).

Operators should not sit with dangling feet but should be able to reach the floor. An adjustable chair must therefore adjust to a low level corresponding to the 5th percentile. The buttock–popliteal depth should be the 5th percentile. If it is longer, a small operator will not be able to reach to the back support with his or her back.

A clearance dimension (*D*) is created under the table. Assuming that the table is height adjustable and can be lowered 10 cm below elbow height, it still needs to be determined if there is enough space for the thighs.

4. Find the anthropometric measures that correspond to the workstation measures. The calculations for the 5th percentile female and the 95th percentile male operator are shown in Figure 8.7. The anthropometric measures are added starting from the floor level. Using the popliteal height and adding 4 cm for shoes, the required range of seat-height adjustability is calculated to be 39.5–52.5 cm. The sitting elbow height for the 5th percentile operator is 18.1 cm and for the 95th percentile operator 29.4

cm. From the sitting elbow height, deduct the thickness of the product (6 cm). This means that the distance from the chair seat to the top of the table is 12.1 cm for the 5th percentile and 23.4 cm for the 95th percentile. Adding these measures to the seat-height adjustability gives a required table height adjustability of 51.6–75.9 cm (or 52–76 cm).

Bearing in mind the thickness of the table top, we find that for the 5th percentile there is 8.1 cm of clearance between the chair seat and the table and for the 95th percentile there is 19.4 cm of clearance. Since the thigh clearance (see Table 8.3) is 10.6 and 17.7 cm, respectively, a small female operator will not have enough space, but a large male will be able to fit his legs under the table.

5. It is sometimes difficult to illustrate a work situation using an anthropometric model. Anthropometric measures are static, and in the real world there are many dynamic elements. Operators reach for tools and parts, and swing around in their chairs. To evaluate the dynamic aspects of a workstation appropriately, one may construct a full-scale mock-up out of cardboard or styrofoam. This can be done in a couple of hours. The purpose is to have people of different sizes testing out the workstation by moving their bodies and simulating the task. Through the full-scale mock-up it may be possible to identify features of the workstation which need to be redesigned.

8.5 NEW DEVELOPMENTS IN THREE-DIMENSIONAL MODELS FOR ANTHROPOMETRIC DESIGN

During the last 10 years there has been a rapid development of three-dimensional data models of the human body. Measures of the full body—or part of the body, such as the face—may be taken using instruments that employ laser scanning. The person is standing on a platform and several laser scanners rotate around the body (see Figure 8.8). Many experts believe that this methodology will eventually replace the traditional measurement of anthropometric data.

An international consortium, Civilian American and European Surface Anthropometry Resource (CAESAR), developed a computer-based methodology for scanning human dimensions in three-dimensional perspective using laser technology. The initiative was supported by the U.S. Air Force and the Society of Automotive Engineering (SAE). In addition, several multinational corporations, in particular manufacturers of vehicles, aircraft, and clothes, participated as members of the consortium (Robinette et al., 2002).

There are several advantages for using this technology. In the first place, there is a great need for anthropometric data which can be used for a variety of applications, including workstation layout, automobile design, apparel sizing, protective equipment design, safety assessment, and cockpit design, to name just a few. By using three-dimensional models rather than specific human measures, one can define the specific measures that are needed for design. It is no longer necessary to add up individual measures.

FIGURE 8.8 Whole body scanning equipment. Courtesy of Cyberware, Monterey, CA.

Measurements are quicker to take than with the traditional methods that use measurement tapes and calipers. Although there is a substantial investment in laser measurement equipment, this methodology may be less expensive in the long run.

One main argument behind a laser model of human dimensions is that nobody is 5th, 50th, or 95th percentile in all measures. This becomes increasingly important for design, which uses several measures of the body. For example, the design of a fighter aircraft cockpit presents a very tight environment, and some 15 to 20 measures must be taken to represent each pilot. The same arguments hold for design of clothes, which any tailor can testify to.

Laser scanning provides detail measures about the surface shape as well as three-dimensional locations of measures relative to each other. The measures are easy to transfer to computer-aided design models. It will take some time until this methodology makes an impact on industry. Other initiatives of whole-body measurement have been undertaken, for example by Wang et al., which may be cheaper than CAESAR. While these technologies are being tested out and further improved, we can also continue to use traditional measures.

The design of workstations is fairly simple and does not require any elaborate measurements. As illustrated above, a standing work station for assembly work can be designed using two to three measures, such as standing elbow height, arm reach, and eye height. For a sitting work station, measures of popliteal height, sitting elbow height, eye height, and thigh clearance are important. These are also the measures that are used in designing adjustable furniture. Other measures such as shoulder breath and buttock breadth do not affect adjustability design, since a chair must be made wide enough to accommodate all users. Unlike clothes which must fit the user, it does not matter if the backseat is a bit too large.

One important drawback of traditional measures is the errors which are obtained when different body measures are added. For example, to design of a height-adjustable table for 5th percentile to 95th percentile users two measures the human body must be added—popliteal height and sitting elbow height. Let's examine the case of the 5th percentile users—the 95th percentile would be similar. This leads to errors, since no user is 5th percentile in all measures. But adding the measures we would put more stringent requirements on the design, and we may find that the resulting measure is for a 3rd percentile user. However, the errors are small; in our case they would be less than 2 cm, which can not be perceived by the user (Helander, Little, and Drury, 2000).

8.6 ACCOMMODATING THE NEEDS OF DISABLED EMPLOYEES

In the U.S. and in several other countries, the needs of the disabled have been recognized in legislation. The Americans with Disabilities Act (ADA, 1990) presented the general framework, and the regulations are interpreted in terms of design recommendations in the ADA Standards for Accessible Design (1994). The standards address the special needs of people who use wheelchairs. The opening of doorways and elevators must be large enough to permit free maneuvering. This is then a problem of the size of the wheelchair (and not the body). There are also recommendations for provisions for employees with limited sight and hearing. Figure 8.9 gives examples of design problems. There are many other design suggestions, as exemplified in Table 8.4.

8.7 ANTHROPOMETRY STANDARDS

Many countries and organizations have developed standards for anthropometric design of office workstations. To date, there are three particularly important standards:

1. International Standards Organization (1998), ISO 9241-5. Ergonomic requirements for office work with visual display terminals (VDTs), Part 5. Workstation layout and postural requirements.
2. Human Factors and Ergonomics Society (2003), BSR/HFES100—Human Factors Engineering of Computer Workstations (Draft Standard for Trial Use). Santa Monica, CA: The Human Factors and Ergonomics Society.
3. CEN European Standard (2000), EN 527. Office furniture: work tables and desks.
   EN 527-1:2000 Dimensions.

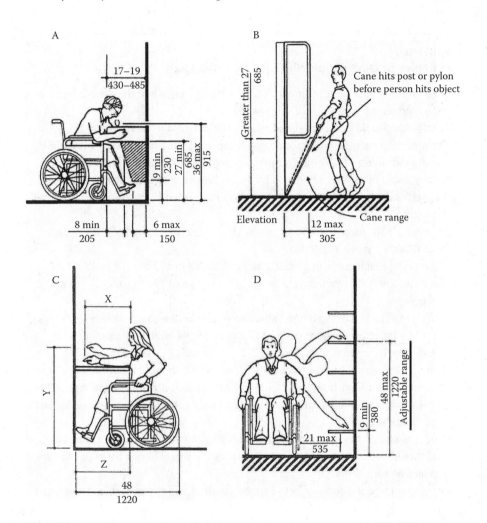

FIGURE 8.9 Examples of regulations from the Americans with Disabilities Act (U.S. Department of Justice, 1994). All measures are in inches and millimeters. (A) Measures for wheelchair access to a water fountain. (B) A pylon is erected against a wall. The cane will hit the pylon or the wall before the person. (C) Measures for forward reach. If $x < 635$ mm, then z must be greater than x. If $x < 510$ mm, then $y < 1220$ mm. If $610 < x < 635$ mm, then $y < 1120$ mm. (D) Measures for wheelchair accessibility of shelves.

These standards determine the design and marketability of workstations. Workstations that do not meet these standards have less chance of entering the local as well as the global marketplaces.

TABLE 8.4
Examples of Design Guidelines for Disabled Users

Door width should be a minimum of 81.5 cm (32 in) when the door is open at 90°; a preferred width is 112 cm (44 in).

Thresholds should be beveled and not more than 1.9 cm (0.75 in) high.

The preferred passage width for a wheelchair is 92 cm (36 in). Two wheelchairs need a minimum of 152 cm (60 in) to pass in an aisle or corridor.

A clear space of 76 by 122 cm (30 by 48 in) is needed to accommodate a person in a stationary wheelchair in front of or to the side of an object, such as a drinking fountain.

To make a 180° turn in a wheelchair, a clear space of 152 cm (60 in) diameter or a T-shaped space is needed.

The forward reach range in a wheelchair is from 38 to 122 cm (15 to 48 in) above the floor. The side reach height range is from 23 to 137 cm (9 to 54 in) above the floor.

A slope of 1:12 is the maximum value recommended for ramps in new buildings. For older buildings, a slope of 1:10 is acceptable when the maximum rise is 15 cm (6 in); if the maximum rise is 7.6 cm (3 in), a 1:8 slope can be used.

On a flight of stairs, the riser depth and tread width should be uniform, with the stair tread width not less than 28 cm (11 in) measured from riser to riser. This should accommodate people with reduced visual capacity when ascending and descending the stairs.

Handrails should be placed on both sides of a stairway, should be continuous, and should extend at least 30 cm (12 in) plus the width of one tread beyond the top and bottom risers.

There should be a clearance between the handrail and the wall of at least 4 cm (1.5 in).

The handrail top should be mounted from 86 to 96 cm (34 to 38 in) above the stair nosing.

Elevator displays should be at least 183 cm (72 in) above the floor.

Visual elements (e.g., floor indicators) should be at least 6.5 cm (2.5 in) high for detection by people with limited vision.

The safety switch to detect a person in the elevator doorway should cover a range of heights of 12 to 74 cm (5 to 29 in) above the floor.

Emergency controls in elevators should be placed no lower than 89 cm (35 in) above the floor. Floor buttons should be mounted no higher than 122 cm (48 in) above the floor.

The sliding force for opening an interior door or a folding partition should not exceed 22 N (2 kg).

EXERCISE: DESIGNING A MICROSCOPE WORKSTATION

Using the setup of the microscope workstation as shown in Figure 8.10, calculate adjustability ranges for seat height, table-top height, and microscope eyepiece eye height (measures A, B, and C in Figure 8.10).

Design for a 5th to 95th percentile female population. There are several assumptions:

1. There is no footrest.
2. The shoes are 4 cm high.
3. In the upper part of the body from the elbow height to the shoulder height there is a postural slump of 2 cm.
4. When looking into the microscope the operators bend the head forward about 30°, which moves the position of the eye downwards by 1.5 cm.
5. The hands are manipulating the focusing controls at elbow height.
6. The arms are horizontal and resting on the granite slab.
7. The table top is 3 cm high. There is also a 4 cm thick granite slab on top of the table to reduce vibration.

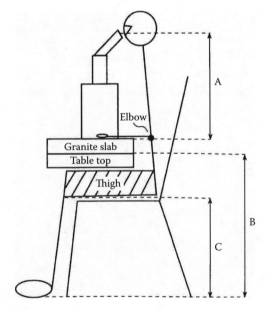

FIGURE 8.10 Example: designing a microscope workstation.

Answer. Use dimensions listed in Table 8.3.

- Measure A: 45.9–46.9 cm
- Measure C: 39.5–48.3 cm
- Measure B: 53.6–72.4 cm

FURTHER READING

Two books dealing with classical anthropometry are recommended:

Pheasant, S.T., 1998, *Bodyspace, Anthropometry, Ergonomics and Design*. London: Taylor & Francis.

Roebuck, J.A., 1995, *Anthropometric Methods: Design to Fit the Human Body*. Santa Monica, CA: The Human Factors and Ergonomics Society.

9 Work Posture

9.1 INTRODUCTION

Engineers who design production processes have a great responsibility. They must consider how the workstation will be laid out and what type of work posture is convenient for the job. Many engineers tend to focus on the engineering aspects; the work station, if even considered, is designed as an afterthought.

In this chapter we present information for evaluation and design of work stations. There is a choice between three types of work postures: standing, sitting, or sit-standing. Criteria and measures for design are presented. Finally we present methods for measuring the ergonomics effect of work posture, including the use of question-naires, Ovako Working Position Analysis System (OWAS), and rapid upper limb assessment (RULA).

9.2 EXAMPLES OF WORK POSTURES

One conventional engineering solution is to use an industrial height workstation with a 92 cm (36 in) high work table. This can accommodate both sitting and standing operators. The working height for the standing operator is about 92 cm (36 in), and a sitting operator can use a high industrial chair with a ring support or a footrest. Such flexibility in a workplace is indeed desirable and Figure 9.1 illustrates how flexibility for sitting or standing can be advantageous for many tasks.

However, the use of a conventional industrial height workstation can also create problems. It is not an appropriate design solution for dedicated seated tasks, which is illustrated by the microscope workstation shown in Figure 9.1(C). Working with a microscope is a dedicated seated task. This work station must support a very precise and static work posture. There is no reason to consider a standing work posture, and hence a regular table should be used. This has one important side benefit in that the operator can put his or her feet on the floor, which improves comfort.

In focusing on the technical aspects of the problem, engineers often forget the critical aspects of human performance. Figure 9.2 illustrates four operators working with very expensive machinery. From these examples it seems that the greater the technological challenge, the greater is the likelihood that the human element will be forgotten in engineering design.

Figure 9.2(A) shows an operator working with a scanning microscope. This was a low magnification microscope with a large exit pupil, so it was fairly easy to look through. (High magnification microscopes have a small exit pupil and cannot tolerate any deviation in eye position.) The first obvious problem was that the operator was standing. The second problem was that the scanning microscope moved back and forth while scanning. Therefore the operator had to move back and forth while

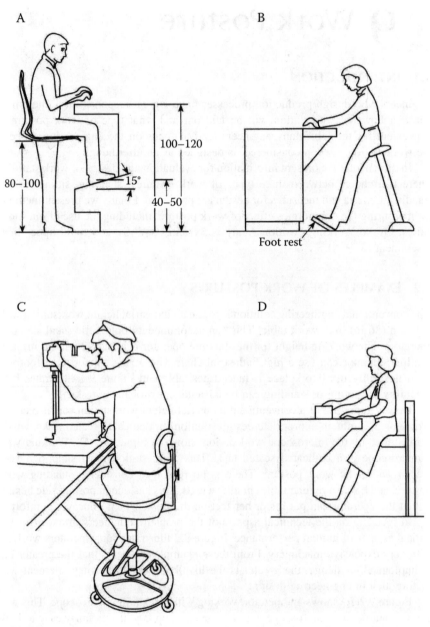

FIGURE 9.1 (A) A worktable for alternately sitting and standing. In this case the table at 110–120 cm is higher than the conventional 92 cm (36 in) table. (B) and (D) Variations of sit-stand arrangements. The operator is free to alternate between standing and sit-standing. (C) Misapplication of a 92 cm (32 in) industrial height table. Working with a microscope is a dedicated seated task, and a regular height table should be used.

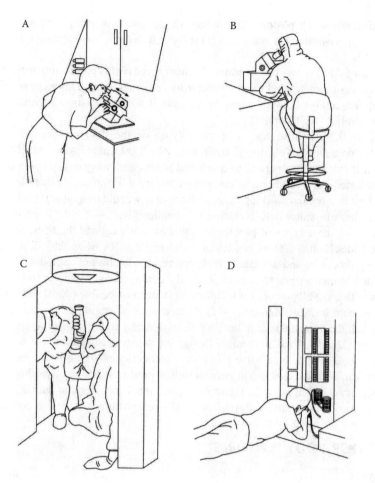

FIGURE 9.2 (A) The microscope is moving, and unfortunately the operator must move along. (B) Automated manufacturing equipment—but the automation never worked fully. (C) Routine maintenance with a flashlight. (D) Final assembly of a very expensive piece of equipment.

looking into the microscope—a very demanding task. Due to the bent-over, standing posture, only very short operators could perform the task. Instead of having the microscope move, the inspected elements should be the ones that moved. Another solution, though less desirable, is to use a moving chair that is synchronized with the microscope movements.

Figure 9.2(B) illustrates a clean-room process which was supposed to be totally automated. However, this very expensive piece of automation never worked out. Visual inspection and quality control had to be performed by a human operator. The main problem was that there was no leg room for the operator sitting at the machine. An armrest was improvised and placed on top of the equipment. The operator could then lean sideways over the top of the machine to perform the task. This was a very uncomfortable work posture. The design of this piece of equipment should have included provisions for a manual workstation. It became too expensive to rebuild afterwards.

Figure 9.2(C) shows two operators in clean-room outfits performing maintenance on process equipment used for manufacturing computer chips. This piece of machinery requires almost constant maintenance, but it was not designed with maintainability in mind (see Chapter 17).

Figure 9.2(D) illustrates an operator lying on the floor completing the final assembly on a piece of electronic equipment which is located inside a steel housing. It was difficult for the operator to reach and to see, and many costly operator errors were reported. As a solution, the company obtained a lifting device that elevated the equipment to a regular working height. The operator could now sit and perform task. The number of quality defects decreased significantly.

To avoid these types of problems, engineers must expand their responsibilities and consider human factors engineering and ergonomics when they design manufacturing processes and workstations. Ergonomic problems are unproductive as they introduce human errors and result in costly quality defects. The examples above illustrate that development of a technical system must be intentionally designed to make it more human. Meister (1971) pointed out that engineers are not unwilling to consider the human operator, but they clearly place a higher priority on engineering problems. This view has to change. Design engineers must bring to the task all the relevant tools and skills to solve problems—technological as well as organizational and human. Just like concurrent engineering of products, design of workplaces also requires many different skills. Design solutions must consider that human operators impose many constraints on design. A good work posture is one such constraint.

9.3 POOR BODY POSTURES

The types of posture that people assume at work can often lead to pain in various parts of the body. Van Wely (1970) identified common complaints for different work postures. Table 9.1 summarizes his observations. This table represents an oversimplification. People usually move around, and it is not easy to characterize a job in terms of a single posture. Nonetheless, the list in Table 9.1 is useful as a checklist for inspections of industrial workstations. For example, if one were to observe an operator who sits with his or her elbows on a high surface, it is a reasonable hypothesis that if the operator has any problems these would be in the upper back or lower neck. If the operator indeed voices such complaints, then the hypothesis has been confirmed, and one should then take measures to improve the work posture by lowering the work height. Similarly, if an operator sits with the head bent back, the common complaint is neck pain. If someone is assuming a cramped work posture, without any possibility of moving around, then the muscles involved may hurt.

A joint that is in an extreme position, either fully flexed or fully extended, may develop biomechanical problems. Rather, joints should be at a mid-range position. For example, arms should not be fully extended or flexed. A few examples are given in Figure 9.3.

TABLE 9.1
Work Postures and Related Complaints
(Van Wely, 1970)

Type of Posture	Location of Complaint
Standing	Feet, lower back
Sitting without lower back support	Lower back
Sitting without back support	Central back
Sitting without proper foot support	Knees, legs, lower back
Sitting with elbows on a high surface	Upper back, lower neck
Unsupported arms or arms reaching up	Shoulders, upper arms
Head bent back	Neck
Trunk bent forward	Lower back, central back
Cramped position	Muscles involved
Joint in extreme position	Joints involved

The recommendations for work posture and the discussions about biomechanical problems are traditional in ergonomics. Yet there are problems that require basic research, as is evident from the following example.

EXAMPLE: SITTING IN INDIA

Sen (1989) explained that industrial workers in India often sit hunched directly on the floor without a chair, or they may sometimes sit on a brick. They develop motion patterns very different from those of industrial workers in Western countries. Sometimes they swing their knees back and forth to manipulate items, at the same time as they work with their hands. Although their knees are flexed in an extreme position, these workers do not have any problems with their knee joints. The reason may be that they have been sitting hunched for their entire lives, and this is a common sitting posture at home and in social gatherings. Sen's statement was indeed surprising since hunch-sitting violates the principle of keeping the joints in a mid-range position. It seems obvious that more basic research in similar cultures (e.g., Indonesia) is necessary in order to analyze this controversy.

9.4 SITTING, STANDING, AND SIT-STANDING

Although our discussions are confined to common work postures, there are additional recommendations in Chapter 8 and Chapter 11. Depending on the type of task, it is advantageous for an operator to stand, sit, or sit-stand (Eastman Kodak, 2004; Michel and Helander, 1994).

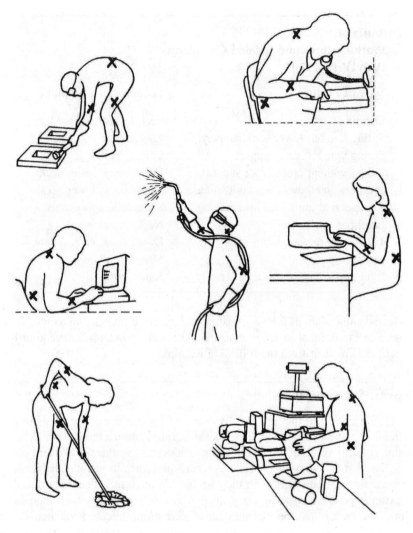

FIGURE 9.3 Examples of work postures where there are problems with extreme joint angle, large muscular force, high degree of repetition, or high contact pressure (from Webb, 1982).

- If there is frequent handling and lifting of heavy objects it is preferable to stand up. However, sit-standing may be an option (see Table 9.2).
- For packaging, or other tasks where objects must be moved vertically below the elbow height, it is preferable to stand or sit-stand. A sitting posture would not be feasible since the hands are reaching downwards and the table cannot be put at a sufficiently low level without interfering with the operator's legs.
- If the task requires extended reaching, it is sometimes preferable to stand or sit-stand, as the operator can then reach further.

TABLE 9.2
Preferred Work Posture for Different Tasks

Type of Task	Preferred Work Posture	
	First Choice	Second Choice
Lifting more than 5 kg (11 lb)	Standing	Sit-standing
Work below elbow height; e.g., packaging	Standing	Sit-standing
Extended horizontal reaching	Standing	Sit-standing
Light assembly with repetitive movements	Sitting	Sit-standing
Fine manipulation and precision tasks	Sitting	Sit-standing
Visual inspection and monitoring	Sitting	Sit-standing
Frequent moving around	Sit-standing	Standing

- Light assembly work with repetitive movements is common in industry, and sitting is preferred. A table is necessary to organize part bins and fixtures and incorporate work aides and supports to relieve local body fatigue due to repetitive movements.
- For fine manipulation and precision tasks the operator must support the underarms. Sitting is definitely preferred.
- Visual inspection and monitoring is best done sitting. The sitting work posture makes it possible to focus one's attention better than if standing.
- If the work task involves a variety of subtasks and also frequent moving around, it may be preferable to sit-stand, since the operator does not have to get in and out of the chair.

The recommendations in Table 9.2 should be used as a first approximation in understanding what the main options are. As we propose in Chapter 15, a task analysis is helpful in understanding the advantages and disadvantages with various design parameters, and how they trade off.

For most of the tasks in Table 9.2 the sit-standing posture is the second choice. Sit-standing workplaces have become increasingly common in industry during the last ten years. Sit-standing is convenient for many tasks, and there are biomechanical advantages since the pressure on the spine and the lower back is about 30% lower for sit-standing and standing as compared to sitting (Andersson and Ortengren, 1974) (see Chapter 14).

EXAMPLE: SIT-STANDING IN MEDIEVAL EUROPE

People have known about the benefit of sit-standing workplaces for over a thousand years. Misericords in medieval churches were used for sit-standing. In old churches in Europe there are altar seats, which can be tipped up, so that the person can sit on the ledge in a sit-standing posture (see Figure 9.4). These are called misericords, a Latin word meaning pity. The first of these were referred to in 1088 in the statutes of the church of Maastricht in the Netherlands (Grössinger, 1996). The monks had to stand up in seven long prayers per day.

This was very taxing, particularly for the older monks. The misericords gave the appearance that the monks were standing, when they were actually sit-standing. A misericord is made out of wood and nicely decorated with wood carvings which are visible when the seat is folded up in its sit-stand configuration (see Figure 9.4).

FIGURE 9.4 Misericord from a church in the U.K. The seat is folded up and exposes the broad ledge which is used for sit-standing.

There are standard recommendations in the ergonomics literature for table (work surface) height for seated and standing workplaces (Ayoub, 1973; Kroemer et al., 2002). Figure 9.5 illustrates that to arrive at a suitable hand position the work table must be put at a lower height for a tall product and a higher height for a short product.

The most advantageous hand position depends on the type of task. For heavy work, it is most convenient to hold the hands about 15 cm (6 in) below elbow height. The arms and the body can then exercise a greater leverage to perform the heavy task more efficiently. For light assembly work the preferred hand height is about 5 cm (2 in) below elbow height.

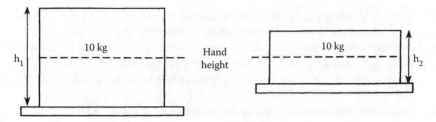

FIGURE 9.5 Appropriate working height for a table is determined by the height of the product and the hand height. In the figure we assume that the products are manipulated or held at an intermediate height. If the product height is h, it would be held at height h/2.

Typing is often performed with the hands about 3 cm (1 in) above elbow height. For precision work with supported elbows and/or supported underarms, the hand height should be about 8 cm (3 in) above elbow height. It is easier to perform precision work with the hands and underarms supported. Another reason is that precision work involves small parts and fine details which can be viewed more easily if the objects are closer to the eyes (at about reading distance).

There are, however, individual preferences in work posture. In typing, for example, some individuals prefer to type with horizontal underarms, but others prefer to raise their hands above the horizontal position and some prefer to lower their hands. Therefore, the values listed in Table 9.3 are intended as guidelines rather than absolute recommendations. In order to calculate a recommended table height we must consider individual preferences for work posture as well as anthropometric measures of 5th to 95th percentile elbow height above the floor. When these two parameters are considered, the result provides a fairly wide range of values, as illustrated in Table 9.3. To arrive at suitable values for table height or work bench height from Table 9.3 the handling height of the product must be deducted from the hand height.

EXERCISE: CALCULATE WORKING HEIGHT

In this industrial task, 25 kg boxes are transported on a conveyor belt. The operator must turn them over to label both sides. The boxes are 50 cm high and are handled at half-height (25 cm). Calculate the preferred height of the conveyor belt using a 5th to 95th percentile range for standing male operators.

Solution: From Table 9.3 the range for hand height over the floor is 91–110 cm. Deducting 25 cm gives a range of 66–85 cm for the height of the conveyor.

Calculate the range of adjustability for a typing table for female 5th percentile to male 95th percentile operators.

Solution: The range for hand height is 63–87 cm. Assuming a 3 cm high home row (center row) of the keyboard, the table top height is 60–84 cm (23.5–33.0 in).

In a manufacturing plant, sitting workstations will be used for light assembly. Assume that there is a female population of workers, that the hand is held at elbow height minus 5 cm, and that the hand height above the floor is 55–73 cm. Assume further that the product has a handling height of $H/2$ cm, where H is the product height. What is the maximum product height if the worktable is 3 cm thick?

Solution: The solution to the problem is shown in Figure 9.6.

Sitting elbow height = 5 + $H/2$ + 3 + Thigh clearance (cm)

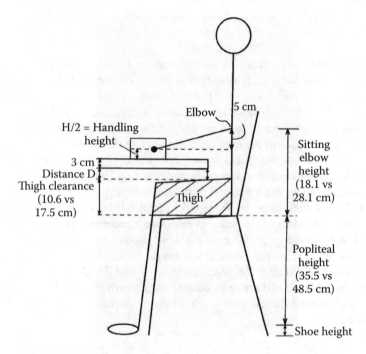

FIGURE 9.6 Calculation of product height. In the calculations assume that D = 0. The numbers given in parentheses are the 5th and 95th female percentiles.

The 5th percentile female operator has a sitting elbow height of 18.1 cm, which is not enough to accommodate the thigh clearance of 10.6 cm, table thickness of 3 m, and hand height of 5 cm below elbow height. In this case $H = -1.0$ cm. Obviously, for small parts assembly this workstation is still acceptable, but if large products are handled we may want to consider a sit-standing or standing work posture. This does not imply that one would disallow products with greater height at a sitting workstation. Operators can adapt to some extent—for example, by gripping the product further down and raising the hands to elbow height. For the 95th percentile female operator, this situation is not so critical because the sitting elbow height is much greater. In this case $H = 5.2$ cm.

TABLE 9.3
Measures (cm) of Preferred Hand Height Over the Floor

Type of Task	Hand Height	Elbow Height (Range)	Preferred Hand Height Over Floor* (cm)			
			Standing (5th to 95th)		Sitting (5th to 95th)	
			Male	Female	Male	Female
Heavy lifting	−15	−20 to −10	91 to 110	85 to 110	Not recommended	Not recommended
Light assembly	−5	−10 to 0	101 to 120	95 to 110	59 to 79	55 to 73
Typing	+3	0 to +6	109 to 128	103 to 118	67 to 87	63 to 81
Precision work	+8	+5 to +10	Not recommended	Not recommended	72 to 92	68 to 91

The ranges are for females and males from 5th to 95th percentile (see Table 3.2) and were obtained by deducting or adding the value for hand height. Shoe height of 3 cm is included. 1 in = 2.54 cm.

9.5 WORK AT CONVEYORS

Conveyors are increasingly being used in manufacturing, not only for transportation, but also at assembly lines and for temporary storage. Often these systems are physically connected. At a workstation this arrangement has the advantage that an operator can push items from a moving conveyor to a storage or an assembly line conveyor and is not paced by the line. The operator can thus work faster or slower, as long as the buffer capacity of the storage conveyor is not exceeded (Konz, 1992a).

There is a common belief in industry that the height of the conveyor line must be fixed and consistent throughout a plant. The commonly preferred height is 92 cm (36 in), which is the same as for industrial standing workstations. This may not always be ideal. Obviously one must avoid downhill and uphill slopes, but there are biomechanical reasons why heights could be different at different locations.

For people working at the conveyors, one should adopt the same rules for determining work height as for regular sitting and standing workstations (see Table 9.3).

The purpose is to make the conveyor height convenient for manual work (not for the engineers who design the plant). Thus, the conveyor height should depend on the size of the object that is being handled. For example, if there are large steel drums transported on the conveyor, and if they are handled by workers, then the conveyor height must be very close to the floor to make such handling convenient. Nagamachi and Yamada (1992) demonstrated that the concept of variable conveyor height worked well in a Japanese plant that manufactured air conditioners. The conveyor line was used for assembly and depending on the height of the work items, the height of the conveyor shifted. They referred to this as a "Panama Canal" conveyor. Productivity and quality improved with this design.

If the work along the conveyor is performed sitting, the hand height should be the same as for other sitting workplaces, that is, for light assembly about 55–79 cm (22–31 in). There must also be leg room and knee room as with other seated workplaces. In addition, to avoid a bad work posture, the conveyor must be thin so that it can fit in the space between the thighs and underarms. A thick conveyor or a tall fixture will force the operator to raise his or her arms, thereby creating a bad work posture.

Sometimes products on a conveyor line create jams. In order to break up the jams, the conveyor must be accessible from both sides so that two people can work together (Eastman Kodak Co., 2004).

Since conveyor lines can extend throughout an entire plant, it is important to provide crossing points or gates where people and material can be brought through. It should not be necessary to crawl under the conveyor line.

Conveyors can help in manual materials handling at workstations. It should be possible to slide assemblies along the conveyor rather than to lift them. This can be achieved by using special rollers or low friction material to connect a moving and a stationary conveyor at a workstation.

Loading, and especially unloading, of conveyors presents hazards and can result in overexertion and back injuries. Typically unloading is much more demanding and there are three times as many injuries as for loading. This is because the operation is often paced by the movement of the conveyor line, and products typically weigh more when they come off the conveyor line after the assembly (Cohen, 1979).

People working at conveyor belts may develop "conveyor sickness" (T. G. and R. L., 1975). This may be true not only for moving conveyors but also for other moving objects such as carousel storage units. If the conveyor speed is greater than 10 m/min (32 ft/min) operators can develop nausea and dizziness. This may be particularly common if a person sits sideways to the conveyor such that the motion is perceived in the peripheral vision.

9.6 MEASUREMENT AND ERGONOMIC IMPLICATIONS OF WORK POSTURE

Poor body posture and forceful working methods can lead to permanent damage of body tissues. It is important to be able to classify body postures and force and draw conclusions that can be used to improve the design of jobs and work places. We must understand that people do not deliberately assume poor work postures; they are forced to do so because of the characteristics of the task and poor ergonomics design of jobs and work stations. Several examples have already been given in this chapter. Below we discuss two types of methods for assessing musculoskeletal problems: subjective assessments performed by employees, and objective assessments performed by an analyst.

SUBJECTIVE METHODS

We describe two methods that can be used by employees to evaluate musculoskeletal problems: a body part discomfort scale and the use of questionnaires.

Body Part Discomfort Scale

Corlett and Bishop (1976) presented a technique for measurement of body part discomfort. They demonstrated that the amount of discomfort (on a five-point scale) is linearly related to the amount and the duration of a particular force. The longer the force is held, the greater discomfort reported. For example, assume that you are carrying a suitcase to the bus station. The weight of the suitcase can be determined, and the time that you carry is can also be measured. The interesting aspect about this scenario is that the amount of discomfort is linearly related to the time (in minutes) that you carry the case. To specify the location of the discomfort the body map in Figure 9.7 is used. The body is divided in segments and the person is asked to rate the amount of comfort on a seven (or five) point scale, where 0 corresponds to no discomfort and 7 to extreme discomfort.

To investigate the effect of hours of work, one can ask the person to assess body part discomfort several times per day. Often the discomfort will be localized to a few areas of the body. It is then possible to ask about the discomfort in those locations and ignore the other locations. Again we would expect that discomfort will increase linearly with the time spent at work.

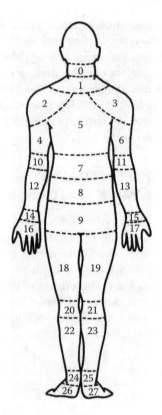

FIGURE 9.7 Dissection of the body in different parts for the measurement of body part discomfort (Corlett and Bishop, 1976).

Questionnaires

Several questionnaire evaluation methods have been developed and validated. One popular method is the Nordic questionnaire for evaluation of musculoskeletal problems (Kuorinka et al., 1987). This is a multiple-page questionnaire. Since it was published it has been validated by Dickinson et al. (1992) and Chaffin and Anderson (1991). This is as close to an international method as one can get, and it is therefore good for cross-cultural evaluations.

Another tool developed at the Swedish National Board of Occupational Safety and Health is a single-sheet analysis for identifying musculoskeletal problems (see Figure 9.8) (Kemmlert and Kihlbom, 1986). This tool is self-explanatory and can be used for example to make before and after comparisons to demonstrate the effectiveness of ergonomic improvements.

For example, assume that you want to measure the effect of a lifting aid at a workstation in a manufacturing plant. You are not quite sure if it helps very much, so you would like to evaluate the effect. By using the questionnaire before and after the installation, you can assess the effectiveness. If the evaluation is favorable, you may want to install some more lifting aids at other work stations.

FIGURE 9.8 One-page questionnaire for self evaluation of musculoskeletal stress factors which may lead to injury (Kemmlert and Kihlbom, 1986).

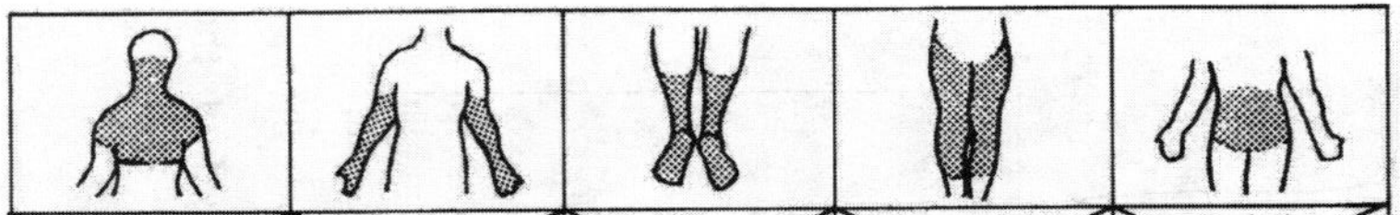

Neck shoulders, upper part of back	Elbows, forearms hands	Feet	Knees and hips	Low back	
		1.	1.	1.	1. Is the walking surface uneven, sloping, slippery or nonresilient?
2.	2.	2.	2.	2.	2. Is the space too limited for work movements or work materials?
3.	3.	3.	3.	3.	3. Are tools and equipment unsuitably designed for the worker or the task?
4.				4.	4. Is the working height incorrectly adjusted?
5.				5.	5. Is the working chair poorly designed or incorrectly adjusted?
		6.	6.	6.	6. (If the work is performed while standing): Is there no possibility to sit and rest?
		7.	7.		7. Is fatiguing foot-pedal work performed?
		8. a b c	8. a b c	8. a b c	8. Is fatiguing leg work performed eg: a) Repeated stepping up on stool, step etc.? b) Repeated jumps, prolonged squatting or kneeling? c) One leg being used more often in supporting the body?
9. a b c d				9. a b c d	9. Is repeated or sustained work performed when the back is: a) Flexed forward, more than 20°? b) Severely flexed forward, more than 60°? c) Bent sideways or twisted, more than 15°? d) Severely twisted, more than 45°?
10. a b c d					10. Is repeated or sustained work performed when the neck is: a) Flexed forward, more than 15°? b) Bent sideways or twisted, more than 15°? c) Severely twisted, more than 45°? d) Extended backwards?
11. a e b f c g d				11. a e b f c g d	11. Are loads lifted manually? Notice factors of importance as: a) Periods of repetitive lifting b) Weight of load c) Awkward grasping of load d) Awkward location of load at onset or end of lifting e) Handling beyond forearm length f) Handling below knee height g) Handling above shoulder height
12.	12.			12.	12. Is repeated, sustained or uncomfortable carrying, pushing or pulling of loads performed?
13.	13.				13. Is sustained work performed when one arm reaches forward or to the side without support?
14. a b	14. a b				14. Is there repetition of: a) Similar work movements? b) Similar work movements beyond comfortable reaching distance?
15. a b	15. a b				14. Is repeated or sustained manual work performed? Notice factors of importance as: a) Weight of working materials or tools b) Awkward grasping of working materials or tools
16.					16. Are there high demands on visual capacity?
	17. a c b d				17. Is repeated work, with forearm and hand, performed with: a) Twisting movements? b) Forceful movements? c) Uncomfortable hand positions? d) Switches or keyboards?

Method of application.

- Find the injured body region
- Follow white fields to the right
- Do the work tasks contain any of the factors described?
- If so, tick where appropriate

Also take these factors into consideration:
a) the possibility to take breaks and pauses
b) the possibility to choose order and type of work tasks or pace of work
c) if the job is performed under time demanded or psychological stress
d) if the work can have unusual or unexpected situations
e) presence of cold, heat, draught, noise or troublesome visual conditions
f) presence of jerks, shakes or vibrations

OBJECTIVE METHODS

Two objective measures of work posture, OWAS and RULA, are described below. Both methods are described in full detail in Corlett (1995).

OWAS Method

The OWAS method was developed in Finland (Karhu et al., 1981). It is now available as software and can be downloaded from <http://turva.me.tut.fi/owas/>. By using OWAS one can code work postures using a three-digit code, to which three more numbers are added to describe the amount of force and the work phase (see Figure 9.9). An experimenter observes the worker, makes an assessment of the posture, and records the result on the data sheet in Figure 9.9. The results from all work phases are then assembled, and an assessment is made whether there is a need to take immediate action to improve the design of the work station or the task. Such immediate action would be prompted if, for example, the person works with a bent and twisted back or with bent knees for more than 70% of the time.

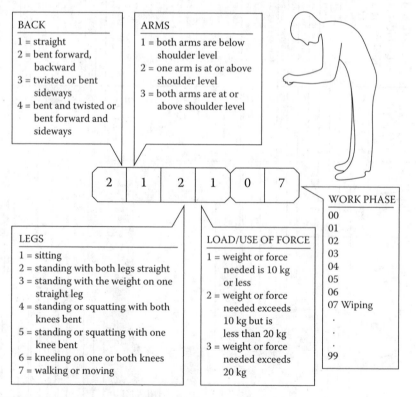

FIGURE 9.9 The first three digits classify the posture and the following digits the force and the work phase.

RULA Method

Rapid upper limb assessment (RULA) is similar to OWAS (McAtamney and Corlett, 1993) (see Figure 9.10). Postures are evaluated using numbers; the greater the number, the worse the posture. Values of force are then estimated.

Figure 9.10 shows the items for assessment when using the RULA method. Group A measures the effect on the arms and hands, and Group B measures the effect on the neck and the trunk. Average values of the postures for Group A and Group B are calculated and the exerted force is added to form an overall score. Just as with the OWAS evaluation, a grand score is calculated, and if the score is high immediate action is required.

Corlett (1995) pointed out that both OWAS and RULA have a deceptive appearance of simplicity and ease of use. It takes, however, much practice to estimate angles and forces.

EXERCISE: EVALUATION OF MUSCULOSKELETAL STRESS FUNCTION

Make copies of the questionnaire in Figure 9.8. Find a workplace where there is much physical workload, such as a construction site. Find 5 experienced workers (about 40 to 50 years of age) and ask them to fill in the questionnaire. Compare the results.

1. Were the results for the five workers similar? If not, try to explain why.
2. Were there problems with the questionnaire? Try to explain why.
3. How can you, as an ergonomist, increase the motivation of workers to take the questionnaire seriously and spend the necessary time to fill in the information?

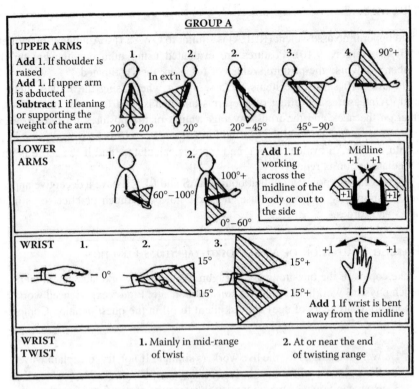

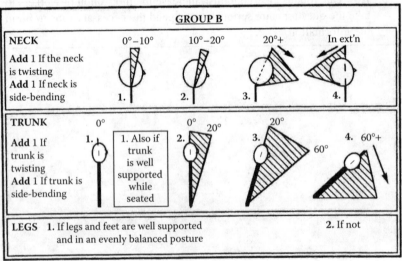

FIGURE 9.10 Items for assessment when using the RULA method.

FURTHER READINGS

The following book is an excellent resource to learn more about industrial ergonomics: Chengalur, N., Rodgers, S.H. and Bernard, T.E., 2004, *Kodak's Ergonomic Design for People at Work*, 2nd ed. New York: Wiley.

10 Manual Materials Handling

10.1 INTRODUCTION

Musculoskeletal disorders account for almost 70 million visits to physicians in the U.S. annually. About 1 million people take time off from work annually to treat and recover from work-related musculoskeletal pain or impairment in the lower back or upper extremities. The economic consequences can be measured by adding compensation costs, lost wages, and lost productivity; the total is usually between U.S. $45 and U.S. $54 billion annually (National Research Council, 2001). There has been considerable debate concerning the causes, nature, severity, and degrees of work-relatedness of musculoskeletal disorders as well as the effectiveness and cost-related benefits of various interventions. None of the common musculoskeletal disorders is uniquely caused by work exposures. They are what the World Health Organization calls "work-related" conditions because they can be caused by work exposures as well as non-work factors.

In the U.K., for the period 2001–2002, 27% of all reported accidents involved manual handling. An estimated 12.3 million workdays were lost (Health and Safety Executive, 2003). Manual handling and lifting are the major causes of work-related back pain (Keyserling and Chaffin, 1986). However, back pain, and in particular low back pain, is also common in other work environments such as seated work, where there is no lifting or manual handling (Lawrence, 1955). In fact, back pain is extremely common. During a lifetime, there is a 70% chance of developing low back pain, and there is a 1 in 7 chance that any individual will be suffering from back pain presently (Pheasant and Stubbs, 1992). Many low back injuries seem to happen spontaneously, and Magora (1974) indicated that lifting and bending were related to only about one third of back injuries. Thus, the prevention of back injuries due to lifting will prevent only a small proportion of such injuries. Liberty Mutual has published yearly information on the cost for workplace injuries in U.S. (see Table 10.1).

In this chapter, we will first present statistics of back injuries associated with lifting. Then, we will analyze correct lifting techniques, and what can be done to help individuals lift correctly through training. A biomechanical model for calculating the compressive force in the lower back is presented. This model has been important in establishing federal guidelines for lifting, such as the current directives for the European Community and the National Institute for Occupational Safety and Health (NIOSH) guidelines in the U.S. Finally, several lifting aids that can be used in manufacturing are described.

TABLE 10.1
Direct Costs and Leading Causes of Serious Workplace Injuries for 2001

Cause	Activities	U.S. $ (billions)	% of Cost
Overexertion	Lifting, carrying, pulling, pushing, etc.	12.5	27.3
Falls on same level	Falling	5.7	12.6
Bodily reaction	Bending, climbing, slipping, tripping without falling	4.7	10.2
Falls to lower level	Falling	4.1	9.0
Struck by object	E.g., tool falling on worker from above	3.9	8.6
Repetitive motion	Repeated stress and strain	2.9	6.3
Highway incident	Driving	2.3	5.1
Struck against object	E.g., worker walking into door frame	1.9	4.1
Caught in equipment	Equipment caught worker	1.7	3.7
Assaults, violent acts		0.4	1.0

Includes payments to injured persons and their medical care providers.

10.2 STATISTICS ON BACK INJURIES ASSOCIATED WITH LIFTING

In 1982, the U.S. Department of Labor published a report of 906 back injuries associated with lifting. The interesting aspect about this study was that only accidents due to manual materials handling were analyzed, and there were no "faking" accidents in the data. From the data it was observed that 42% of all back injuries due to lifting occurred in manufacturing. This was three times as much as for any other industry. Back injuries are therefore frequent in manufacturing and it is important to analyze their causes.

The 906 workers were asked what they were doing when they injured their backs (Table 10.1). The percentage values in Table 10.1 add up to more than 100% because many workers reported engaging in more than one activity. In the table we report the number and percentage of accidents rather than the accident rate. The accident rate would be obtained if the number of accidents were divided by the amount of time engaged in each activity. Accident rate would be a much more informative measure. For example, if the individuals who engage in, say, pushing (4% of the accidents) took only 1% of the total time, then the accident rate for pushing would be 4 times greater than the average. Unfortunately, there is no information on the amount of time that workers spend on each of the different activities. We cannot therefore only analyze the data from a frequency point of view; that is, the number of accidents per hour of work. From Table 10.2, we conclude that it makes sense to focus on lifting, since a reduction in the number of lifting accidents would have the greatest impact on overall safety.

TABLE 10.2
Workers' Accounts of What Activities
They Were Engaged in While Lifting

Activity	No. of Accidents	Activity %
Carrying	133	15
Holding	96	11
Lifting	692	77
Lowering	107	12
Placing	145	16
Pulling	65	7
Pushing	39	4
Shoveling	14	2
Other	25	3
Total	906	

Some workers reported more than one activity.

Workers were then asked what types of movements they were doing when their backs were injured (Table 10.3). Again, many workers reported several simultaneous activities. The most dangerous activity was bending, followed by twisting and turning. This verifies what has been pointed out by many researchers: a combination

TABLE 10.3
Common Movements Undertaken When the Back
Was Injured

Activity	No. of Accidents	% of Accidents
Bending	505	56
Climbing	16	2
Squatting	107	12
Standing	243	27
Stretching	141	16
Suddenly changing position	159	18
Twisting/turning	299	33
Walking	72	8
Other	26	3
Total	894	

TABLE 10.4
Workers' Responses to the Question "Was Lifting Equipment Available?"

Response	Number of Accidents	% of Accidents
Equipment not available	434	60
Equipment available but not used because:		
Did not think it was necessary	61	9
It was not practical to use	121	17
It was not working	11	2
It takes too long	16	2
Injury occurred while using equipment	41	6
Other	34	5
Total	708	101

of bending and twisting/turning puts a torque on the spine, and the likelihood for back injuries therefore increases.

Workers were asked if lifting equipment was available. The responses are shown in Table 10.4. In the majority of accidents, lifting equipment was not available. However, the availability of lifting equipment does not necessarily mean that it will be used. In many cases existing equipment was not used because it was not practical (17%) or workers did not think its use was necessary (9%). The practicality of lifting aids is crucial. If workers find that lifting aids slow down work, the chances are that they will not be used. A slow lifting aid reduces productivity and a worker's sense of accomplishment. It is essential that before ordering lifting equipment, an analysis is made of the practicality of the equipment with regard to the task and the effect on productivity.

10.3 BASIC BIOMECHANICS

The basic problem in lifting is that the force from a lifted load becomes ten times larger in the spine. We will explain this phenomenon below. The human spine is a flexible column of 24 vertebrae with a large wedge-shaped bone at the bottom, which is called the sacrum (Figure 10.1). Between each pair of vertebrae there are discs which act as shock absorbers. On top of the sacrum are five lumbar vertebrae referred to as Ll to L5. The bottom disc L4/L5 is the site of most back injuries.

A disc has a fibrous outer layer and is filled with fluid. With increasing age, and also with increasing exposure to manual material handling, cracks develop in the disc, and if there is a great amount of pressure, there is a risk of disc herneation (Michel and Helander, 1994). The fluid of the disc will press through the fibrous outer layer and put pressure on the nerves adjacent to the spine. A graphic but not quite accurate analogy is squeezing a jelly doughnut. Most medical experts now

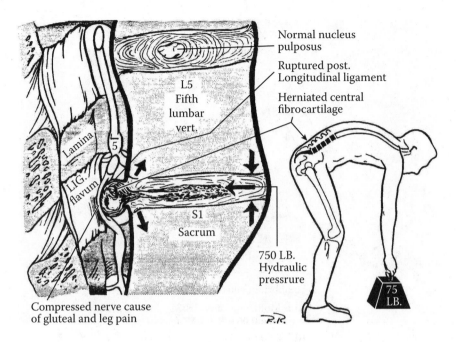

Normal nucleus
pulposus

Ruptured post.
Longitudinal ligament

Herniated central
fibrocartilage

L5
Fifth
lumbar
vert.

Lamina

5

LIG.
flavum

S1
Sacrum

750 LB.
Hydraulic
pressrure

75
LB.

Compressed nerve cause
of gluteal and leg pain

FIGURE 10.1 Illustration of the L4/L5 and the L5/S1 discs. The L5/S1 disc shows herniation. From the classic drawings of Keegan (1953).

believe that only about 5% of back injuries involve damage to the discs. However, when they occur, these injuries tend to be more serious and long lasting. Fracture of the vertebrae is very rare in lifting accidents, unless the bones have become softened, as in osteoporosis.

Since disc injuries occur in either the L4/L5 disc or the L5/S1 disc, the biomechanical calculations are done for these discs. To calculate the compressive force on disc L5/S1 we make several assumptions. We assume that the individual weighs 75 kg, and that 65% of the body mass is in the upper part of the body, denoted by the vector B (Lindh, 1980). The length of the moment arm from the erector spinae muscle to the disc is 6 cm. The calculations are illustrated in Figure 10.2.

Let us apply this model to the two different cases of lifting shown in Figure 10.3. Assume that for the case of lifting with a bent back (A) the moment arms are w = 40 cm and b = 26 cm. For lifting with a straight back (B) the moment arms are somewhat reduced: w = 35 cm and b = 18 cm.

Assuming that B = 75 × 0.65 × g = 75 × 0.65 × 9.81 = 478 N and that W = 250 N, we can use Equation 10.1 in Figure 10.2 to calculate that for case (A), ES = 3658 N. Assuming a body inclination of 30°, the disc compressive force is calculated using Equation 10.2: F = 3658 + 478 × 0.89 + 2500.89 = 4306 N. Similarly, for case (B), ES is calculated to 2892 N and, assuming a body inclination of 30°, F = 3540 N. This corresponds to a reduction in disc compressive force by 18% for the case with bent knees (B).

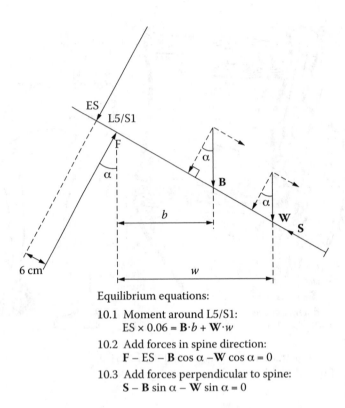

Equilibrium equations:

10.1　Moment around L5/S1:
$$ES \times 0.06 = \mathbf{B} \cdot b + \mathbf{W} \cdot w$$

10.2　Add forces in spine direction:
$$\mathbf{F} - ES - \mathbf{B} \cos \alpha - \mathbf{W} \cos \alpha = 0$$

10.3　Add forces perpendicular to spine:
$$\mathbf{S} - \mathbf{B} \sin \alpha - \mathbf{W} \sin \alpha = 0$$

FIGURE 10.2 Calculation of the disc compressive force *F* and disc shear force *S*. *ES*: erector spinae muscle force; *B*: force from upper body weight; *W*: force from lifted weight.

This model makes many simplifying assumptions. In the first place, lifting is analyzed as a static activity, whereas in reality it is very much a dynamic activity. Dynamic lifting models have been developed, and these give compressive disc forces that are 20–200% of the static case (e.g. Garg et al., 1982; McGill and Norman, 1986). These models are still under development, and are not yet practical to use. Chaffin (1969) assumed that abdominal pressure will also affect the lifting model. This additional assumption may, however, not affect the calculations very much (Waters et al., 1993).

10.4　THE SO-CALLED CORRECT LIFTING TECHNIQUE

In many organizations, courses are given to train employees in correct lifting technique. This entails lifting with a straight back and bent knees, which, as we have seen earlier, can reduce the disc compressive forces. The International Labor Organization (1972) published several kinetic methods which build upon this technique, as illustrated in Figure 10.4.

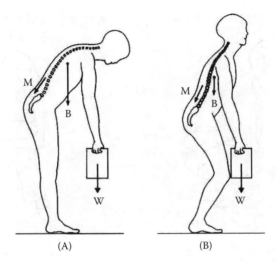

FIGURE 10.3 An individual weighing 75 kg is lifting an object of 25 kg. (A) Lifting with a bent back. (B) Lifting with bent knees and a straight back.

The guidelines for correct lifting techniques—straight back, bent knees—have become quite controversial in recent years. The first observation is that this technique only applies to small compact objects that can be held between the legs while lifting. Larger boxes, for example, are too large to lift with the straight back, bent knees technique. To clear the knees it will be necessary to hold the box at some distance. In addition, there is much more load on the leg muscles. For many lifts, this technique is simply difficult and awkward and it will not be used by workers. Furthermore, Garg and Herrin (1979) calculated disc compressive forces and concluded that stoop lifting (bent back, straight knees) is sometimes superior to squat lifting (straight back, bent knees).

There have been many studies investigating the effect of training in manual materials handling. Unfortunately very few studies have used control groups, so it is not possible to draw any firm conclusions. (For a complete discussion of these issues, see Kroemer et al., 2001.) The problem with not having a control group is that other simultaneous changes in a company, or among those who are tested, could affect the outcome of training. Let us assume, for example, that training in manual materials handling is introduced in a company. As this is done, not only is awareness of good lifting techniques promoted, but there are also several other simultaneous effects.

- Understanding that low back pain is the cause of elevated workmen's compensation premiums in the company
- Understanding that the company expects fewer accidents to be reported in the future
- Understanding the priority expressed by management of reducing low back injuries

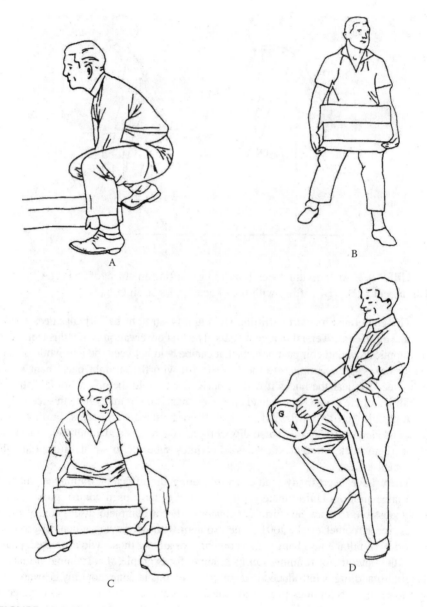

FIGURE 10.4 Illustration of "correct" lifting techniques. (A) Squatting while lifting. The angle of the knee of the front leg is approximately 90°, the arms are held close to the body, and the back is straight. Before raising the load the chin is tucked in, which tends to further straighten the back while lifting. After raising the load the lifter is immediately ready to move horizontally by using the momentum of the body weight. (B) A weight should be carried with straight arms. This reduces the tension in the upper arm and shoulder muscles. (C) The arms should remain straight while lifting. The feet are placed apart to prepare for forward movement. (D) This illustrates how a load can be raised to bench height by using leg muscles, thus reducing the risk of back strain (International Labor Organization, 1972).

- Feeling better about the personal concern shown by the company compared with the past
- Experiencing greater job satisfaction and cohesion with coworkers

These secondary factors are likely to affect the reporting of accidents. Thus, there are reasons why the reported injury rate could be affected by parameters other than the incidence of low back pain. In a similar fashion, a worker's decision to return to work after low back pain treatment is affected by management attitudes and psychological factors (Snook, 1988).

Assuming one can control for these motivational effects, it could then be possible to make a fair comparison. Scholey and Hair (1989) investigated the incidence of back pain among 212 physical therapists who were involved in back-care education. One would think that physical therapists would be careful to report all occurring back pain and that they would be less biased by secondary factors. Their incidence rate was compared with that in a carefully matched group consisting of individuals who were not physical therapists. There was no difference in the incidence rate reported in the two groups.

- Lifting skills: correct body positioning, posture, and movement
- Awareness and attitudes: physics and biomechanics of lifting
- Fitness and strength, "Chinese" style

In developing a training program one must first consider what to train. There are many problems with training programs (Kroemer et al., 2001). There is usually a limited time effect of training. During the immediate time following training, trainees have a sense of enthusiasm and relevance. After a few weeks the information sinks in and is perceived as secondary to many other problems. People tend to revert to previous habits if training is not reinforced.

One of the problems in teaching correct lifting techniques is the lack of feedback from the body itself while lifting. There are no nerve endings in the discs, which means that the lifter is not aware of differences in disc pressure due to lifting technique. The trainee must rely on feedback from the training instructor, peers, and managers.

Emergency situations which lead to back injury are difficult for an individual to control. As with other accidents, several different things occur simultaneously. The individual must make quick decisions, and body movements cannot be controlled. The situation is quite different from planned, deliberate lifting, which can be controlled. Therefore, if job requirements are basically stressful, behavior modification through training may not be successful. It is better to design safe jobs where manual handling is less frequent.

10.5 NIOSH GUIDELINES AND STANDARDS FOR LIFTING

In many countries there are guidelines and standards which limit lifting in the workplace. The purpose of these guidelines is to reduce the amount of low back pain as well as to reduce work injuries. The rationale is that manual lifting poses a risk of low back pain, and low back pain is more likely to occur if the load exceeds the worker's physical capabilities. In addition, the physical capabilities of workers vary extensively, and in designing workplaces and tasks one must consider that some workers are less capable than others.

Below we look at three important sets of guidelines: the NIOSH guidelines (Waters et al., 1993), the European Community guidelines, and the Safety and Health Commission Guidelines from the U.K. Each takes a different approach to the determination of acceptable weights. Mital et al. (1993) provide detailed information on these and other regulations pertinent to manual materials handling.

1991 NIOSH EQUATION FOR EVALUATION OF MANUAL LIFTING

This new equation replaces the former NIOSH lifting equation published in 1981. In developing the present guidelines, three criteria of lifting were considered: biomechanical, physiological, and psychophysical (Waters et al., 1993).

The biomechanical criterion was based on calculating the compressive forces in the L5/S1 disc. Several studies have indicated that, during lifting, the largest moments are created in the trunk area and the L5/S1 disc is at greatest risk. This criterion is most important to delimit the weight of infrequent, heavy lifts in the lifting equation. Based on studies of human cadavers it was concluded that a maximum disc compressive force is 3.4 kN, although for some individuals it may be twice as much.

The physiological criterion evaluates the metabolic stress and muscle fatigue that may develop during lifting. This criterion is most important for frequent lifting. To limit muscle fatigue the maximum aerobic work was set to 9.5 kcal/min. This corresponds to the average, 50th percentile female work capacity. A single lifting task should not impose greater demands than 70% of the maximum aerobic capacity. For long work periods such as 1 hour, 1–2 hours, and 2–8 hours, the maximum work rate must be lowered to 50%, 40%, and 33% of the maximum aerobic capacity, respectively (see Figure 12.2).

In developing the equation it was considered that working at waist level, at a height of 75 cm, is the most comfortable. Lifts above waist level involve both the shoulder and the arm, whereas lifts below waist level involve the whole body.

The third criterion, the psychophysical criterion, took into consideration the acceptability of lifts to workers. This type of criterion is based on subjective judgment among workers; the chosen limit for lifting should be acceptable to 75% of female workers and 99% of male workers. The calculations are based on experimental studies where subjects are asked over the course of an experiment to rate the acceptability of a lifting task.

The NIOSH equation for calculating the recommended weight limit (RWL) represents a compromise between the three different criteria discussed above. It is a multiplicative model and several task variables are included as weighting functions (Waters et al., 1993) (see Table 10.5):

$$RWL = LC \times HM \times VM \times DM \times AM \times FM \times CM$$

The calculations are performed twice—once for the point of origin and once for the point of destination. If the point of destination does not involve controlled lifting—for example, when the lifter drops the object in place—the latter calculation is excluded.

Example 1: Loading Punch Press Stock

The normal job of a punch press operator is to feed small parts into a press and remove them (Putz-Anderson and Waters, 1991). Once per shift the operator is required to load a heavy reel of supply stock from the floor to the machine (i.e., to a height of 160 cm), as illustrated in Figure 10.5. The reel is 75 cm in diameter and weighs 20 kg. Assume that the operator lifts the reel in the sagittal plane (in front of the body) as shown, and that to load the reel the operator must exercise significant control at the destination of the lift.

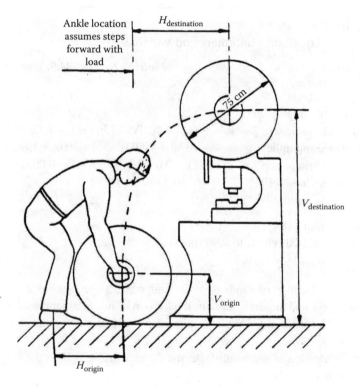

FIGURE 10.5 Calculation of NIOSH lifting limits: loading punch-press stock.

Solution:

H (origin)	H (destination)	V (origin)	V (destination)	F
57.5 cm	57.5 cm	38 cm	160 cm	<0.2

For the origin:

RWL = 23 × HM × VM × DM × AM × FM × CM

   = 23 × (25/57.5) × (1 − 0.003|38 − 75|) × (0.82 + 4.5/122) × 1.0 × 1.0 × 1.0

   = 7.6kg

For the destination:

RWL = 23 × (25/57.5) × (1 − 0.003|160 − 75|) × (0.82 + 4.5/122) × 1.0 × 1.0 × 1.0

   = 6.4 kg

TABLE 10.5
The NIOSH Equation: Multipliers and Variables

Multipliers	Metric	U.S. Customary
LC: Load constant	23 kg	51 lb
HM: Horizontal multiplier	(25/H)	(10/H)
VM: Vertical multiplier	(1 − 0.003|V − 75|)	(1 − 0. 0075|V − 75|)
DM: Distance multiplier	(0.82 + 4.5/D)	(0.82 + 1.8/D)
AM: Asymmetric multiplier	(1 − 0.0032A)	(1 − 0.0032A)
FM: Frequency multiplier*		
CM: Coupling multiplier**		

* Obtained from Table 10.6.
** Varies from 1.00 (good) to 0.90 (poor).

Variables

H Horizontal location of hands from the midpoint between the ankles. Measure at the origin and the destination of the lift (cm or in). *H* is between 25 cm (10 in) and 63 cm (25 in). Most objects cannot be lifted closer than 25 cm from the ankles.

V Vertical location of the hands from the floor. Measure at both the origin and the end-point of the lift.

D Vertical travel distance between the origin and the destination of the lift (cm or in).

A Angle of asymmetry—angular displacement of the load from the sagittal plane. Measure at the origin and at the destination of the lift (degrees).

TABLE 10.6
Frequency Multipliers (CM)

Frequency (lifts/min)	Work Duration (h)					
	<1		<2		<8	
	V < 75	V > 75	V < 75	V > 75	V < 75	V > 75
0.2	1.00	1.00	0.95	0.95	0.85	0.85
0.5	0.97	0.97	0.92	0.92	0.81	0.81
1	0.94	0.94	0.88	0.88	0.75	0.75
2	0.91	0.91	0.84	0.84	0.65	0.65
3	0.88	0.88	0.79	0.79	0.55	0.55
4	0.84	0.84	0.72	0.72	0.45	0.45
5	0.80	0.80	0.60	0.60	0.35	0.35
6	0.75	0.75	0.50	0.50	0.27	0.27
7	0.70	0.70	0.42	0.42	0.22	0.22
8	0.60	0.60	0.35	0.35	0.18	0.18
9	0.52	0.52	0.30	0.30	0.00	0.15
10	0.45	0.45	0.26	0.26	0.00	0.13
11	0.41	0.41	0.00	0.23	0.00	0.00
12	0.37	0.37	0.00	0.21	0.00	0.00
13	0.00	0.34	0.00	0.00	0.00	0.00
14	0.00	0.31	0.00	0.00	0.00	0.00
15	0.00	0.28	0.00	0.00	0.00	0.00
>15	0.00	0.00	0.00	0.00	0.00	0.00

From Waters et al. (1993). 75 cm = 30 in.

Because the operator must exercise significant control to load the reel, the calculation for the destination is required. The most protective of the two RWL values is used to estimate the job demands: the RWL for the destination is 6.4 kg, which is smaller than the RWL at the origin (7.6 kg). According to the lifting index formulated below, RWL may be multiplied by a factor of 3 which brings the load to about 20 kg.

Example 2: Product Packaging

In this example, products arrive via a conveyor at a rate of 1 per minute (Putz-Anderson and Waters, 1991). The worker packages the product in a cardboard box and then slides the packaged box to a conveyor behind table B, as illustrated in Figure 10.6. The product weighs 7 kg (16 lb), and the job is performed for an 8-hour shift. For this example, assume that significant control of the object is not required at the destination. Workers twist their bodies to pick up the product. Furthermore, assume that workers can flex the fingers to the desired 90° angle to grasp the container. The job is performed for a normal 8-hour shift, including regular rest breaks.

Solution: The task data are as follows:

H (origin)	H (destination)	V (origin)	V (destination)	F	Asymmetry (origin)	Asymmetry (destination)	Coupling
35 cm	33 cm	60 cm	100 cm	1/min	90°	0°	Fair

Since the worker can grasp the object with the fingers flexed at 90°, the couplings are classified as "fair" (Waters et al., 1993). In this example, the RWL is only computed at the origin of the lift, since significant control is not required at the destination.

For the origin:

$$RWL = 23 \times HM \times VM \times DM \times AM \times FM \times CM$$
$$= 23 \times (25/H) \times (1 - 0.003|V - 75|) \times (0.82 + 4.5/D) \times (1 - 0.0032A) \times 0.75 \times 0.95$$
$$= 23 \times (25/35) \times (1 - 0.003|60 - 75|) \times (0.82 + 4.5/40) \times (1 - 0.0032 \times 90) \times 0.75 \times 0.95$$
$$= 7.4 \text{ kg}$$

Thus the recommended weight limit is 7.4 kg, which is about the same as the actual product weight of 7 kg.

Lifting Index

The lifting index (LI) provides a simple estimate of the hazard of an overexertion injury for a manual lifting job:

$$LI = (\text{Load of weight } L)/(\text{Recommended weight limit RWL}) = L/RWL$$

where L is the weight of the object lifted (lb or kg). In their discussion of the lifting index, NIOSH conceded that lifts are often greater than RWL (Waters et al., 1993).

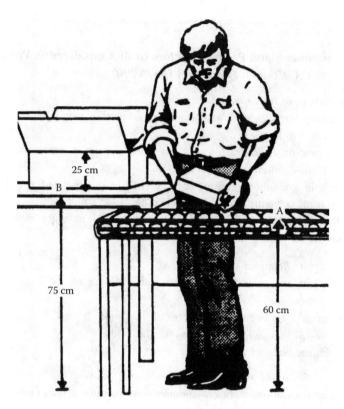

25 cm

B

75 cm

A

60 cm

FIGURE 10.6 Calculation of NIOSH lifting limits: product packaging.

Several experts agree that the lifting index should not exceed 3.0, because many individuals would be at a great risk.

10.6 GUIDELINES IN EUROPE

The Council of the European Communities has also formulated qualitative requirements for manual handling of loads "where there is a risk of back injury to workers" (EC Council Directive L156, 1990). This directive mandates employers to organize workstations to make manual handling a safe activity. Several factors are listed in Table 10.7.

GUIDELINES FOR MANUAL LIFTING IN THE U.K.

The Health and Safety Commission (1991) in the U.K. developed consultative guidelines for materials handling (Table 10.8). The criterion for the development of the guidelines was to consider a boundary "beyond which the risk of injury is sufficiently great to warrant a more detailed assessment of the work system." The guidelines are for lifts performed less than once per minute "under relatively favorable conditions." This implies a stable load which is easy to grasp and an upright

TABLE 10.7
Work, Environment, and Personal Factors to Be Considered in Workstation Organization (EC Council Directive L156, 1990)

1. Characteristics of the Load
The manual handling of a load may present a risk, particularly of back injury, if it is
- too heavy or too large
- unwieldy or difficult to grasp
- unstable or has contents likely to shift
- positioned in a manner requiring it to be held or manipulated at a distance from the trunk, or with a bending or twisting of the trunk
- likely that its contents and/or consistency would result in injury to workers in the event of a collision.

2. Physical Effort Required
A physical effort may present a risk, particularly of back injury, if it is
- too strenuous
- only achieved by a twisting movement of the trunk
- likely to result in a sudden movement of the load
- made with the body in an unstable posture.

3. Characteristics of the Working Environment
The characteristics of the work environment may increase a risk, particularly of back injury, if
- there is not enough room, in particular vertically, to carry out the activity
- the floor is uneven, thus presenting tripping hazards, or is slippery in relation to the worker's footwear
- the place of work or the working environment presents the handling of loads at a safe height or with good posture by the worker
- there are variations in the level of the floor or the working surface, requiring the load to be manipulated on different levels
- the floor or foot rest is unstable.

4. Requirements of the Activity
The activity may present a risk particularly of back injury if it entails one or more of the following requirements:
- over-frequent or over-prolonged physical effort involving in particular the spine
- an insufficient bodily rest or recovery period
- excessive lifting, lowering, or carrying distances
- a rate of work imposed by a process which cannot be altered by the worker.

TABLE 10.7
Work, Environment, and Personal Factors to Be Considered in Workstation Organization (EC Council Directive L156, 1990)

5. Individual Risk Factors

The worker may be at risk if he/she

- is physically unsuited to carry out the task in question
- is wearing unsuitable clothing, footwear, or other personal effects
- does not have adequate or appropriate knowledge or training.

TABLE 10.8
Guidelines for Lifting According to the Health and Safety Commission (1991)

Height	Less than Half Arm's Length (kg)	Between Half Arm's Length and Full Arm's Length (kg)
Below knee height	10	5
Knee height–knuckle height	20	10
Knuckle height–elbow height	25	15
Elbow height–shoulder height	20	10
Shoulder height–full length	10	5

work posture with a nontwisted trunk. Under such circumstances the guideline figures are assumed to provide reasonable protection to nearly all men and between one-half and two-thirds of women. There are also correction factors for stooping and twisting the body. For example, for 90° stooping the weight should be reduced by 50%, and for 90° twisting it should be reduced by 20%. One major advantage of these guidelines is that they are very easy to use.

10.7 MATERIAL HANDLINGS AIDS

In an industrial facility there are many different needs for materials handling: transportation of goods to and from the facility; unloading of materials at the receiving department; transportation of materials to workstations until the product has been assembled, tested, and inspected; and transportation of the product to packaging and to a warehouse for final distribution to customers. In addition to these primary transportation needs, there are also secondary transportation requirements, such as removal of waste products and housekeeping. Transportation and materials handling in manufacturing constitute a major expense. We therefore have a dual interest in designing an effective materials handling system:

- To reduce manufacturing costs.
- To reduce ergonomic costs and injuries.

The planning for materials handling and smooth transportation should start at the product design stage (Grossmith, 1992). One important aspect of product design is "design for ease of handling and transportability." Thus a product could have a smooth bottom, which makes it easier to transport on conveyor belts. The product can also be equipped with handholds (permanent or temporary) to simplify manual lifting.

Product design is also important because, by virtue of the design features, certain manufacturing processes will become necessary. The process equipment may be available in only one part of the plant, and a transportation need is then created. It may be possible to move the process equipment, so that it is practical for the manufacture of a specific product. However, there is usually a mix of products, and expensive process machinery must be used for many different products. Such issues are then important for the design and layout of a manufacturing facility.

The purpose of just-in-time (JIT) manufacturing is to structure the transportation activities. According to this philosophy, smaller quantities of parts are delivered to a manufacturing plant and then distributed to workstations, just in time for processing and assembly. The JIT philosophy has an interesting effect in that the need for storage of parts and products is reduced. Therefore, the manufacturing plant can be made smaller, and the cost of buying the land is also reduced (in Japan the cost of land is very high, which favors JIT).

10.8 MATERIALS HANDLING DEVICES

Table 10.9 presents a list of materials handling devices. There are many possible usages of materials handling devices in receiving, at workstations, and between workstations, testing, packaging, and warehousing. The usage depends entirely on the application and the task at hand. We cannot suggest any fixed formula; it depends on the creativity of the designer. Several of the aids are illustrated in Figure 10.7. Some devices are used for horizontal transportation in the plant and some of them for vertical transportation. From the ergonomics points of view it is particularly important to minimize vertical transportation, particularly if manual lifting is involved. "Don't put it on the floor, so you won't have to pick it up again."

It is difficult to avoid horizontal transportation in the plant, although one can try to minimize the transportation distance by optimizing the layout of the facilities. For horizontal transportation, conveyors have generic applicability and can be used for all the different manufacturing stages, including storage in the warehouse. Carts

TABLE 10.9
A List of Manual Materials Handling Devices and Their Possible Uses in Manufacturing

	Horizontal (H) or Vertical (V) Transportation	Receiving	At Workstation	Between Workstations	Testing	Packaging	Warehousing
Conveyor	H	x	x	x	x	x	x
Snake conveyors (easily movable)	H	x					x
Ball transfer table	H				x	x	
Carts	H	x	x	x	x		
Carousels	H		x	x	x		
Turntables	H			x	x		
Cranes	V	x	x	x	x	x	x
Hand trucks	H	x		x			x
Forklift trucks	H, V	x	x			x	x
Gravity feed conveyors/slides	H, V				x	x	
Automatic storage and retrieval	V, H	x			x		x
Stackers	V, H	x					x
Lift/tilt table	V		x		x	x	
Levelators	V		x		x	x	
Scissor table	V		x		x	x	
Vacuum lifting devices	V	x	x		x	x	x
Self-leveling table	V		x		x		
Adjustable table	V		x		x		
Overhead balancer	V		x		x		

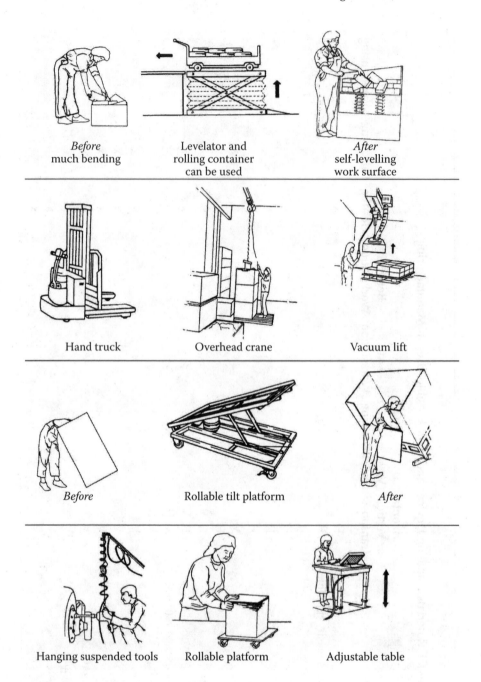

Before
much bending

Levelator and
rolling container
can be used

After
self-levelling
work surface

Hand truck

Overhead crane

Vacuum lift

Before

Rollable tilt platform

After

Hanging suspended tools

Rollable platform

Adjustable table

FIGURE 10.7 Illustration of some aids for lifting and materials handling (courtesy of Swedish Work Environment Fund, 1985).

and carousels are also fairly generic, and can be used at several sequential processes. A cart can be used as a moveable workstation that is passed down the line. It can be designed so that an operator can work conveniently at the cart.

Horizontal transportation is continuous and connects the different manufacturing functions. Vertical transportation is mostly local and discrete, and does not connect the different systems. An interesting maxim for the design of a plant would be "Minimize the vertical movement." This can be done, for example, by removing the top and the bottom shelves in storage. For JIT, with its minimal requirement for storage, this is not an unrealistic scenario.

Three of the vertical devices listed in Table 10.9 are automatic: self-leveling tables, gravity feed conveyors, and overhead balancers. These devices are particularly interesting because they do not require any action by the worker.

EXERCISE: MATERIALS HANDLING DEVICES

Discuss the materials handling devices in Table 10.9. If possible, make a study visit to a local manufacturing company. Make a map of the facilities and indicate which handling devices are used. Analyze the materials handlings solutions and transportation, and propose a redesign. Discuss how the devices listed in Table 10.9 and Figure 10.7 can be used to rationalize transportation and improve ergonomics.

RECOMMENDED READINGS

The reader is referred to the following books:

Chaffin, D.B., Andersson, G.B.J., and Martin, B.J., 1991. *Occupational Biomechanics*, 3rd ed., New York: Wiley.

Kroemer, K., Kroemer, H. and Kroemer-Elbert, K., 2001, *Ergonomics: How to Design for Ease and Efficiency*, 2nd Ed., Englewood Cliffs, NJ: Prentice-Hall.

11 Repetitive Motion Injury and Design of Hand Tools

11.1 INTRODUCTION

Repetitive motion injury (RMI) or cumulative trauma disorder (CTD) has been increasingly recognized in ergonomics during the last 20 years. But going back in history, RMI had already been recognized in 1717 by Ramazzini in Italy. He described RMI among office clerks and believed that these events were caused by repetitive motions of the hand, by constrained body postures, and by excessive mental stress (Franco and Fusetti, 2004).

There are many other terms, such as overuse disorder, musculoskeletal disorder, work-related disorder, repetitive distress or strain, and "motion injury". In this chapter we use the terms RMI and CTD interchangeably. Typically these injuries are caused by repetitive motions, such as of a hand, and there is a cumulative affect so that RMI may develop after an extended period of time (Putz-Anderson, 2005).

Liberty Mutual (2004) publishes annual statistics for the U.S. on causes and costs for workplace injuries. For 2001 the estimated costs for repetitive motion injuries was $2.9 billion. This is a significant amount, but less than overexertion (lifting, pulling, carrying, etc.), which cost $12.5 billion.

11.2 COMMON REPETITIVE MOTION INJURIES

Table 11.1 lists several different types of syndromes with both their medical and popular names. Below we will describe some of the most common syndromes.

CARPAL TUNNEL SYNDROME

The carpal tunnel is enclosed in the wrist and delimited by the bones of the hand and the carpal tunnel ligament (Figure 11.1). The carpal tunnel is a tight space containing several tendons, some blood vessels, and the median nerve. This crowded space is reduced in size even further when the hand or fingers are flexed or extended or bent to the side—ulnar deviation and radial deviation. Ulnar deviation is illustrated in Figure 11.1(C). The hand is bent outward—a common hand posture for keying and piano playing. Radial deviation is the opposite direction, when the hand is bent inward.

The median nerve enervates the index and middle fingers and the radial side of the ring finger. If there is a swelling inside the carpal tunnel, such as would incur

TABLE 11.1
Common Repetitive Motion Injuries

Scientific Disorder Name	Popular Names
Carpal tunnel syndrome	Telegraphist's wrists
Cubital tunnel syndrome	Clothes wringing disease
De Quervain's disease	Tennis elbow
Epicondylitis	Golfer's elbow
Ganglion	Bible bump
Shoulder tendonitis	Space invader's wrist
Tendonitis	Slot-machine tendinitis
Tenosynovitis	Pizza palsy
Thoracic outlet syndrome	Trigger finger
Ulnar nerve entrapment	

if a tendon was inflamed, or if there is external pressure, the median nerve can get squeezed and nerve conduction is no longer efficient. The symptoms of carpal tunnel syndrome are numbness, tingling, pain, and clumsiness of the hand—very much the same as when a foot falls asleep.

Carpal tunnel syndrome has been reported for many occupations in manufacturing (Silverstein et al., 1987). It is particularly significant for meat packers (Brogmus and Marko, 1990) and automobile workers (White and Samuelson, 1990). But it has also been observed among supermarket cashiers (Margolis and Kraus, 1987) and a variety of occupations in manufacturing (Table 11.2).

CUBITAL TUNNEL SYNDROME

This is a compression of the ulnar nerve in the elbow. The ulnar nerve enervates the little finger and the ulnar side of the ring finger, and this is where tingling and numbness will occur. It is believed that cubital tunnel syndrome can be caused by resting the elbow on a hard surface or a sharp edge.

TENDONITIS OR TENDINITIS

This is inflammation of a tendon. The symptoms are pain, burning sensation, and swelling. One special case is shoulder tendonitis or bursitis at the rotator cuff (see Figure 11.2). This entails irritation and swelling of the tendon or of the bursa, and it will sometimes occur when the arm is frequently elevated or raised (Kroemer et al., 1994).

TENOSYNOVITIS

This is an inflammation of tendons and tendon sheaths. It frequently occurs in the wrist and ankle where tendons cross tight ligaments. The tendon sheath swells, which

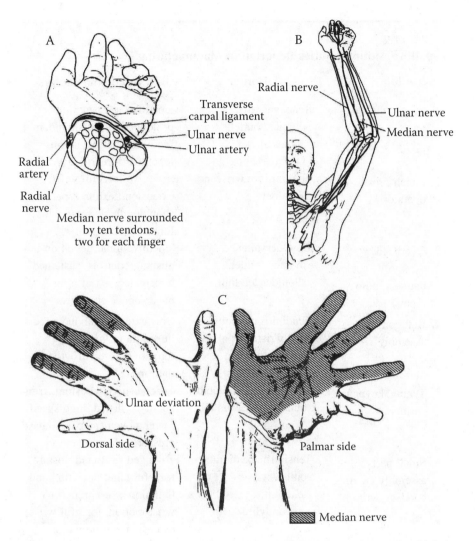

FIGURE 11.1 (A) Cross-section of the wrist showing the carpel tunnel, which is formed by the five bones on the one side and the transverse carpal ligament on the other. (B) Pathway of the three major nerves that originate in the neck and feed into the arm. (C) Enervation of the hand of the median nerve. The shaded areas indicate where numbness would occur in carpal tunnel syndrome. The amount of ulnar deviation can be measured as indicated in the figure. (Adapted from Putz-Anderson, 1988.)

makes it more difficult for the tendon to move back and forth inside the sheath. Like any inflammation, the symptoms are pain, burning sensation, and swelling.

There are many special cases of tenosynovitis, such as De Quervain's disease. This is tenosynovitis of the tendons of the thumb at the wrist. It may occur due to forceful gripping and twisting of the hand, such as when using a screwdriver. It has

TABLE 11.2
Repetitive Motion Injuries Reported in Manufacturing

Type of Job	Disorder	Occupational Factors
1. Buffing/grinding	Tenosynovitis, thoracic outlet, carpal tunnel, De Quervain's	Repetitive wrist motions, prolonged flexed shoulders, , forceful ulnar deviation, repetitive forearm pronation
2. Punch press operators	Tendinitis of wrist and shoulder	Repetitive forceful wrist extension/flexion, repetitive shoulder abduction/flexion, forearm supination
3. Overhead assembly (welders, painters, auto repair)	De Quervain's, thoracic outlet, shoulder tendinitis	Repetitive ulnar deviation in pushing controls, sustained hyperextension of arms, hands above shoulders
4. Belt conveyor assembly	Tendinitis of shoulder and wrist, carpal tunnel, thoracic outlet	Arms extended, abducted, or flexed more than 60°, repetitive, forceful wrist motions
5. Typing, keypunch, cashier	Tension neck, thoracic outlet, carpel tunnel	Static, restricted posture, arms abducted/flexed, high speed finger movement, palmar base pressure, ulnar deviation
6. Small parts assembly (wiring, bandage wrap)	Tension neck, thoracic outlet, wrist tendonitis, epicondylitis	Prolonged restricted posture, forceful ulnar deviation and thumb pressure, repetitive wrist motion, forceful wrist extension and pronation
7. Bench work (glass cutters, phone operators)	Ulnar nerve entrapment	Sustained elbow flexion with pressure on ulnar groove
8. Packing	Tendinitis of shoulder or wrist, tension neck, carpal tunnel, De Quervain's	Prolonged load on shoulders, repetitive wrist motions, overexertion, forceful ulnar deviation
9. Truck driver	Thoracic outlet	Prolonged shoulder abduction and flexion
10. Core making	Tendinitis of the wrist	Prolonged shoulder abduction and flexion, repetitive wrist motions

TABLE 11.2 (continued)
Repetitive Motion Injuries Reported in Manufacturing

Type of Job	Disorder	Occupational Factors
11. Stockroom, shipping	Thoracic outlet, shoulder tendinitis	Reaching overhead, prolonged load on shoulder in unnatural position
12. Material handling	Thoracic outlet, shoulder tendinitis	Carrying heavy load on shoulders

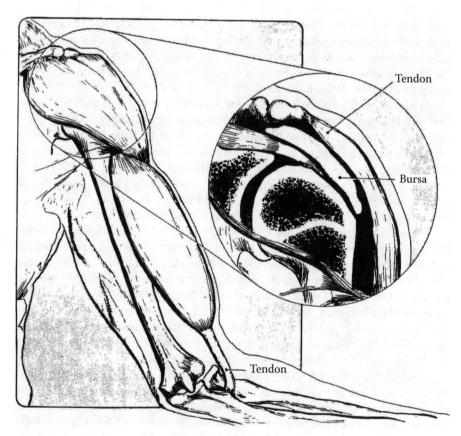

FIGURE 11.2 A view of the muscle–tendon–bone unit illustrating the relationship between a bursa and a tendon in the shoulder (Source: Putz-Anderson, 1988).

also been called "clothes wringing disease." Another special case of tenosynovitis is "trigger finger," which occurs in the flexor tendons of the finger. The tendon can become nearly locked up so that the movement of the finger is sudden and jerky.

THORACIC OUTLET SYNDROME

This is a disorder that results from compression of the three nerves of the arm and the blood vessels (see Figure 11.1[B]). This bundle of nerves and blood vessels, which is located between the clavicle and first and second ribs, can be compressed by the pectoralis minor muscle which leads to a reduction of the blood flow to and from the arm. The arm becomes numb and difficult to move.

11.3 CAUSES OF REPETITIVE MOTION INJURY

There are several different factors that may play a part in causing cumulative trauma disorder. For the individual case, it is often impossible to pinpoint a primary cause. One must take a comprehensive look at all the various manual activities that may have contributed to the RMI. It is not just a matter of inappropriate or aggressive work methods, but also what type of activities are performed when one is off work. Leisure activities such as knitting, carpentry, and tennis playing will also impact the likelihood of developing RMI. Some of these factors are listed in Table 11.3 (Armstrong and Chaffin, 1979; Eastman Kodak Co., 1986; Putz-Anderson, 2005).

In addition, there may be psychological "causes" of cumulative trauma disorder. One well-known incidence is the so-called RSI epidemic in Australia. During 1984, the repetitive motion injury rate increased by a factor of 15 (from 50 to 670) among employees of the Australian Telecom. However, the injury rate decreased, and by the beginning of 1987 it was back to normal (Hadler, 1986; Hocking, 1987). This sudden increase and subsequent drop in injury rate could be attributed to psychological factors. Some operators may have heard that colleagues were having problems and would interpret their own symptoms as being serious manifestations of RMI.

In the last couple of years, the RMI rate has increased in the U.S. and in Europe, and it would be natural to assume that some of the reported injuries are psychological in nature. But there is also a real problem, and the increased injury rate may be due partly to the situation where it has become accepted in society to report RMI, whereas this was not an accepted work injury in the past. Indeed, Hadler (1989) documented the types of back injuries reported in Switzerland, Germany, and Holland. The legal definitions of back injuries are different in these countries, and as a result different types of back problems are reported. Society norms and acceptance seems to affect greatly the type of occupational injuries that are reported.

Another example is for VDT workers. In the Scandinavian countries there were frequent complaints of pain in the neck and shoulder in the end of the 1990s (Hagberg and Sundelin, 1986), but RMI was rare (Winkel, 1990). In the U.S. the situation was different, and carpal tunnel syndrome was frequently reported among VDT operators (National Institute for Occupational Safety and Health, 1992). The shoulders and hands are connected by the three nerves—see Figure 11.1(B)—and there may be a possibility that the etiology of the injuries is the same, although the manifestation of complaints are different, so as to conform to the local norms. Whatever reason employees may have (physical or psychological), one must take

TABLE 11.3
Causes of Cumulative Trauma Disorders

Inappropriate work methods
- Repetitive hand movements with high force
- Flexion and extension of hand
- High force pinch grip
- Uncomfortable work postures
- Lack of experience of manual work
- New job
- Back from vacation

Inappropriate leisure activities
- Insufficient rest due to working in a second job
- Knitting, playing musical instruments, playing tennis, bowling, home
- improvement work

Pre-existing conditions
- Arthritis, bursitis, other joint pain
- Nerve damage
- Circulatory disorders
- Reduced estrogen level
- Small hand/wrist size

Note that many of the listed causes have not been confirmed by research, since they are difficult to investigate, and it takes a long time to accumulate epidemiological data.

complaints seriously. There are often simple modifications and additions to workstations that can alleviate some of the problems. For example, VDT operators often ask for a soft wrist rest, a split keyboard, a lower typing surface, or a footrest. These are inexpensive modifications, and one should not question the utility of such measures.

11.4 DESIGN GUIDELINES TO MINIMIZE REPETITIVE MOTION INJURY

Table 11.4 illustrates several engineering guidelines that can be used to minimize RMI. The assumption for presenting these guidelines is that the working environment, the task, and the workstation can be improved or redesigned by using various measures.

TABLE 11.4
Guidelines for Reducing RMI through Product Design, Process Engineering, Workstation Design and Use of Appropriate Handtools

Guidelines for hand posture
- Watch out for sudden flexion or extension of the hand or fingers
- Avoid extreme ulnar deviation and radial deviation
- Avoid operations that require more than 90° wrist rotation
- Keep forces low during rotation or flexion of the wrist
- For operations that require finger pinches keep the forces below 10 N; this represents 20% of the weaker operators' maximum pinch strength

Guidelines for handtools
- Cylindrical grips should not exceed 5 cm (2 in) in diameter
- Avoid gripping that spreads the fingers and thumbs apart by more than 6 cm (2.5 in)
- Use hand tools that make it possible to maintain the wrist in a neutral position (see Figure 8.2)
- Guidelines for workstation design
- Keep the work surface low to permit the operator to work with elbows to the side and wrists in a neutral position
- Avoid sharp edges on the work table and part bins that may irritate the wrists when the parts are procured; keep reaches within 50 cm (20 in) from the work surface so that the elbow is not fully extended

Guidelines for process engineering
- Allow machinery to do repetitive tasks and leave variable tasks to human operators
- Provide fixtures that hold parts together during assembly, and which can present the assembly task at a convenient angle to the operator
- Minimize time pressure or pacing pressure by allowing operators to work at their own paces

Guidelines for product design
- Minimize the number of screws and fasteners used in the assembly
- Minimize the torque required for screws
- Locate fasteners and screws at "natural" angles so they are easy for the operator to insert
- Design a product with large parts to permit gripping with fingers and palm instead of pinching

11.5 HAND TOOL DESIGN

Hand tool design affects the incidence of musculoskeletal disorders. Below we will explain some of the issues in designing and selecting good hand tools.

Hand tools have been used since the beginning of humankind, and ergonomics was always a concern. Tools concentrate and deliver power, and aid the human in tasks such as cutting, smashing, scraping, and piercing. Various hand tools have been developed since the Stone Age, and the interest in ergonomic design can be traced back in history (Childe, 1944; Braidwood, 1951).

During the last century there has been one important modification: many hand tools are now powered. The forces are greater, and thus the opportunities for injuries are also greater. In this chapter we give some guidelines for designing hand tools. There are several issues. A hand tool must

- Fit the task
- Fit the user and hand
- Not create injuries

There are two basic grips: the power grip and the precision grip (Figure 11.3). In the power grip, the hand makes a fist with the forefingers on one side and the thumb reaching around. There are three different categories of power grip that are differentiated by the direction of the force: (1) force parallel to the forearm, e.g., a saw; (2) force at an angle to the forearm, e.g., a hammer; and (3) torque about the forearm, e.g., a screwdriver (Konz, and Johnson, 2004).

For precision grips there are two subcategories: (1) the internal precision grip where the tool is held inside the hand, e.g., a table knife; and (2) the external precision grip where the tool is pinched by the thumb against the index finger and middle finger, e.g., a pen.

A hand tool can often be designed in different ways, since there are different ways of exerting power on the tool and the task. An electric screwdriver can have a pistol grip or an inline grip (Figure 11.4), and a surgical knife can be handled with an internal precision grip or an external precision grip. The option chosen should depend on how the task is organized and what is convenient for the operator.

There are many special-purpose hand tools. An accomplished chef has at least a dozen different knives for different purposes. Some of them are handled with a power grip and some with a precision grip and, depending upon the task, they are small or large, flexible or stiff. Likewise, for manufacturing one can design special purpose hand tools to fit specific tasks. Sometimes it is also possible to combine several hand tools into one—for example, a hammer with an extension for pulling nails, which makes it convenient for carpentry. A combination hand tool will save time because the operator can use one tool rather than two.

One particular concern is the size of the hand and handedness of the person—whether left- or right-handedness. As demonstrated in Table 8.2, there are few other dimensions of the human body where the differences between genders are as great

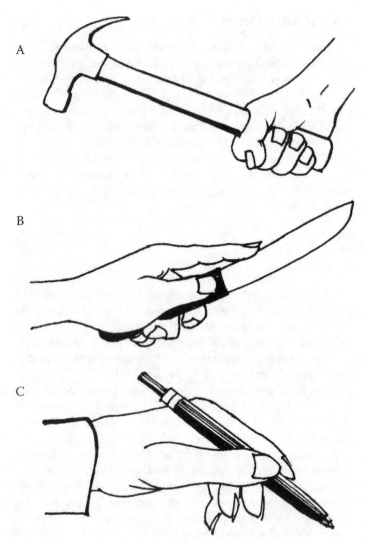

FIGURE 11.3 (A) Power grip; (B) internal precision grip; (C) external precision grip.

as for the size of the hand (Ducharme, 1973). Typically, the hand circumference for a 5th percentile male is the same as that for a 50th percentile female. Several organizations in the U.S., such as General Motors and the U.S. Navy, have a large number of female operators. They now supply hand tools appropriate for the female hand. Figure 11.5 shows the difference in the maximum grip strength for the average male and the average female. The maximum grip force for a female is about half that of a male operator.

Right-handed tools for left-handed users create awkward situations. The left-handed person can try to use the tool with the right hand but his or her dexterity

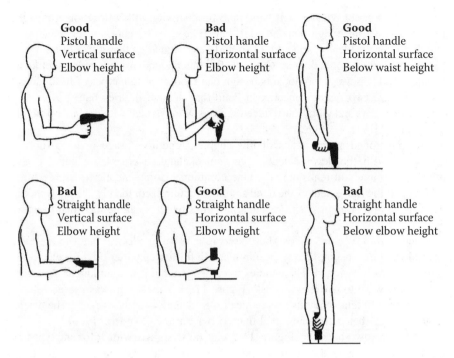

FIGURE 11.4 A hand tool should be selected so that it is possible to operate with a straight wrist.

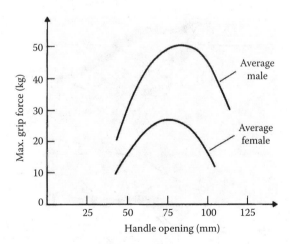

FIGURE 11.5 Maximum grip strength for male and female employees in electronics manufacturing (adapted from Greenburg and Chaffin, 1977).

and power is better with the left hand, and productivity will suffer. Sometimes the left hand can grip a right-handed tool, but there may be cut-outs for the fingers which do not fit. Ideally, a hand tool should be designed so that it can fit both left-handed and right-handed user. Cut-outs for the fingers, for example, should be avoided. Today, several tool makers design hand tools for both types of handedness.

There are two major concerns in hand-tool design: injuries due to musculo-skeletal disorders and vibration-induced injuries. We will deal with the latter problem in Chapter 13.

As mentioned at the outset of this chapter, repetitive usage of hand tools is associated with the development of musculoskeletal disorders, including carpal tunnel syndrome and tenosynovitis. One common recommendation for preventing CTD is that the movement of the hand should be minimized. Ideally, the hand should be in its neutral straight position, and sometimes handles can be modified to better fit a task.

Tichauer's (1966) study of pliers used in a Western Electric plant is a classic example that has inspired many ergonomists. In this case, a plier used for electronic assembly was redesigned. The handles were bent so that it was no longer necessary to bend the wrist to perform the task (Figure 11.6). This design was successful and the incidence of tenosynovitis was reduced significantly among workers. The design motto is, "It is better to bend metal than to twist arms" (Sanders, 1980).

The hacksaw shown in Figure 11.7 can be designed with different types of handles. In (A) the hand close to the body would be in ulnar deviation, and the hand at the far end would be in dorsiflexion. For case (B), both hands would be perfectly straight and aligned with the tool, so this is a much better design.

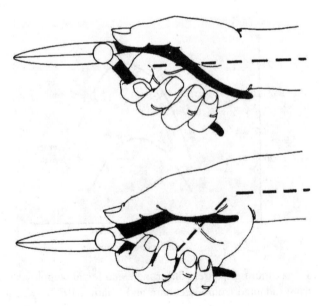

FIGURE 11.6 The handles of the tool are bent so that the wrist can remain straight.

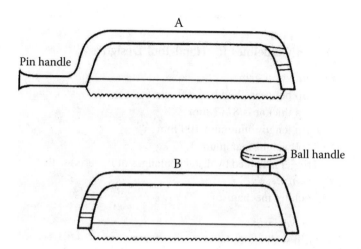

FIGURE 11.7 Two design options for a hacksaw. Case (B) is clearly better as both hands can operate with straight wrists.

John Bennett, an enterprising ergonomist, obtained a patent for the so-called Bennett's bend. This implies using bent handles for a variety of different tools (hammers, knives, broom handles, and tennis rackets). An appropriate amount of bending is 19 ± 0.5 degrees. Investigations by Schoenmarklin and Marras (1989) and Krohn and Konz (1982) verified that a bent hammer handle had the effect of reducing ulnar deviation and did not hamper performance, compared with straight handle hammers. However, there is nothing magic about 19 degrees—what is wrong with 25 degrees? Some skepticism would be appropriate; the optimum angle really depends on the task.

11.6 DESIGN GUIDELINES FOR HAND TOOLS

Table 11.5 summarizes several design guidelines for hand tools. The aim of these guidelines is to increase operator comfort, convenience, and controllability of hand tools.

TABLE 11.5
Design Guidelines for Hand-Tool Design

For precision grip
- Grip between thumb and finger
- Grip thickness 8–13 mm
- Grip length minimum 100 mm
- Tool weight maximum 1.75 kg
- Trigger activated by distal phalanges of finger(s) with fast-release
- Locking mechanism

For power grip
- Grip with entire hand
- Grip thickness 50–60 mm
- Grip length minimum 125 mm
- Grip force maximum 100 N
- Grip shape noncylindrical, preferably triangular with 110 mm periphery
- Tool weight maximum 2.3 kg, preferably about 1.2 kg
- Trigger activated by thumb with locking mechanism

General guidelines
- Grip surface smooth, slightly compressible and non-conductive
- Avoid vibration, particularly in the range of 50–100 Hz
- Design handles for use by either hand
- Keep the wrist straight in handshake orientation
- Tool weight balanced about the grip axis
- Eliminate pinching hazards

EXERCISE

The purpose of this exercise is to analyze kitchen knives. Kitchen knives are used for a variety of tasks (see Figure 11.8). For each of the knives, describe the following:

1. The cutting task. How is the knife used? What is the supporting surface, such as a cutting board, a plate, or unsupported (in the air)?
2. Are there differences in dimension of the knife, the friction of cutting, the force applied?
3. Are there differences in the way the knives are gripped?

Discuss the results of your analysis and provide reasons why it is important to have so many types of knives.

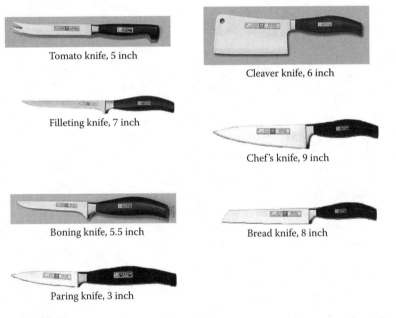

Tomato knife, 5 inch

Cleaver knife, 6 inch

Filleting knife, 7 inch

Chef's knife, 9 inch

Boning knife, 5.5 inch   Bread knife, 8 inch

Paring knife, 3 inch

FIGURE 11.8 Different types of knives that are used in preparation of meals.

RECOMMENDED READING

Putz-Anderson, V. (Ed.), 2005, *Cumulative Trauma Disorders: A Manual for Musculoskeletal Diseases of the Upper Limbs*, London: Taylor & Francis.

12 Physical Workload and Heat Stress

12.1 INTRODUCTION

In most Western countries, physical workload is no longer as common as it used to be. In manufacturing, hard physical labor has been taken over by materials handling aids, mechanized processes, and automation. Legislation has also put limits to the amount of workload that employees can be exposed to. Yet, in some occupations such as construction work, commercial fish netting, and logging, workers still perform much physical work. Such work generally involves less structured tasks, and they are difficult to mechanize.

Physical work activities are still common in industrially developing countries where mechanization has yet to pay off in the light of easily available labor. For example, in the construction industry materials are typically carried by workers. Eriksson (1976) estimated 30 years ago that 200 workers at a road construction site in Bangladesh could move as much dirt manually as one Caterpillar, and the costs were equivalent. Under such circumstances, the national economy, as well as the workers' personal economy, will gain by using manual labour.

Although the physical work demands in manufacturing have been substantially reduced, there are still many situations which require ergonomic analysis. Many individuals are less capable of physical work, and in this chapter we are particularly interested in individual differences due to gender and age.

12.2 METABOLISM

Metabolism is defined as the conversion of foodstuffs into mechanical work and heat (Astrand, Rodahl, Dahl, and Stromme, 2003). In order to be useful to the body, the foodstuff is converted into a high-energy compound *adenosine triphosphate* (ATP). ATP serves as a fuel transport mechanism. It can release chemical energy to fuel internal work in the various body organs. The phosphate bond can easily be broken down to *adenosine diphosphate* (ADP) according to the following formula:

$$ATP + H_2O = ADP + \text{energy release}$$

This basic reaction supplies the energy for the muscle cells. After the energy has been delivered the ADP is restored again to ATP using a combination of foodstuffs. At first glucose is used (if available), then glycogen, and finally fats and protein.

During continuous work there is oxygen available in the blood. This oxygen is used for the conversion, so that each molecule of glucose will generate 36 molecules of ATP. This is an energy-efficient process, and it is called an *aerobic* process (one that employs oxygen).

When there is a sudden burst of energy demand, the body will not have the time to use oxygen to resynthesize ATP. This is the case in a 400 meter dash. During the first 100 meters there is enough ATP available which can be broken down into ADP. After that ADP must be converted into ATP. Since the remaining 300 meters require much energy and there is not enough oxygen available, ATP will be recreated without oxygen. This is called an *anaerobic* (without oxygen) process. It is much less energy efficient than the aerobic process. In this case one molecule of glucose will generate only two molecules of ATP. In the anaerobic process lactic acid is produced as a byproduct. This is what makes the muscles burn with fatigue.

Lactic acid will hence accumulate in the working muscles rather than being carried away by the blood. Eventually lack of available energy, lack of fuel, and accumulation of lactic acid in the muscles lead to fatigue and cessation of work. After the 400 meter run, the muscles will ache and it will take a minute or so to recover.

The same phenomenon is also noted for static work with continuous contraction of some muscles. In this situation (such as carrying a suitcase) the static contraction of muscles results in a swelling, which may block the arteries, so that oxygen cannot be transported to the muscles. This will then create an anaerobic process and since the blood is not circulating and cannot remove the waste products, lactic acid is produced. This again produces local muscle fatigue and aching muscles.

The aerobic ATP conversion process is only about 50% efficient, so that half of the total food energy is lost as heat before it can be used. This is because the ATP energy is used to support three different processes. First, it maintains chemical processes, such as the synthesis and maintenance of high energy bonds in chemical compounds. Second, it is used to fuel neural processes and muscular contractions to maintain the body functions, such as blood flow and breathing. Finally, some of the ATP energy is used for muscular work. At most, 25% of the energy that enters the body in the form of food can be used for muscular work. This is the upper limit of the energy efficiency for the human body, and it is typically achieved only for the large muscles in the body, such as the leg muscles. The 25% efficiency exceeds that of a steam engine and is about equal to the efficiency of a combustion engine (Brown and Brengelmann, 1965). For the smaller muscles in the arms and shoulders an efficiency of about 10–15% is typical. Therefore one should try to use the large muscles for work rather than the small muscles. Figure 12.1 shows a water pump powered by leg movements and the body weight.

The amount of energy expenditure associated with a task can be assessed by measuring the amount of oxygen used. The oxygen uptake is calculated by measuring the volume and oxygen content of exhaled and inhaled air. This analysis is performed using special instruments. The oxygen uptake is then converted into kilocalories (kcal) of energy expenditure; one liter of oxygen generates 4.83 kcal of energy. Measurement of oxygen uptake therefore provides an exact assessment of energy expenditure, but it is quite tedious to measure. A much easier, but more approximate, method is to measure heart rate. Heart rate gives a fair estimate of energy expenditure

FIGURE 12.1 It has long been understood that leg movements are more energy efficient that arm movements. This leg driven water pump was invented in France in 1660.

in the intermediate range. Heart rate is less suitable for assessing small and very high rates of physical work.

Maintaining the basic body functions at rest requires about 1200 kcal/day. This is referred to as the basic metabolic rate (BMR). It includes functions such as the heart (215 kcal/day), brain (360 kcal/day), kidney (210 kcal/day), and muscles at rest (360 kcal/day). On top of maintaining the basic body functions, people usually engage in some minimal activity. This is referred to as leisure activity and does not include work activities. Together the BMR and leisure activities give an average energy consumption of 2500 kcal/day.

Different occupations incur different energy consumption rates. For an 8-hour work day the following values are typical:

- Seated office work: 800 kcal/day
- Light assembly work: 1680 kcal/day
- Ocean fish netting: 4800 kcal/day
- Lumberjacking: 6000 kcal/day

Ocean fish netting and lumberjacking are unusual because of their very high energy requirements.

Total energy requirements are obtained by adding BMR, leisure activities, and occupational rates. A total energy requirement of less than 4000 kcal/day is considered moderate, between 4000 and 4500 kcal/day is considered heavy, and above 4500 kcal/day is considered severe.

12.3 INDIVIDUAL DIFFERENCES

One of the main reasons for taking an interest in work physiology is to consider variations in work capacity between individuals. One important difference is physical condition (Figure 12.2). A highly trained individual (such as a marathon runner) can sustain 50% of the maximal aerobic capacity for an 8-hour work day; an average individual can sustain 35%, and an untrained individual 25% (Michael et al., 1961).

Chronological age is a fairly poor determinant of work capacity. A definite conclusion is that the variability between individuals increases with age. Figure 12.3 shows the maximal oxygen uptake for 2 individuals from the age of 35 years onwards (Miller and Horvath, 1981). The two curves represent two male professors of work physiology. Besides them, who else would have their maximal oxygen uptake tested so frequently? From the figure we observe that by the age of 65 years, individual A was as fit as ever, while individual B had a maximal oxygen uptake of 65% of his high value at the age of 35 years.

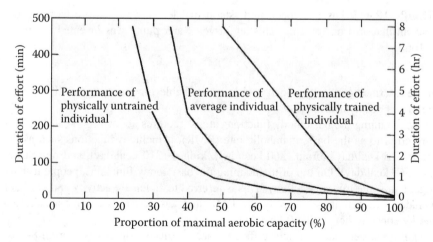

FIGURE 12.2 The capacity for sustained physical work depends on the amount of physical conditioning.

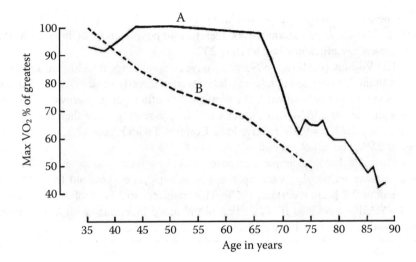

FIGURE 12.3 Volume of maximal oxygen uptake VO_2 as a function of age for two individuals. The oxygen uptake is given as a percentage of the greatest value attained for that individual.

12.4 METABOLISM AT WORK

Once work has begun, it takes some time for the metabolism to catch up with the energy expenditure of the muscles that are engaged in work. In fact, metabolism does not reach a stable level until several minutes after work has begun. The amount of time taken depends upon how hard the work is, but is typically about 5 minutes. Thus, the metabolic activity (or oxygen uptake) does not increase suddenly at the onset of work. Rather, there is a gradual, smooth increase in oxygen uptake (Figure 12.4). As mentioned, during the first few minutes of work, the muscles use energy that does not require oxygen.

As the oxygen uptake increases, the body can use the aerobic, or oxygen-requiring, fuel ATP. Returning to Figure 12.4 it can be seen that the metabolic rate eventually stabilizes. This steady-state level represents the body's aerobic response to the demands of increased workload. When the work ceases, the oxygen uptake returns slowly to the resting level prior to work. During this slow return after work the oxygen debt incurred during the onset of work (area A) is repaid (area B).

EXAMPLE: CALCULATION OF RELATIVE WORKLOAD

With a general understanding of the internal energy conversion processes, an example of the calculation of human work efficiency can be discussed. A 30-year-old man of average height (173 cm) and average weight (68 kg) is employed in packaging. This task imposes 23 watts (W) of external work. His resting metabolic rate just prior to work is about 93 W. The steady-state energy expenditure for this task is 209 W. (Both values can be calculated by measuring his oxygen consumption.)

The increase in oxygen uptake due to the imposed task is as follows: 209 – 93 = 116 W. The 23 W of external work therefore imposes 116 W of internal work, and the energy efficiency is 23/116 = 20%.

The VO$_2$max (maximum volume of oxygen uptake) for this 30-year-old man is 3.5 l/min. The oxygen uptake can be converted directly to work, and 3.5 l/min corresponds to 1179 W of work. Assuming a 20% efficiency in energy conversion, this translates to 236 W of external work. The assembly work therefore corresponds to a 23/226 = 9.7% relative load. Compared with Figure 12.2, this is much below 25%, and is not excessive.

This calculation example can be expanded by analyzing other individuals with a lower maximal oxygen uptake. For example, a 60-year-old female has a VO$_2$max of 2.2 l/min (Åstrand, 1969). This translates to 134 W of external work and a relative workload of 17%. For an untrained individual with a maximum workload of 25% (see Figure 12.2) this value would be on the high side.

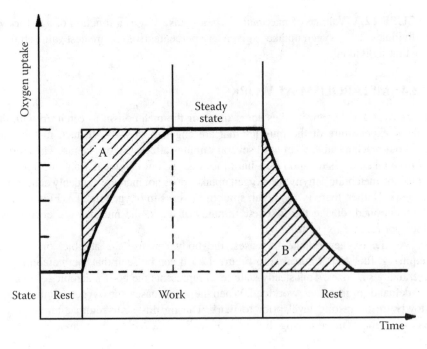

FIGURE 12.4 Oxygen uptake at the onset of, during, and after work. (A) Oxygen debt; (B) repayment of oxygen debt during rest. A = B.

12.5 MEASUREMENT OF PHYSICAL WORKLOAD

As we have previously noted, it is mostly impractical to use oxygen uptake to assess workload in a manufacturing situation. Heart rate (pulse rate) is a far easier measure.

However, heart rate is a good predictor only of workloads of intermediate intensity (about 100–140 beats/min). Simple measurements of heart rate can be useful to estimate if there are any problems with the current level of physical workload. This is illustrated by the following example.

The author once visited an automobile assembly plant. There was a female assembly worker who seemed physically exhausted. She was about 45 years of age and of small stature (about 150 cm [5 ft]). The type of work did not seem to put overly great demands on any of her coworkers. However, I stepped up and asked if I could take her pulse rate. It was running at about 130 beats/min, clearly excessive for an 8-hour work day. She was moved to another less physically demanding task.

12.6 HEAT STRESS

Heat stress is often a serious problem in hot climates, especially in industrially developing countries, where work is conducted outdoors, or where manufacturing facilities lack insulation and/or cooling. Surprisingly, it is also a problem in southern Europe and the U.S. In this section I will also briefly review some of the many standards on heat stress that have been issued by the International Standards Organization.

THERMOREGULATION

There are several physiological mechanisms for regulating body temperature. These are under involuntary control by nerve cells in the hypothalamus (a structure in the lower brain), and they maintain the body temperature within a narrow range (about $37 \pm 0.5°C$). This process is known as thermoregulation. As illustrated in Chapter 16, the body temperature exhibits daily variations. It peaks in the late afternoon and reaches its lowest level in the early morning. In order to keep the body temperature within a narrow regulated range, the amount of heat gained and lost by the body over the short span of time must be equivalent. If the body gains an excessive amount of heat, there could be excessive sweating, dehydration, and heat stroke; finally, death may occur.

There are two major ways of adapting to a hot environment: through *acclimation* and through *acclimatization* (Parsons, 2003). Acclimation refers to physiological changes, such as sweating, in response to temperature. Acclimatization refers to more enduring, long-term changes in physiological mechanisms that enable an individual to work in extremely hot environments. Repeated exposure to hot environments leads to an improved tolerance to the heat load. During acclimatization there are progressive increases in body temperature, working heart rate, and sweat rate. These processes can be completed in 1–10 days of exposure to a hot environment. The time required for acclimatization is reduced when people actually perform physical work in the heat. However, acclimatization to a hot environment can be lost over a period as short as a weekend. People who work outdoors and spend the weekend in an air-conditioned environment will have to acclimatize again. Recovery to the prior level will take about a day. Acclimatization is usually completely lost after 3–4 weeks in a cool environment.

12.7 MEASUREMENT OF HEAT STRESS

In addition to the ambient temperature, there are several other factors that effect heat exposure. In order to calculate their effect, the thermal balance of the body may be expressed in the thermal balance equation (Barnard, 2002). A somewhat simplified version of this equation is as follows (in W/m²):

$$M - W = C + R + E + S$$

where M is the metabolic power, W is the effective mechanical power, C is the heat exchange by convection, R is the heat flow by radiation at the skin surface, E is the heat flow by evaporation at the skin surface, and S is the heat storage.

As explained above, the metabolic processes are only partially effective. For the most effective muscles only about 25% of the metabolism (M) can be used for work. The rest is used to produce heat and maintain the basic metabolic processes. By expressing the metabolic power in watts per square meter, it is possible to compensate for the body size of individuals. Thereby individuals can be compared. For the calculation of an average individual, one can assume a body area of 1.8 m².

Heat transfer by convection (C) refers to the temperature exchange produced by moving air. The amount of convection depends on the difference between skin temperature and air temperature. The radiated heat (R) may be heat radiated by the human body (in the infrared light spectrum). The human body can also absorb radiated heat from external sources.

The evaporated heat loss (E) occurs primarily at the skin surface. Moisture is present on the skin because of sweating, and when the moisture evaporates heat is taken from the body surface. The evaporation is a function of air speed and the difference in water vapor pressure between the sweat (at skin temperature) and the air. In hot, moist environments, evaporated heat loss is limited, since the air cannot accept or absorb more water. If the air has maximum water pressure (100% humidity) there can be no further evaporation of sweat, and therefore no cooling of the body (ASHRAE, 1997). In a hot, dry environment, however, evaporated heat loss is limited only by the amount of perspiration that can be produced by the worker. The sweat disappears immediately from the skin and is absorbed by the air. The maximum sweat production that can be maintained by an average man throughout a day is about 1 l/h.

The heat storage (S) in the heat equation should in essence balance around zero. If S becomes large there is a risk of heat stroke. There are obviously many ways to reduce S. Stopping work is one way. Several additional methods are mentioned below in Table 12.3. The metabolic rate for different tasks can now be classified as in Table 12.1 (International Standards Organization, 1989a).

WET BULB GLOBE TEMPERATURE

One common method of evaluating heat stress is to record the wet bulb globe temperature (WBGT) (International Standards Organization, 1989b). This index takes into account four basic parameters: air temperature, mean radiant temperature, air speed, and absolute humidity. There are two different formulations for WBGT.

TABLE 12.1
Classification of Industrial Activities in Terms
of Workload and Metabolic Rate (ISO, 1998c)

Activity	Workload	Metabolic Rate (W/m²)
Seated, relaxed	Resting	58
Standing, light industry	Low	93
Standing, machine work	Low	116
Heavy machine work	Moderate	165
Carrying heavy material	High	230

(1) Inside buildings and outside buildings where there is no sunshine:

$$WBGT = 0.7\, T_{NW} + 0.3\, T_G$$

(2) Outside buildings with solar load:

$$WBGT = 0.7\, T_{NW} + 0.2\, T_G + 0.1\, T_A$$

where T_{NW} is the natural wet bulb temperature, T_G is the globe temperature, and T_A is the dry bulb temperature.

These measurements are easy to obtain using three different types of temperature measurements, as illustrated in Figure 12.5. The values of WBGT are used to classify if a certain work activity is advisable and to suggest limits for exposure to heat stress (see Table 12.2).

There are many ways of reducing heat stress or the effects of heat stress in work environments. This is referred to as heat stress management (see Table 12.3).

During the past 20 years there has been a debate concerning the maintenance of a pleasant climate in office environments. In order to measure the thermocomfort under these circumstances, an index called the predictive mean vote (PMV) is used (Fanger, 1970; Webb and Parsons, 1998). The PMV is an index that predicts the mean value of the votes that would be obtained if a large group of persons were asked to evaluate the climate. The following seven-point thermal sensation scale is used:

+3 Hot
+2 Warm
+1 Slightly warm
 0 Neutral
−1 Slightly cool
−2 Cool
−3 Cold

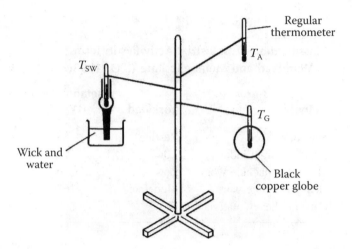

FIGURE 12.5 The globe temperature T_G measures the radiated heat using a thermometer inside a black painted copper globe. The more sun, the hotter the globe, so T_G will increase. The capacity of the air to absorb evaporated water is measured by the wet bulb temperature T_{NW}. This uses a thermometer which is covered by a cloth wick, the lower part of which is immersed in a reservoir of water. In dry air the water will evaporate. This will draw energy (calories) from the thermometer, and temperature reading will decrease. The dry bulb air temperature T_A is measured using an ordinary thermometer.

TABLE 12.2
Reference Values of the WBGT Heat Stress Index

		Reference Value of WBGT (Celsius)	
Workload	Metabolism (M) (W/m²)	Acclimatized	Not Acclimatized
Resting	<65	33	32
Low	65–130	30	29
Moderate	130–200	28	26
High	200–260	26	23
Very high	>260	24	19

Adapted from International Standards Organization (1989).

The PMV can be used to predict the percentage of dissatisfied (PPD) office users (Figure 12.6). The results of this research gives credence to the saying, "You can't please everybody." Regardless of the temperature setting in an office at least 5% of the office workers will be dissatisfied. The International Standards Organization suggests that the temperature be chosen so that the PPD is less than 10%. This means

TABLE 12.3
Measures to Reduce Heat Stress in Hot
and Humid Environments

Reduce the relative humidity by using dehumidifiers
Increase air movement by using fans
Lower the temperature by using air conditioners
Remove heavy clothing; permit loose-fitting wide clothing
Provide for lower energy expenditure levels
Schedule frequent rest pauses; rotate personnel
Schedule outside work so as to avoid high-temperature periods
Select personnel who can tolerate extreme heat
Permit gradual acclimatization to outdoor heat (2 weeks)
Supply cool, refrigerated vests (containing cooling elements)
Install local cold spots; e.g., refrigerated rooms for rest breaks
Maintain hydration by drinking water and taking salt tablets

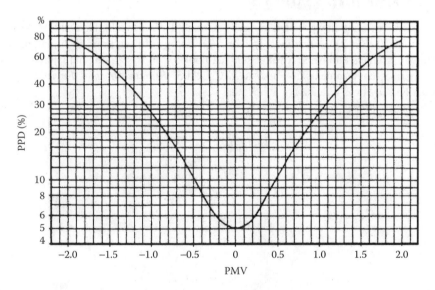

FIGURE 12.6 The predicted percentage of dissatisfied (PPD) users as a function
of the predictive mean vote (PMV) (International Standards Organization, 1994).

that 90% of office users will like the climate. During the winter season this translates
to an indoor temperature of 20–24°C, and during the summer season to an indoor
temperature of 23–26°C. Both these temperature ranges assume sedentary activities,
common in an office environment. The reason for the lower temperature range during
the winter is that people wear thicker, more insulating clothes during the winter

time. The ISO 7730 points out that there is insufficient information available to establish comfort limits for activities that are more physically demanding than seated office work (International Standards Organization, 1994).

EXERCISE: DISCUSSION OF HEAT STRESS MANAGEMENT

Discuss the effect of each of the measures in Table 12.2. Analyze the effect of each measure in the equation $M - W = C + R + E$.

Some of these measures may be practical for some environments but impractical in other environments. Discuss each of the measures using the examples given at the end of this chapter. What measures would be practical to reduce heat stress in the following:

1. An underground metal mine. This mine is very deep and the temperature is 90°F (32°C), and the humidity is 99%.
2. An office at 90°F and a humidity of 50%.
3. An outdoor tennis game at 90°F and a humidity of 50%.

RECOMMENDED READINGS

A classic textbook in work physiology is Åstrand, P.-O., Rodahl, K., Dahl, H.A. and Stromme, S.B., 2003, *Textbook of Work Physiology*, 4th edition, Champaign, IL: Human Kinetics Publisher. An excellent textbook for understanding human reactions to thermal environments is Parsons, K.C., 2003, *Human Thermal Environments*, London: Taylor & Francis.

13 Noise and Vibration

Noise is the most impertinent of all forms of interruption. It is not only an interruption, but is also a disruption of thought.

Arthur Schopenhauer (1788-1860)

13.1 INTRODUCTION

This chapter deals with the effects of noise and vibration on human performance and comfort. From an engineering design perspective, noise and vibration are close cousins; vibration of a steel plate will cause noise. Many engineering measures that reduce vibration will also cut noise. From a human performance and health perspective, they are, however, quite different.

Noise is very physical and very noticeable to most employees. Questionnaire investigations in industrial plants show that workers usually single out noise as the most important ergonomics problem in factories (Karlsson, 1989). This is not totally unexpected, because compared to many other ergonomic problems noise is very obvious and concrete.

There are four different aspects that can make noise unacceptable in the working environment:

1. Noise can cause hearing loss
2. Noise can affect performance and productivity
3. Noise can be annoying
4. Noise can interfere with spoken communication

In this chapter we first discuss several different methods for assessing the effects of exposure to noise. We then examine some performance effects of noise that are likely to affect an industrial worker, and we will discuss engineering methods for reducing noise in the workplace. Table 13.1 gives some examples of typical noise levels.

13.2 MEASUREMENT OF SOUND

A sound-level meter is used to measure sound. It consists of a microphone, an amplifier, and a meter that gives a visible reading in decibels (dB) on a scale. Most meters incorporate three different types of weighting of a sound. These are known as the A, B, and C scales (Figure 13.1). In particular, the dBA scale has achieved widespread use in work environments. This scale (or weighting function) approximates

TABLE 13.1
Examples of Activities and Corresponding Noise Levels

Activity	Typical Noise Level (dBA)
Near jet aircraft at take-off	125
Punch press at 1 m	105
Lathe	90
Quiet manufacturing (e.g., electronics)	75
Automobile at 20 m	65
Conversation at 1 m	50
Inside quiet home	42
Public library	20
Recording studio (threshold of hearing)	0

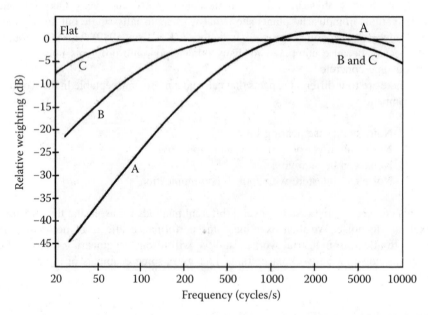

FIGURE 13.1 Weighting curves A, B, and C for sound-level meters. A is less responsive to lower frequencies and gives the best approximation of the sensitivity of the human ear.

the sensitivity of the human ear. The dBA scale is referenced to a sound pressure level of 0.00002 N/m², which corresponds to the threshold of hearing. Thus 0 dBA = 0.00002 N/m².

TABLE 13.2
Permissible Noise Exposures in the U.S.

Duration per Day (hours)	Sound Level (dBA)
8	90
6	92
4	95
3	97
2	100
1.5	102
1	105
0.5	110
0.25 or less	115

To calculate the sound pressure level (L_P, in decibels) the following formula may be used:

$$L_P = 20 \log P/P_0 \text{ dB}$$

where P is the root mean square (rms) sound pressure and P_0 is the reference sound pressure (0.00002 N/m^2). From the formula it can be derived that doubling the sound pressure would lead to an increase of 6 dB.

In most countries there are laws that regulate the amount of noise that employees can be exposed to. In the U.S., the maximum noise exposure throughout a working day of 8 hours is 90 dBA (OSHA, 1999). Note that other countries have similar laws in place; the permissible noise exposure limit varies between 85 and 90 dBA. When employees are subjected to sounds exceeding those listed in Table 13.2, administrative or engineering controls will be utilized. If such controls fail, personal protective equipment will be used to reduce sound levels to the levels in the table.

For every 5 dBA increase beyond 90 dBA, the exposure time is reduced by half. For example, if the noise is 95 dBA, then the maximum exposure time is 4 hours, and for 100 dBA it is 2 hours. According to Occupational Safety and Health Administration (OSHA) regulations in the U.S., noise exposure of different intensity can be added according to the formula:

$$D = C_i/T_i$$

where D is the allowable noise dose (should be 1), C_i is the number of hours of exposure to a noise level i, and T_i is the permissible number of hours of exposure to noise level i.

EXAMPLE: CALCULATION OF NOISE DOSE

A machine subjects its operator to 90 dBA when it is idle and to 95 dBA when it is used at full power. Assume 7 hours of use per day, with 2.1 hours at 90 dBA and 4.9 hours at 95 dBA. The total noise dose is calculated accordingly:

$$D = 2.1/8 + 4.9/4 = 1.487$$

Since the noise dose is greater than 1.0, this work exposes its operator to excessive noise that is not permissible.

13.3 NOISE EXPOSURE AND HEARING LOSS

The major concern in the manufacturing industry is that noise exposure will lead to loss of hearing. There are two major types of hearing loss: *conductive hearing loss* and *neural hearing loss*. Conductive hearing loss can be caused by mechanical rupture or dislocation of the eardrum and the bones in the middle ear (Figure 13.2). This may be due to a sudden intense pressure wave, such as produced by an explosion or a blow to the external ear. As a result there may be physical damage to the middle ear, for example by dislocation of the stirrup. The hearing loss may be partial or total, temporary, or permanent (Kryter, 1985).

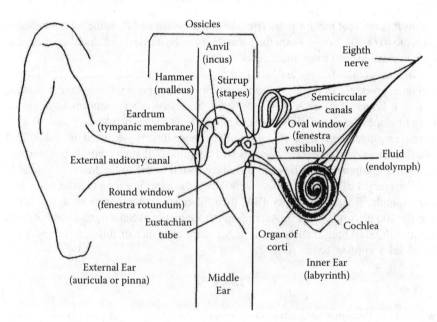

FIGURE 13.2 The structure of the ear. Some of the mechanisms are exaggerated in size to illustrate their functionality.

Prolonged noise exposure can cause hearing loss due to auditory nerve damage, also called neural hearing loss. In this case the intensity, frequency, and duration of exposure must be considered. For example, noise levels at about 130 dB may cause swelling of the hair cells of the organ of Corti on short exposure, and destruction of these cells on longer exposure. These changes are usually localized and involve only part of the organ of Corti, corresponding to certain (high) frequencies. The destruction of the hair cells in the organ of Corti is an irreversible process, and the resultant hearing loss is permanent. However, if the noise exposure time is short there may be only temporary swelling of the organ, which is reversible and causes only temporary hearing impairment, called temporary threshold shift (TTS) (Ward, 1976).

A person with auditory nerve damage first loses hearing of the higher frequencies at around 4000 Hz (Loeb, 1986). It is then difficult to hear a woman's voice but it is easier to hear a man's lower pitch. The person affected soon begins to speak louder and in a monotone voice, since the modulating effect of hearing is impaired. Because low tones are heard better than higher ones, it becomes difficult to understand words and sentences. Low pitch noise seems unduly loud and conversation becomes difficult in a noisy environment. Amplification of the sound through a hearing aid may not solve any problems, since high frequencies will still not be heard.

In contrast, a person with conductive deafness will complain that others in conversations do not speak loud enough. Understanding is not impaired if the sound level is sufficiently high, and such a person can benefit from the use of a hearing aid.

As mentioned, the first and most notable damage caused by excessive noise is to hearing, in particular frequencies at about 4000 Hz. However, there is extreme variability in the individual reactions to noise. Similar loss of hearing may also occur because of aging (presbycusis).

Loss of hearing may also be caused by ear infections, several diseases (mumps, measles, scarlet fever), and by common colds. Helander (1992) suggested that presbycusis may actually be caused by the cumulative effect of common colds over a lifetime. These viral infections can destroy auditory nerve cells.

13.4 HEARING PROTECTORS

There are two types of hearing protectors that are commonly used in industry: ear plugs and ear muffs (Berger and Casali, 1997). The plugs are designed to occlude the ear canal and are available in many types of material. Cotton has traditionally been used, but unfortunately, and contrary to popular belief, it affords no protection. Ear plugs made out of rubber, neoprene, glass down, and plastics offer good protection. Custom-molded ear plugs are also available (Casali and Park, 1990). They are made individually to fit the ear canal and offer excellent protection. Ear muffs are designed to cover the entire external ear. They consist of ear cushions made of soft spongy material or specially filled pads to ensure a snug fit.

Ear plugs provide a sound attenuation of between 15 dB for low frequency sounds and 35 dB for higher frequencies. At frequencies above 1000 Hz, muffs provide about the same protection as plugs. At frequencies below 1000 Hz, certain muffs provide more protection than plugs. Ear plugs and ear muffs may be worn together in intense noise situations. This combination provides an additional attenuation of approximately 5 dB.

Workers who regularly wear ear protection report that they actually hear conversations better. Cutting down the noise level that reaches the ear helps to decrease the distortion in the ear so that speech and warning signals are actually heard more clearly. An analogy can be drawn with the wearing of sun glasses to reduce excessive glare, thereby improving vision.

13.5 ANALYSIS AND REDUCTION OF NOISE

There are two main methods for measuring noise: the use of dosimeters and the use of sound level meters. Workers' exposure to noise can be quantified using a noise dosimeter (ANSI, 1991). A dosimeter is attached to the worker's body, e.g., on the chest. It summarizes the noise exposure over one working day, providing a measure for assessing whether an individual, with his or her particular work habits, has been overexposed to noise.

The other method is to use a sound level meter to analyze the working environment and obtain readings of the noise produced by various machines. A sound level meter can be set at different frequencies and a curve is constructed (Figure 13.3).

There are two common types of analysis: octave-band analysis and third-octave-band analysis. In an octave-band analysis the noise is measured at each octave. The preferred practice is to divide the audible range into 10 bands having the central frequencies 31.5, 63, 125, 250, 500, 1000, 2000, 4000, 8000 and 16000 Hz (ANSI, 1986). However, this may not give sufficient resolution for a detailed analysis of the noise. With a third-octave-band analysis there are three readings for every octave, which increases the resolution considerably. Figure 13.3 illustrates the frequency spectrum of a wood planer machine, where the noise spectrum was recorded using both an octave-band and a third-octave-band analysis. There are two peaks at about 125 Hz and about 1000 Hz. These peaks are due to the rotating elements in the wood planer machine. Since we now understand where the noise comes from, it is possible to take engineering measures to reduce these two peaks.

Through engineering change, noise energy can sometimes be moved in frequency to solve a noise problem, as has been shown by the U.S. Department of Labor (1980). A large diesel engine in a ship was designed to operate at a 125 revolutions per minute (rpm) with a direct drive connection to the ship's propeller. Noise of 125 Hz from the propeller would have been extremely disturbing to the crew. The solution was to add a differential gear between the engine and propeller in order to gear down the propeller's speed to 75 rpm. A larger propeller was also required. Shifting the noise to a lower and mostly inaudible frequency resulted in much less disturbance.

Only a detailed analysis of the frequency spectrum of the noise source can reveal such possibilities. In the case of the wood planer it may be possible to modify the cutting speed of the machine, which could possibly reduce the noise level to a legal level of 85 dBA.

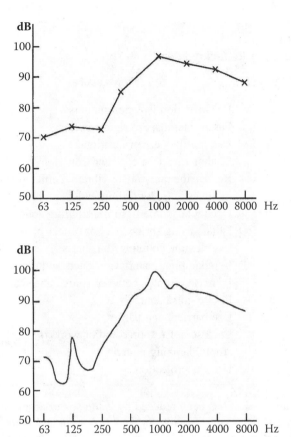

FIGURE 13.3 Octave-band and third-octave-band analyses for a wood planer machine. Note how the increased resolution makes it possible to identify 125 Hz and 1000 Hz as critical frequencies.

13.6 REDUCTION OF NOISE IN A MANUFACTURING PLANT

In a manufacturing plant one can take many different measures to reduce the noise (US Department of Labor, 1980). The noise can be controlled at the noise source, by reducing the structure-borne transmissions of noise, and by reducing the airborne transmissions of noise. A summary of some common measures is given in Table 13.3.

Many of the common noise sources in a plant (from manufacturing processes and machinery, air intake, and other equipment) are illustrated in Figure 13.4. Several measures have been taken to reduce noise, including the following: using vibration isolation mounts, placing heavy vibrating equipment on a separate rigid structure, and using an air intake muffler with laminar flow of air. The structure-borne transmissions have been reduced, for example, by use of flexible pipe on the air intake and sound isolating joints between the vibrating equipment and the floor. Finally,

TABLE 13.3
Approaches to Reducing Noise

Control Target	Measures
Noise source	Use vibration isolation mounts
	Fasten members to rigid structures
	Use mufflers on exhaust/intake
	Change direction of sound emission
	Reduce the radiating or vibrating efficiency of sound sources; e.g., by drilling holes in plates or covers
Structure-borne transmissions	Decouple source from transmitting solid
	Isolate using spring steel or rubber plate
	Use flexible couplings on shafts
	Use damping materials in ducts and conveyors
Air-borne transmissions	Increase distance between source and worker
	Rotate noise source
	Use barriers and baffles
	Enclose noise source and/or workers
	Apply damping material
	Use ear protection

the airborne transmission has been reduced by using sound absorbing ceilings and shields, and by enclosing noise sources in a control room and in the basement.

Most of the engineering measures listed in Table 13.3 are equally effective in reducing vibrations as well as noise. In fact, vibrations and noise are concomitant; noise and sound are vibrations of the air mass introduced by compressions and rarefactions of the density of air molecules. A vibrating plate will vibrate air masses and produce noise.

13.7 EFFECTS OF NOISE ON HUMAN PERFORMANCE

There are no clear-cut effects of noise on performance. In fact, this has been a much debated topic among researchers (Broadbent, 1978; Poulton, 1978; Kryter, 1985). Gawron (1982) reviewed 58 noise experiments and found that 29 showed a reduction in performance, 22 showed no effect, and 7 showed that noise improved task performance. Part of the problem in research is to provide a theory of the effects of noise on performance. If a viable theory exists, experiments could be undertaken and the theory tested. A problem in formulating a theory is that there are many types of noise and many types of task. Noise can be anything from intermittent to continuous and from music to white noise. The task can be skill-based (manual automatic behavior), rule-based (if scenario A, then do X; if B, then do Y, etc.), or knowledge-based (requiring deep thinking and pondering of alternatives) (Rasmussen, 1986).

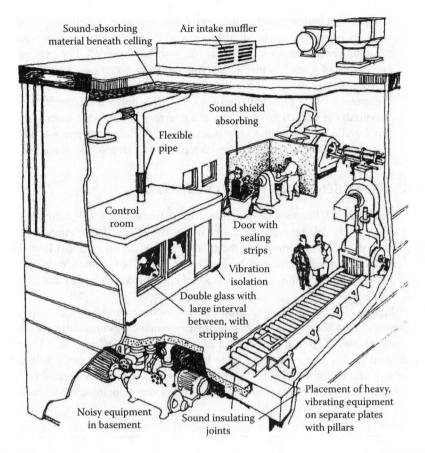

FIGURE 13.4 Example of noise control measures that can be implemented in an industrial building (U.S. Department of Labor, 1980).

The existing studies simply do not cover a sufficient range of noise and task conditions to be able to draw firm conclusions and formulate a viable theory. In reviewing the literature on noise, one can draw a few guarded conclusions:

1. Visual functions, such as visual acuity, eye focusing, and eye movements are little affected, if at all, by noise.
2. Motor (manual) performance is rarely affected by noise.
3. For the performance of simple, skill-based, routine tasks, noise may have no effect.
4. For rule-based tasks where the individual makes quick choices between different alternatives, noise may have some effect, particularly if the noise is louder than 95 dBA.
5. The detrimental effects of noise seem to be associated primarily with knowledge-based tasks, where operators must apply their knowledge of

different scenarios, think hard, and make tentative conclusions. This involves heavy use of the short-term as well as long-term memory, and the short-term memory capacity is likely to be exceeded. For example, processing of verbal, semantic information (which can be a knowledge-based task), suffers in noise well below the legal requirement of 85 dBA. Weinstein (1977) reported that a 68–70 dBA noise level significantly impaired the detection of grammatical errors (knowledge-based) in a proofreading task, but the same amount of noise did not appear to have any adverse effects on the ability to detect spelling errors (rule-based task).

13.8 BROADBENT AND POULTON THEORIES

At the end of the 1970s, two famous researchers, Broadbent and Poulton, became engaged in a lively debate on the effects of noise (Broadbent, 1978; Poulton, 1978). They had very different theories about the effects of noise on performance (Table 13.4). Both researchers made reference to Yerkes-Dodson's law, which postulates an inverted U-curve relationship between stress and performance (Yerkes and Dodson, 1908; Figure 13.5).

In Figure 13.5, an increase in arousal (A) can have probability of improving performance, as in situation (1), or hampering performance, as in situation (2). There is an optimal level of arousal (or stress) at which an operator performs best. If arousal is increased further (e.g., the task gets to be too stressful) performance will suffer. Conversely, if the arousal level is very low (a typical task with low arousal is visual inspection), then people have problems staying awake or being alert enough and

---

TABLE 13.4
Poulton's and Broadbent's Theories on the Effects of Noise on Performance

Poulton
- Noise masks acoustic task-related cues and inner speech. People cannot hear what they think.
- Noise is distracting.
- There is a beneficial increase in physiological arousal when noise is first introduced, but this beneficial increase lessens over time.

Broadbent
- The detrimental effects of noise are due to over-arousal (Figure 13.5) and not to the masking of inner speech.
- At high noise levels there is a funneling of attention (due to over-arousal). People cannot focus attention on a wide variety of information, but tend to lock on the most important information. As a result, errors are committed, but the operators may not be aware of these errors.

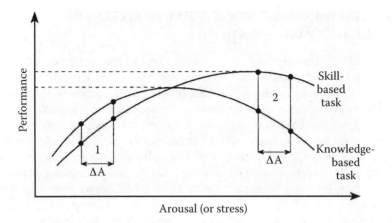

FIGURE 13.5 Yerkes-Dodson's law, formulated in 1908, postulates a relationship between arousal (or stress) and performance. The curves are original, but the classifications into skill-based and knowledge-based tasks were added by the author.

performance may suffer due to under-arousal. Note that in Figure 13.5, skill-based (easy) tasks suffer less from over-arousal than do knowledge-based (difficult) tasks.

Note that Yerkes-Dodson's law does not apply in the individual case. Some people thrive under high stress and some cannot perform. However, there is evidence that, from a probabilistic point of view, if we were to examine the behavior of a large group of people, they would on average behave as in Figure 13.5. Kahneman (1973) made reference to much research in support of Yerkes-Dodson'a law.

So who is right, Poulton or Broadbent? As with many theories of human behavior, the truth may have elements of both Poulton's and Broadbent's theories (Sanders and McCormick, 1993). More research is necessary to answer these intriguing problems.

EXERCISE: DISCUSSION OF THEORIES

Within the frameworks outlined by Poulton, Broadbent, Yerkes-Dodson, and Rasmussen, discuss the following:

1. Noise may facilitate certain tasks such as repetitive assembly.
2. Noise may degrade performance on tasks requiring information processing, such as working on manufacturing orders and calculations of pricing, billing, and shipping information.

Discuss the positive effects of noise on the following tasks:

3. Can noise degrade repetitive assembly?
4. Can noise improve performance in problem solving?

13.9 ANNOYANCE OF NOISE AND INTERFERENCE WITH COMMUNICATION

There are also psychological effects of noise; people reportedly become irritated and annoyed. But the amount of irritation depends on the circumstances. Much research has gone into assessing the effect of noise (e.g., traffic noise) on communities. Sperry (1978) noted that there are many acoustic as well as nonacoustic factors which influence the reaction to traffic noise. Among the nonacoustic factors are the time of day, the source of noise, and the attitude of the exposed person. The nighttime tolerance level for noise is about 10 dB lower than the daytime tolerance. Noise from aircraft is perceived as more annoying than the noise from automobiles and trucks. In fact, vehicular noise needs to be about 10 dB higher than aircraft noise to be equally annoying. Finally, the attitude to noise is very important. Comparative studies have demonstrated that individuals living in Rome, Italy, tolerated a 10 dB greater noise level than did people in Stockholm, Sweden. Is this a case of stiff Swedes and *laissez faire* Italians?

Surveys in industry have shown that noise is the primary source of dissatisfaction or annoyance (Karlsson, 1989). Perhaps this is because noise is so physical and clearly evident that people complain about it. Certainly it is easier to complain about noise than to formulate complaints about abstractions, such as the presentation of information on displays, even though the latter may be far more important to the task. The author once visited an air traffic control tower to make a survey of ergonomic problems. The air traffic operators' first complaints were of uncomfortable chairs. Later we found severe problems with the information that was presented. For example, the design of displays that illustrated how airplanes were taxiing and lining up on the ground for take-off was relatively complex. The modification of the information displays was clearly the most important ergonomic problem. But the issue is somewhat abstract, difficult to think of, and difficult to talk about.

13.10 INTERFERENCE OF NOISE WITH SPOKEN COMMUNICATION

Noise is a well-qualified problem because it disrupts communication, and some ergonomics standards have postulated that the noise level should be no greater than 55 dBA in office environments, in order to facilitate communication (Human Factors and Ergonomics Society, 2003). There are two common methods for evaluating the effect of noise on communication: preferred noise criteria (PNC) curves, and preferred speech interference level (PSIL).

PREFERRED NOISE CRITERIA (PNC) CURVES

This methodology was developed by Beranek et al. (1971) (Figure 13.6). The curves are based on office workers' subjective ratings of noise. The ratings were given during several experiments done to investigate which frequencies in the noise were particularly disturbing to speech communication. The PNC curves in Figure 13.6 represent equal-sensitivity (iso-sensitivity) curves to noise of different frequencies.

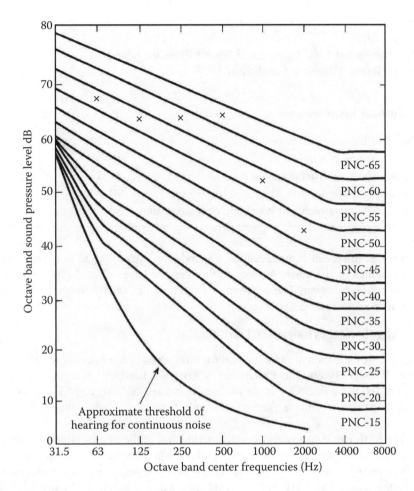

FIGURE 13.6 Preferred noise criteria (PNC) curves are iso-sensitivity curves for noise of different frequencies (the number after PNC is the sound pressure in decibels at 1000 Hz). To evaluate a noisy environment, octave-band readings are obtained and plotted on top of the PNC curves.

For the higher frequencies the curves come down. The human sensitivity to high frequency noise is greater and less sound pressure is needed to perceive the noise, but for the lower frequencies the human ear is less sensitive and much greater sound pressure is needed to perceive the noise.

The PNC curves are used to evaluate the acoustical requirements for different tasks. Some recommended PNC curves and the approximate sound pressure levels are presented in Table 13.5.

At higher PNC values (around 50 and 60) it becomes very difficult to communicate with other individuals. The PNC values are different from dBA, because dBA is an average weighting across the entire sound spectrum, whereas the PNC provides an evaluation throughout the noise spectrum. As an example of using PNC curves,

TABLE 13.5
**Recommended PNC Curves and Sound Pressure Levels
for Different Listening Conditions**

Acoustical Requirements	PNC	Approximate dBA
Excellent listening conditions	5–20	5–30
Good listening conditions	20–35	30–42
Moderately good listening conditions	35–45	42–52
Fair listening conditions	40–50	47–56
Just acceptable speech and telephone communication	50–60	56–66

assume that in speech communication a sound-level meter is used to obtain octave band readings of the noise. Assume further that we have selected PNC = 60 as a criterion for evaluation. The criterion is exceeded for 500 Hz, but otherwise the noise level is acceptable (see Figure 13.6).

PREFERRED SPEECH INTERFERENCE LEVEL (PSIL)

The PSIL is the most common method for rating the speech interference effects of noise (Webster, 1969). The PSIL value is first calculated by averaging the sound pressure levels (in decibels) of octave bands centered on 500, 1000, and 2000 Hz. Thus, if the levels of noise were 65, 70, and 75 dB, respectively, the PSIL would be 70 dB.

The PSIL can give a fairly good approximation of the impact of noise having a flat spectrum. However, if there are irregularities in the spectrum, it loses some of its usefulness because the simple average of the three octave bands cannot characterize the noise. The PSIL value is evaluated using a graph. In Figure 13.7 the

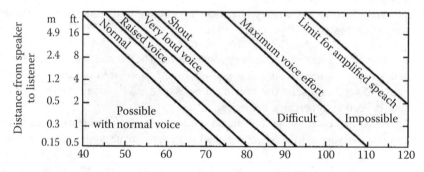

FIGURE 13.7 Voice level and the distance between the speaker and the listener as a function of PSIL noise level.

distance from a speaker to a listener is given as a function of the PSIL value. The necessary speech level is then characterized as normal, raised voice, very loud voice, shout, maximum vocal effort, or limit for amplified speech. Thus, for example, if the PSIL value is 65 dB and the distance to a listener is 8 ft, the speaker would have to talk with a very loud voice. PSIL has also been used to characterize office communication in private offices and secretarial offices (Beranek and Newman, 1950). Of particular interest is the effect of noise on telephone use. For a PSIL value greater than 60 dB it is difficult to use a telephone, and for a value greater than 76 dB it is impossible to talk on the telephone.

Exercise: How to Use PSIL

To evaluate the ease of communication in an industrial plant, the noise was measured for three octave-bands: 500, 1000, and 2000 Hz. The recorded noise levels were 75, 80, and 82 dBA, respectively.

1. Calculate the PSIL value.
2. Using the values in Figure 13.7, what is the maximum distance at which two individuals can communicate without raising their voices?
3. Given the social unacceptability of a very close distance, what would be the necessary speech level if the distance was 1.0 m?

13.11 WHOLE-BODY VIBRATION

In today's work environment machines can often cause vibration, which in some circumstances may pose a health hazard (Griffin, 1997). There are two major kinds of vibration: whole-body vibration and hand vibration. The latter is commonly referred to as segmental vibration, implying vibration of the extremities. In addition to these, there is a third phenomenon, sea sickness, which involves exposure to slow vibrations in the range 0–1 Hz. A common source of whole-body vibration is vehicles of all types including forklift trucks, long-haul trucks, earth-moving equipment, and other industrial moving machines. Hand vibration or segmental vibration is often induced by hand-held tools such as power drills, saws, jack hammers, concrete vibrators, and chain saws. These are dealt with in the Chapter 8 [check –correct?]. In this chapter we give an overview of the most common problems related to vibration.

13.12 SOURCES OF VIBRATION DISCOMFORT

A common source of whole-body vibration is from transportation vehicles where drivers are exposed to a vibration generated by the vehicle and the roadway. Figure 13.8 illustrates that different parts of a driver's body have different resonant frequencies. For the shoulder and the stomach the resonant frequency is 3–5 Hz. This perhaps explains why this particular frequency range produces the greatest reported discomfort.

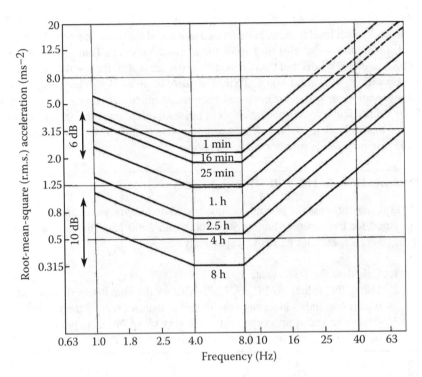

FIGURE 13.8 ISO Standard 2631 prescribes exposure limits of vibration for 8 hours of work and less than 8 hours. The figure illustrates exposure limits for vertical (y direction) vibration. There are similar regulations in the x and z directions. To obtain exposure limits for reduced comfort, subtract 10 dB. For exposure limits to avoid tissue damage, add 6 dB.

Laboratory studies have confirmed that vibrations between 3 and 5 Hz are likely to be physically uncomfortable at an acceleration level of approximately 0.1 g, to be painful and distressing at intensities of about 1 g, and to cause injuries if the acceleration exceeds 2 g. These findings form the main background for the present ISO standards for vibration (see Figure 13.9) (Mackie et al., 1974; Gruber, 1976).

Hansson et al. (1976) studied exposure to whole-body vibrations of 44 industrial truck drivers. He found that using ISO Standard 2631 for exposure limits (ISO, 1997), 6 of the industrial truck drivers presented a risk to health if exposure lasted for 8 hours. Vibration was fatiguing and reduced the work capacity of the drivers in two thirds of the trucks studied (according to ISO standards).

Large and heavy trucks exposed the drivers to lower frequencies than did small and light trucks. Obviously, the vibration characteristics of similar machines vary considerably, depending upon the design. Hansson et al. (1976) concluded that manufacturers and designers of trucks are not always well informed about the implications of different design alternatives on whole-body vibration.

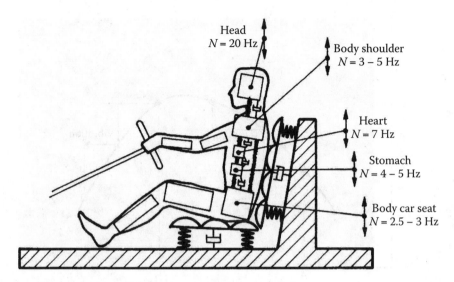

Head
$N = 20$ Hz

Body shoulder
$N = 3 - 5$ Hz

Heart
$N = 7$ Hz

Stomach
$N = 4 - 5$ Hz

Body car seat
$N = 2.5 - 3$ Hz

FIGURE 13.9 The resonant frequencies of different parts of the body of a seated driver.

Mackie (1974) pointed out that it is not only the amount of physical energy that can greatly influence the discomfort of vibrations; there are also several psychological factors, including the following:

1. The nature of the task. For example, riding in a recreational boat is usually associated with pleasure, although the same magnitude vibrations would be perceived as very stressful in an industrial environment.
2. The person's degree of training or familiarity with the task. For example, a skilled horseback rider can compensate for much of the vibration by rhythmically contracting certain muscles. Likewise, an industrial worker can compensate for some of the movements of a vibrating forklift truck or piece of industrial machinery. There are also individual differences in sensitivity to vibration; heavy individuals suffer more from vibration than do light individuals.
3. The presence of other stressors acting in combination. For example, vibration in combination with noise produces a greater level of stress than vibration alone or noise alone (Poulton, 1979). This will affect the physiological arousal of the individual, which in turn has implications for the performance level (Figure 13.10).

In addition to the discomfort effects of vibration, there are several reputed health effects, such as various spinal, anal–rectal, and gastrointestinal disorders (Fothergill and Griffin, 1977). However, these have been difficult to verify in research. Most of the evidence comes from epidemiological investigations of truck drivers and heavy equipment operators (Seidl and Heide, 1986). A large U.S. study of truck drivers

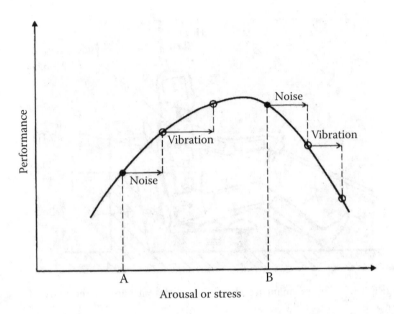

FIGURE 13.10 Yerkes-Dodson's law can be used to illustrate the additive effect of two stressors on performance. The initial arousal level (A or B) is crucial to performance.

reported that drivers complain about these problems, but there are other possible factors that could also contribute, such as extended sitting and poor eating habits.

Exposure to vibration also induces physiological responses. The most basic physiological response to a moderate level of vibration is an increase in heart rate, about 10–15 beats/min above the resting level. The heart rate returns to its normal level after the vibration ceases. Blood pressure can also increase, particularly for vibration frequencies around 5 Hz. Some investigations have revealed a slightly increased breathing rate and oxygen consumption. These changes may be related to increased muscular activity which is induced by vibration. As mentioned above, people exposed to vibration will often contract their muscles to tighten up the body and avoid vibration in some tissues.

Another notable finding is that at vibrations of about 10–25 Hz, the visual acuity level decreases. This frequency range is thought to represent the resonant frequency of the eyes, and as a result there is often a reduction in the operator's performance level (Grether, 1971; Collins, 1973).

Whole-body vibration also effects both cognitive and motor performance. Sherwood and Griffin (1992) noted that whole-body vibration impairs learning. From a review of the literature, Hornick (1973) concluded that for tracking experiments (with a joystick) the tracking errors could increase by 40% in a vibrating environment. By supplying an arm support, the errors were reduced to about half.

13.13 SEGMENTAL VIBRATION

Hand-tool vibration can cause vibration injuries. There are two common types of vibration injury: Reynaud's disease (or white finger disease) and Dart's disease. Reynaud's disease is caused by hand-tool vibration in the frequency range 50–100 Hz. Examples of such hand tools are pneumatic drills, jackhammers, and concrete vibrators (Gemme et al., 1993). The white fingers are caused by a reduction in blood flow to the hand and to the fingers, which is due to constriction of the smooth muscles of the blood vessels in the hand and fingers. Both the nerves and the blood vessels in the hand are permanently damaged (National Institute for Occupational Safety and Health, 1989).

The reduction in blood flow causes stiffness and numbness of the fingers and gradual loss of muscle control of the hand. Workers have difficulty in holding, grasping, and manipulating items. White finger disease is aggravated by other conditions that cause vasoconstriction of the hand, such as cold weather and smoking. The feeling in the hand is the same as when a foot falls asleep, and there are complaints of tingling, numbness, and pain.

Dart's disease is less common. This disease is caused by vibration frequencies around 100 Hz. The symptoms are the opposite to those of white finger disease. In Dart's disease, blood pools in the hands, which become blue, swollen and painful.

One way of reducing the transmission of vibration is to use a vibration-attenuating handle. Andersson (1990) used a handle that consisted of a hand grip and a rubber element which acted as a universal joint. This handle effectively reduced transmitted vibration by about 70%. Soft handles such as foam grips do not seem to work. A study by Fellows and Freivalds (1991) demonstrated that grip force was greater when using a foam grip, since the deformation of the foam led subjects to feel as though they were losing control. Due to the increased grip force more vibration energy was transferred to the hand.

RECOMMENDED READINGS

Crooker, M.J., 1997, Noise, In: Salvendy, G. (Ed.), *Handbook of Human Factors and Ergonomics*, New York: Wiley.

Griffin, M., 1996, *Handbook of Human Vibration*, Amsterdam: Elsevier.

Part III

Organization/Management-Centered Human Factors

14 Ergonomics of Computer Workstations

The computers are up—we are down.
We have lots of information technology.
We just don't have any information.

14.1 INTRODUCTION

Today computers are used in all types of environments—in offices, manufacturing, farming, construction, military operations, and at home. About 50% of the households in Western countries in 2004 had a computer at home; the Scandinavian countries led with 80% of homes having computers. In the early writings about computer terminal workplaces, they were referred to as visual display terminal (VDT) workplaces. In this chapter we refer to computer workplaces.

As the quotation above reveals, the problems with information technology have today moved from hardware issues (which dominated the 1970s and the 1980s) to software issues. Nonetheless, many office users worry about ergonomics problems, including visual fatigue, poor sitting posture, repetitive motion injuries, exposure to radiation, and poor job satisfaction.

In this chapter we will highlight some of the concerns, and the common myths about the deleterious effects of computers on office work. This area has been well researched, and there are clear recommendations concerning the design of computer workstations. There is also an increasing awareness among office employees that ergonomics is beneficial, and today, many organizations hire ergonomics consultants to propose ergonomics design solutions for offices. Some of issues have already been dealt with in other chapters: sitting work posture (Chapters 8 and 9), illumination (Chapter 4), and input devices (Chapter 6).

The ergonomics of computer workstations has been researched for over 30 years. Some very influential publications spread the interest in Europe (Östberg, 1976; Cakir et al., 1978). Many of the problems with computer workstations were due to the poor visibility of cathode ray tube (CRT) screens. Over the years the CRT technology has improved significantly. Today, with the introduction of liquid crystal displays (LCD) and other flat panel technologies, most of the visual problems have disappeared. At the time of writing this chapter CRTs still dominate, although the move to LCD displays is rapid.

Since the introduction of computers in the workplace, there have been tremendous developments in technology and the ergonomics of design. This is illustrated in Figure 14.1, a photograph which was taken in 1982. The figure shows the former

FIGURE 14.1 Most of the ergonomics problems you could think of: (1) facing a window; (2) keyboard not detachable; (3) no document holder; (4) glare on the screen; (5) chair too low; (6) table too high; (7) arm rests interfere with keying; (8) inadequate leg clearance under the table; and (9) a very uncomfortable chair. Courtesy of United Press International.

President of the U.S., Jimmy Carter, in his home in Plains, Georgia. It illustrates several ergonomics problems, which today have disappeared due to improved technology.

14.2 SITTING WORK POSTURE

In designing a computer workstation one must understand the ergonomics requirements and how they are related. Figure 14.2 defines the important design elements

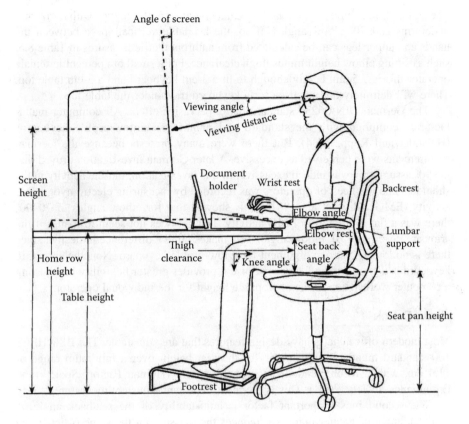

FIGURE 14.2 Definition of computer terminal workstation terms.

of a computer workstation. Some of these design elements have been standardized through national ergonomics standards (Human Factors and Ergonomics Society, 2003) as well as international standards (International Standards Organization, ISO Series 9241, 2004). Below we comment on some of the more important design concepts.

VIEWING ANGLE

The center of the screen should be lower than the eye height, so as to obtain a viewing angle of about 25–35° below the horizontal (Hill and Kroemer, 1989). People who sit in an upright posture prefer to look down rather than look up or look straight ahead. In particular, looking up with the head bent back is a common cause of strain and muscle fatigue in the neck.

THIGH CLEARANCE AND LOW-PROFILE KEYBOARDS

A person sitting at a desk has limited space for the keyboard and the table top (see Figure 14.2). In 1981, the early German DIN 66234 standard mandated the use of low-profile keyboards and thin table tops (Deutsches Institut für Normung, 1981).

The assumption behind this standard was that operators prefer to type with horizontal underarms and 90° elbow angles. If so, the available vertical space between the hands and upper legs can be calculated from anthropometric measures in Table 8.3, such as sitting elbow height minus thigh clearance. For a small 5th percentile female operator this is 7.5 cm, barely enough to fit a 3-cm keyboard and a 3-cm table top. There will definitely not be space for a keyboard tray under the table top.

The German DIN 66234 standard had a pervasive effect. All computer manufacturers complied with the standard and manufactured low-profile keyboards (Helander and Rupp, 1984). But there were many protests because the German requirements were perceived as excessive. A later German investigation proved that the 90° assumption was indeed excessive. In this investigation, muscle activity in the shoulders and the neck of operators was recorded by measuring electromyographic activity (EMG) (Zipp et al., 1981). This showed that for elbow angles of 70–90° there was a flat minimum in EMG activity. Thus, it does not seem to matter if the arms are raised to a 70° elbow angle. This makes quite a difference in design, since there is no longer a strong argument for a low-profile keyboard. Nonetheless a thin keyboard is a good design feature because it provides greater flexibility in adjusting a computer workstation to the appropriate height for the individual operator.

Chair Design

Most modern office chairs have design features that are adjustable. The BSR/HFES 100 standard mandates adjustability of the seat height over a minimum range of 11.4 cm, with a recommended range of 38–56 cm (Human Factors Society and Ergonomics Society, 2004). This is the most important adjustability feature.

The second most important factor is adjustability of the seatback angle. A seatback angle of greater than 110° reduces the pressure on the spine (Michel and Helander, 1994). As a person moves from a straight standing posture to a straight sitting posture, the hip joint angle goes from 180° to about 90°. The last 30° of movement from 120° to 90° are absorbed by the pelvis, which rotates forward. This biomechanical change reduces the length of the leverage arm from erector spinae muscles (back muscles) to the spine. As a result the disk pressure is about 30% greater while sitting as compared with standing (Andersson and Ortengren, 1974).

The third most important adjustability factor is the lumbar support. This design feature may have been oversold. Lumbar supports are often not used since chair users do not sit straight and usually do not press their back all the way into the backrest. In fact, many chair users prefer a more relaxed sitting posture (Grandjean, 1986). However, individuals with bad backs are sensitized to the effect of lumbar supports, and have a tendency to use them much more than persons with healthy backs. The lumbar support can become very uncomfortable if it puts pressure on the wrong spot on the back, and lumbar supports must therefore be adjustable (Branton, 1984). Modern aircrafts have introduced lumbar support in seats, but airlines exaggerate their contribution to sitting comfort. Some of them rotate, and many of them are not adjustable so they do not contact the lumbar.

A classic study by Anderson and Örtengren (1974) was instrumental in understanding how disc pressure increases as a function of chair angle and use of lumbar

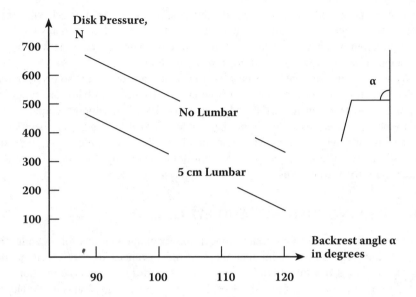

FIGURE 14.3 Normalized disc pressure in N in the lumbar disk as a function of backrest angle in degrees and protrusion of the lumbar support. Two situations are depicted: 5 cm lumbar support versus no lumbar support.

support. They inserted pressure transducers into the L3/L4 disc of experimental subjects sitting on a chair. Then they varied the backrest angle of the chair, from 120° to 90°. The backrest angle was measured from the horizontal seat pan to the back rest. At the same time they varied the protrusion of the lumbar support (see Figure 14.3). For a 90° backrest angle and no lumbar support, the disk pressure was close to 670 N. As the backrest angle opened up to 120°, the pressure dropped to about 370 N. With the lumbar support the pressures were lowered to 470 N and 120 N respectively.

This research then gives much support for the notion that it is not good to sit straight up. A leaned back work posture is better. At the same time the lumbar support may have a significant impact. Unfortunately most chair users don't put their back all the way back into the chair and can therefore not take advantage of the extra reduction in pressure. This is, in a way, expected! There are no nerve endings or pressure sensors in the disks, so chair users do not get any feedback from the increased pressure, unless they happen to have a bad back, in which case there may be feedback from pressure on adjacent nerves. The implication for design of chairs is that an adjustable back rest angle is important so that users can fold the chair back.

SUPPORTS FOR THE HANDS, ARMS, AND FEET

A footrest can be helpful for short operators, so that they can support their feet. However, footrests should not be used out of convention. In Figure 14.2 the footrest is unnecessary since the operator can put her feet on the floor.

Arm rests should not interfere with the table or desk top. For a keying task, where the operator must pull the chair close to the table, short arm rests (elbow rests) are often preferred over long arm rests. Today, many chairs come with height adjustable arm rests that can be used in many different ways.

Wrist rests are optional. Because typing habits are different, some operators prefer wrist rests and some do not. Soft wrist rests (rather than hard) are supposed to put less pressure on the wrist and reduce the risk of developing carpal tunnel syndrome. However, research has not been able to prove the significant benefits of wrist rests, and whether soft wrist rests really make a difference. But footrests, armrests and wrist rests are inexpensive, and should be readily available to operators who require them (Sauter et al., 1985).

14.3 VIEWING DISTANCE AND EYE GLASSES

There are a few controversial issues concerning recommendations for viewing distance. Some researchers claim that the viewing distances to the screen, to the documents on the document holder, and to the keyboard should all be the same, so that it is not necessary to refocus the eyes. Refocusing takes time and is unproductive (Cakir et al., 1980). In addition, for older operators with presbyopia (inflexible lens in the eye) the uniform distance is helpful since it is easier to focus.

Other researchers claim that it is important to keep exercising the focusing mechanism (accommodation) of the eye, and that thereby visual fatigue and temporary myopia can be avoided (National Research Council, 1983). The term temporary myopia implies that the accommodation or focusing of the eyes adjusts to the somewhat closer viewing distance which is imposed by a close working task. Thereby the range of clear vision is moved closer to the eye, and it is difficult to focus on distant objects. This phenomenon is not unique to computer work. Every close work task may cause temporary myopia, which typically disappears an hour after work. Nonetheless, many individuals notice these effects and are overly concerned. For example, when driving home after work during darkness, temporary myopia, combined with dilated eye pupils, makes it difficult to read traffic signs. Some individuals may misinterpret this and obtain eye glasses to correct a condition which hardly needs any correction.

Several investigations have addressed the long-term effect of screen viewing on vision. Researchers generally agree that there are no adverse effects (Bergqvist, 1986). The eyes do not become more myopic, hyperopic, or presbyopic than in any other close work task. Most of the changes in eyesight that have been reported by computer operators are normal, and due to aging of the eye; they would have happened with any close work. Nonetheless, CRT viewing is visually more exacting than other visual tasks. Compared with printed characters on a paper, CRT characters are more blurred and there is less luminance contrast between the characters and the screen background. This has been shown to decrease the speed of reading CRT screens as compared with paper (Gould and Grischkowsky, 1984). However, high-resolution screens improve character definition and are easier to read.

Another more severe aspect of computer screen viewing is that many individuals lack eyeglasses with appropriate correction (Sauter et al., 1985). This is particularly true for older operators who use bifocal lenses. The lower part of the bifocal lens is typically ground for a viewing distance of about 30 cm and the upper part for a far viewing distance of about 400 cm. The distance from the eyes to the monitor may be around 60 cm, and to maximize visibility many operators who wear bifocals will bend their heads back to read the screen through the lower part of the lens, and at the same time move the head closer. This causes neck strain, and operators often complain about neck and shoulder pain (Sauter et al., 1985).

Many companies now supply special glasses known as terminal glasses. An optometrist will go to the workplace and measure the viewing distance between the operator's eyes and the screen, and prescribe lenses which are ground for the exact viewing distance to the screen.

Visual Fatigue

Many operators complain about visual fatigue. Usually, visual fatigue does not have anything to do with the CRT screen but rather the type of work that people undertake. Computer terminal work can indeed be very intense and fatiguing (Helander et al., 1984). For example, a data entry operator may input as many as 20,000 characters per hour for 8 hours a day. Typically the operator looks at the source document and glances at the screen only occasionally to check the data. After such an intense work day it should not be surprising that operators are fatigued in their entire bodies as well as in their heads. Thus, visual fatigue is just another aspect of general fatigue. Several studies have indeed confirmed that data input operators complain the most about visual fatigue, although this type of work involves comparatively little screen viewing (Helander et al., 1984).

14.4 EFFECT OF RADIATION

The myth about screen radiation is a popular debate in news media and sales. Clearly there is now solid evidence that CRT screens do not generate hazardous radiation (National Research Council, 1983; Bergqvist, 1986). In fact, the amounts of X radiation, ultraviolet radiation, and infrared radiation are at such low levels that they are difficult to measure, and they are not considered a health risk. I recently consulted on office ergonomics to a large multinational corporation. Surprisingly many office workers and even a top manager were concerned about radiation. May employees used screen filters to "offset" the radiation.

Many computer terminal workers expressed concern that the exposure to radiation emitted by CRT screens might lead to the formation of cataracts. Available data indicate that the threshold dose of X radiation that induces cataracts in humans is between 200 and 500 rad for a single exposure and around 1000 rad for exposure spread over a period of several months (National Research Council, 1983). In comparison, an operator exposed to 0.01 mrad/h would absorb less than 1 rad in 40 years of work at a CRT screen. Likewise, the level of ionizing radiation generally believed to increase significantly the risk of birth defects is more than 1 rad for acute exposure. In contrast a worker exposed to CRT screen work would absorb 14 mrad over a period of 9 months. There are indeed many other items in our daily lives which generate more X radiation than CRTs, including brick walls and self-illuminating dials on wristwatches.

One remaining concern is the effect of electromagnetic radiation on CRT screen operators. There is a lack of basic research to prove the effect of electromagnetic radiation on laboratory animals and organisms in general. However, some research indicates that railway engineers, who are exposed to a large amount of electromagnetic radiation from overhead power lines as well as train engines, may have an increased health risk of leukemia and pituitary cancer (Floderus et al., 1993). It is however unlikely that CRT screen operators, who experience much lower radiation levels, would be at any risk.

14.5 REDUCING REFLECTIONS AND GLARE ON CRT SCREENS

A special problem with CRT screens is that glare and reflections on the screen can make the text difficult to read. Several national and international ergonomic standards have proposed guidelines for designing workplaces so as to maximize the visibility of the screen (ISO, 1998).

As the illumination level in a workplace increases, so will the amount of glare on the screen. The ideal working environment for screen viewing is a pitch-black room. The absence of illumination will enhance the screen contrast and make characters very visible. However, this is not very practical since there are other important tasks which do require ambient illumination, including communication with coworkers, and the need to see characters on the keyboard. There are several ways of reducing reflections on a CRT screen (Table 14.1).

TABLE 14.1
Seven Ways of Reducing Screen Reflections

Location	Measure
At the source	1. Cover windows partially
	2. Place light fixtures strategically
	3. Use directional lighting
At the workstation	4. Move the workstation
	5. Tilt the screen
	6. Use screen filters or coatings
	7. Use partitions between the source and workstation

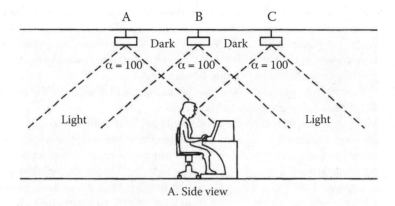

A. Side view

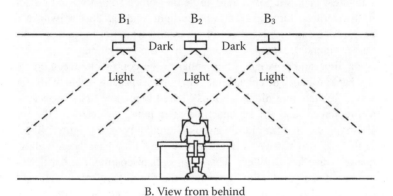

B. View from behind

FIGURE 14.4 From these two figures one can conclude that the ideal location of luminaires is to the side of the operator.

1. Cover windows completely or partially by using draperies, vertical louvres, horizontal louvres, or a gray plastic film. Generally, vertical louvres are preferred over horizontal louvres, because they can be positioned to block the sun, yet permit most workers to look outside. Horizontal louvres often shield off the outside view totally. Windows can also be covered with a neutral density film (usually a gray sheet of plastic) to reduce the transmittance of light from the outside.

2. Place light fixtures strategically. Figure 14.4 provides a side view and a view from behind of an operator at a workstation. In the figure we assume that the light fixtures have a restricted light angle of about 100°. This may be typical for "egg crate" types of luminaires. The operator in Figure 14.4(A) sits at the borderline location of luminaires A and C where there is no direct glare from luminaire C and no reflected glare from luminaire A. Figure 14.4(B) illustrates that luminaire B2, which is closer than the

other luminaires, will cause more veiling reflections and wash out more contrast on the screen. Locations 131 and B3 are better. In summarizing the points made in Figure 14.4, luminaires should be placed to the side of operators and not at the front or the back, where they cause more direct glare, indirect glare, and veiling luminance.

3. Use directional lighting. The examples in Figure 14.4 illustrate the use of directional lighting or task lights.

4. Move the workstation. An operator should not face a bright window, since the large contrast between the dark screen and the bright window may cause discomfort due to glare. Nor should an operator work with his or her back against a window, as screen reflections from the window are inevitable. Rather, the screen should be positioned at 90° to the window. Workstations can also be moved from a bright area to a darker area in an office. This will reduce veiling screen luminance and wash-out of contrast.

5. Tilt the screen. The tilting mechanism, which is mandatory in many standards, makes it possible to angle the screen so as to avoid reflections from overhead luminaires and other light sources. Just as with a tilted mirror, one can decide what to look at and what not to look at!

6. Screen filters or coatings. Filters, such as the neutral density (gray) filter, color filter, and polarized filter, enhance the contrast between characters and background. The enhancement in contrast is achieved in the following way. The incoming illumination is filtered twice: the first time on its way to the screen, and the second time after being reflected by the screen. However, the character luminance is filtered only once (Figure 14.5). Most of these filters were used in the past. Today, they have been replaced by quarter-wave length filters. This is an optical coating, similar to what is used on a camera. As the light enters the filter, there are reflections from the first and the third surface. Since the difference in traveled distance between these two reflections is one half wavelengths, the two reflections are in counter phase and the reflected light is therefore extinguished. The quarter wavelength filter can be bought separately and mounted on top of the display. This will reduce reflections considerably. However, one must make sure that there is not already a filter attached to the display, since this would lower the character luminance, so that the characters become difficult to read.

7. Hang or erect partitions. By hanging partitions from the ceiling or standing them on the floor, it is possible to block off illumination from light sources in an open-plan office or plant.

Application of an etching or frosting to the screen surface reduces specular (mirror-like) reflections. It is no longer possible to see clearly any reflections of one's clothes or face or overhead luminaires, since the reflections become fuzzy. Unfortunately, the screen characters also become a bit fuzzy.

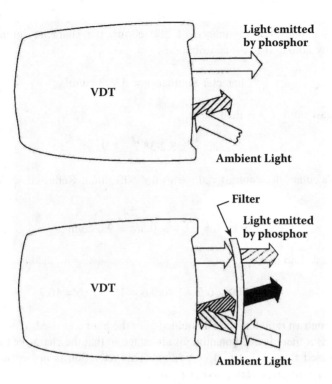

FIGURE 14.5 A filter increases the contrast between the characters and the background.

EXERCISE: CALCULATING THE EFFECT OF A SCREEN FILTER ON THE
DISPLAY CONTRAST RATIO

The purpose of this exercise is to understand how illuminance, luminance, and contrast ratio are related, and to gain an appreciation of how screen filters work. We use Figure 14.5 as a basis for calculating how a screen filter can improve the contrast ratio between the luminance of the characters and the screen background. The ambient light is reflected by the phosphor coating on the back of the screen (this is why the screen surface becomes lighter in a lighter environment). We assume that the screen phosphor has a reflectance of 60%, and the neutral density filter has a 50% transmittance.

(1) Calculate the contrast ratio without the filter. If the incident ambient illumination is 200 lux, the reflected screen luminance can be calculated. Reflected screen luminance:

$$200 \times 0.6/\pi = 38.20 \text{ cd/m}^2$$

Assuming a phosphor luminance of 300 cd/m², the character luminance is obtained by adding the two contributions:

$$\text{Character luminance} = 338.2 \text{ cd/m}^2$$

The contrast ratio (C_r) is then:

$$C_r = 338.2/38.2 = 8.9$$

(2) Calculate the contrast ratio with the 50% filter. Reflected screen luminance:

$$200 \times 0.5 \times 0.6 \times 0.5/\pi = 9.6 \text{ cd/m}^2$$

The contrast ratio is then:

$$C_r = (300 \times 0.5 + 9.6)/9.6 = 159.6/9.6 = 16.7$$

Thus the contrast ratio has almost doubled after the filter was used. This enhances visibility. But there is one potential disadvantage in that the character luminance has decreased from 338.2 to 159.9 cd/m². This reduction is not critical, since 159.9 cd/m² still gives very good visibility.

14.6 FROM CRTs TO LCDs

We have arrived at a technology crossroad. In many countries, LCDs are now replacing CRT displays at a rapid rate. LCD displays as well as other flat panel technology have many interesting properties, and several of the problems with CRT displays do not apply to LCD displays. For example, there is no electromagnetic radiation from LCD displays.

Several comparisons between CRT and LCD displays are given in Table 14.2. (Menozzi et al., 2001; Noro, 2002; Nylén, 2002, and Ziefle, 2001). As the price of LCD displays keeps coming down, we predict that they soon will be generally adopted.

TABLE 14.2
Comparison between CRT and LCD Technology

Issues	CRT Technology	LCD Technology
Price	Inexpensive	Moderate
Footprint	Large	Small
Weight	Heavy	Light
Workplace arrangements	Heavy monitor rarely moved	Easy to move
Viewing distance	The large footprint of the display forces the display closer	Long
Energy consumption	High	Low
Character definition	Has improved over the years	Excellent
Color rendering	Very good	Good
Image distortion	Some	None
Flicker	Depends on refresh rate	None
Specular reflections	Quarter wavelength coating solves problems	Less reflections
Heat emission	High	Low
Electromagnetic emission	Some	None

14.7 SUMMARY

This chapter has focused on specific design issues concerning computer use in offices with the main purpose of dispelling myths about visual fatigue and radiation and emphasizing the real concerns, which are based on concrete research findings. These include the design of workspaces and work postures as well as illumination.

FURTHER READINGS

Kroemer. K.H.E. and Kroemer, H. (2001), *Office Ergonomics*, London: Taylor & Francis.
Raymond, S. and Cunliffe, R., 2000, *Tomorrow's Office: Creating Effective and Human Interiors*, London: Taylor & Francis.

15 Training, Skills, and Cognitive Task Analysis

15.1 INTRODUCTION

Due to the introduction of computers and automated tools, manufacturing and production systems have become increasingly complex (Sheridan, 2002). Although the automation of manufacturing systems may have the effect of removing some employees from the shop floor, those who remain get greater responsibilities. They supervise a production system, they participate in the planning and scheduling of production, they exercise quality control, and sometimes they take responsibility for ordering and deliveries. In this complex environment, human errors can have serious consequences.

In referring to human errors, we do not mean to imply that an operator was truly at fault and should be blamed. Rather, the reason for error may be poor design of the production environment, poor management, or lack of training. In an unexpected crisis situation, there will not be much time to act, and the human operator will use intuition to make decisions. The decision may be entirely logical and rational, except that sometimes the production system is not designed for logical or rational input, so a "human" error is committed. The issue of transfer of training is often brought up in this context. In situations of emergency, individuals act according to previously learned stereotypes—except that the old stereotypes may not fit a new environment. One example is a pilot who can fly several airplanes. In an emergency situation, the pilot may decide on instinct. The response was appropriate for the first aircraft the pilot learned to fly, but not the present aircraft.

In previous chapters we have addressed the issues of design of work systems; in this chapter we focus on training and development of programs that can modify how people work together, solve work-related problems, and actively fulfill their roles in ergonomics implementation. We then introduce tasks analysis, which has a long tradition in HFE as a tool for design of artifacts and for training programs. Finally we give an overview of cognitive task analysis, which is very different from the regular task analysis. The purpose in this case is to understand expertise and how to support the development of experts in a work system.

15.2 ESTABLISHING THE NEED FOR TRAINING

To establish the need for training, one must understand what skills are required to perform a task and the characteristics of the trainees (Goldstein, 1980; Patrick, 1992). This is traditionally done in a systems approach to defining training needs (see Figure 15.1). First the training objectives are defined. These explain the purpose of the

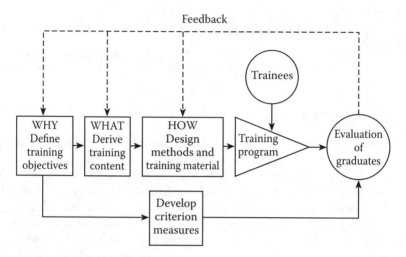

FIGURE 15.1 The development and evaluation of a training program.

training. Obviously the training should be based on perceived needs for training, and these needs must be clarified and formalized. Using the training objectives, criterion measures are developed, which express what the students should master after the course. The training program must be effective in improving the trainees. To understand if this is the case, the effectiveness of the training program is formally evaluated after the training course. The criterion measures are used for the evaluation. They should be stated in terms of concrete goals that can be measured after the training course.

Examples of such criteria are as follows: for manufacturing assembly, being able to assemble a widget in 5 minutes (skill based); for safety, being able to specify the safety procedures for several scenarios (rule based); and for manufacturing production, being able to produce a manufacturing production schedule for a mix of orders (knowledge based) (Rasmussen, 1986).

After the training objectives and the criterion measures have been clarified, the content of the training must be defined. For skill-based and rule-based scenarios, a task analysis is often used. For the knowledge-based situation, where individuals have to use their problem-solving skills, a cognitive task analysis is appropriate.

15.3 DETERMINING TRAINING CONTENT AND TRAINING METHODS

The type of training depends on the type of task. In Table 15.1 we use Rasmussen's (1986) distinction between tasks that are skill-based, rule-based, and knowledge-based. The model is presented in section 5.8. The skill-based scenario refers to manual or perceptual motor skills such as are used in driving a car. Over time these skills become so well learned that they become automatic. Rule-based skills are used for procedural tasks. Here the worker must first recognize a scenario and then

TABLE 15.1
Skills and Options for Training Skills in Manufacturing

Type of Skill	Example of Tasks	Training Options
Skill based	Manual skills; e.g., assembly	On the job, coaching
Rule based	Procedures; e.g., workplace organization, housekeeping, safety procedures	On the job, coaching, classroom
Knowledge based	Problem solving: e.g., production scheduling	Classroom, problem solving at work

decide what to do: if condition A, then action 1; if condition B, then action 2; etc. Knowledge-based tasks are those that require deep thinking and the pondering of alternatives, as in complex problem solving.

As we suggest in Table 15.1, these skills should be trained for differently. On-the-job training is clearly appropriate for skill-based jobs (Holding, 1986). The job site has better ecological validity than the classroom. At the job one finds the real scenarios that remain difficult to simulate in a classroom. Classroom training may be appropriate for knowledge-based jobs, particularly if there are theoretical components which may be useful in problem solving (e.g., mathematics and physics). But problem solving at work is essential, since it provides the necessary context of real task cues.

The options for training listed in Table 15.1 are quite traditional. Many researchers have expressed reservations against the use of fashionable training tools such as computer-aided instruction, multimedia, and e-learning. There is nothing wrong in using such tools, but the availability of tools should not dictate the strategy for training; rather, the tools must be adapted to the particular instructional needs. Gilbert (1974) expressed great caution:

"If you don't have a gadget called a teaching machine—don't get one. Don't buy one; don't borrow one; don't steal one. If you have such a gadget—get rid of it. Don't give it away, for someone else might use it. This is a most practical rule based on empirical facts from considerable observation. If you begin with a device of any kind you will try to develop the teaching program to fit that device."

As illustrated in Table 15.2, there are two basic scenarios: training to learn a new job, and training to improve job performance. In a manufacturing plant, workers, supervisors, and management may assess the need for training by analyzing task requirements. For existing production systems, one can analyze production reports and quality reports, since these may give important hints of what employees need to learn. As illustrated in Figure 15.1, there are three stages in training development:

TABLE 15.2
The WHY? WHAT? and HOW? of Training Development

Development of Objectives, Content, Materials	Training	
	To Learn a New Job	To Improve Job Performance in an Existing Job
WHY?		
Defining training objectives		
• Production reports		x
• Quality reports		x
• Customer feedback		x
• Employee feedback		x
WHAT?		
Developing training content		
• Employee information	x	x
• Experience with similar case	x	x
• Expert option	x	x
HOW?		
Designing methods and training materials		
• Task analysis	x	x
• Discussion with employees	x	x
• Experience with similar case	x	x

defining training objectives, developing training contents, and designing training methods and materials. These stages correspond to the questions why, what, and how in Table 15.2.

Training needs are diagnosed by using feedback from customers and employees and reports of production, quality, and yield. Note here that feedback requires an existing scenario. For new jobs there is no direct feedback, since there are no production reports or customer complaints.

For a new job, the analysis must be based on past experience, conventions, and trial and error. Feedback is apparently missing. Often, however, one can draw from experiences of similar cases within the company, or one can ask a consultant.

In the third stage, the training program is designed in detail based on familiarity with the work scenario. Further information is derived from task analyses and discussions with employees.

15.4 USE OF TASK ANALYSIS

Task analysis has been used extensively as a tool to develop training programs (Luczak, 1997). The main purpose of task analysis is to obtain a thorough understanding of the task, and thereby capture what is important in training. Task analysis has several additional purposes, such as the design of products, safety systems, and workplaces. We do not supply examples of such analyses here, but the procedures are quite similar to those illustrated below.

In task analysis a job is first broken down into its various components. The U.S. Air Force has suggested a hierarchical breakdown:

- Job
- Duty
- Task
- Subtask
- Activity

The purpose of the hierarchical breakdown is to provide a logical description of the various activities that constitute the job; an example is given in Figure 15.2. In this example, the job of a car mechanic can be regarded as being composed of three major duties: relining the brakes, tuning the engine, and servicing the cooling system. These duties are broken down into tasks. In tuning an engine there are three major tasks: repair the distributor, replace the plugs, and repair the carburetor. For some

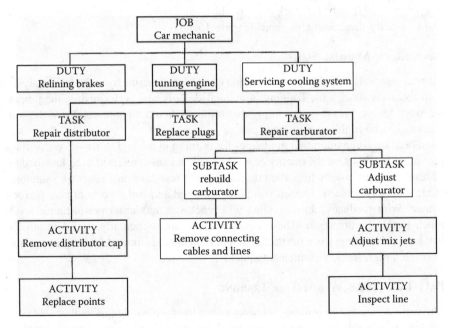

FIGURE 15.2 An example illustrating the hierarchical breakdown of a task (U.S. Air Force).

of these tasks a further breakdown of activities may be necessary, e.g., remove the distributor cap and replace the points.

The extent to which a task is broken down depends on the purpose of the task analysis; one could stop at the activity level, or one could go further and study very minute details, such as eye movement and finger movements. This is not warranted for training development. But in designing a cockpit, it is useful to analyze eye movements because such information can suggest where displays should be located.

Depending upon the purpose of the analysis, the analyst will have to define what information needs to be collected and how best to represent the information so that it is easy to use. The results of a task analysis can be presented in many different ways. The most common is a table with columns (Table 15.3). These tables were developed to describe the task of a process control operator (Drury, 1983; Swain and Guttman, 1980). The definition of the column headings and the description of the task is up to the analyst. For every analysis one must define what information is necessary and how to present it, so that it is useful for design. There are no set formats or general procedures, for a task analysis—as Montemerlo and Eddower (1978) in their well-known report explained—task analysis is an art, not a science.

The training program is defined based on the results of the task analysis. Some aspects of a task will be more important to train than others, and in Table 15.3 the criticality of operators' actions is analyzed. Although it is important to train the entire task, actions that are highly critical should be emphasized.

After the completion of training it is important to evaluate the effectiveness of the training, using the criterion measures developed previously. The results of these evaluations can then be fed back to refine the training objectives and training content and to modify the training methods (Figure 15.1).

Training of Manual Skills

Manual assembly is highly procedural and is easier to describe by using task analysis than are many other jobs. Training of manual skills is best performed on the job, at the workstation. This does not imply that training should be haphazard. On-the-job training can be highly structured.

There are some important guidelines for training in manual skills—in particular, providing feedback on the quality of work. Feedback, also referred to as knowledge of results, is particularly important during the first few days on a new task (Salmoni et al., 1984). An effective coach will point out good and bad aspects of task performance. With feedback, an individual will reach the maximum performance level much faster (Figure 15.3). These considerations are particularly useful in small-batch manufacturing, since the production of each batch is limited, and it is important to reach a high level of proficiency quickly.

Part-Task versus Whole-Task Training

The main advantage in training subtasks rather than the whole task is that subtasks are easier to learn. After one subtask has been trained, one can continue training similar tasks, thereby using the transfer-of-training effects. The disadvantage is that one may lose dynamic or contextual cues that are available only if the entire job is

TABLE 15.3
Two Different Column Formats Used to Describe the Task of a Process Control Operator

Task	1. Hardware interface, software interface	2. Visual and manual cues	3. Required response	4. Operator feedback	5. Performance Criteria, e.g. time and accuracy	6. Criticality of action
1						
1.1						
1.2						
2						
2.1						

TASK DESCRIPTION / TASK ANALYSIS

Task or step number	Instrument or control	Activity	Cue for initiation	Remarks	Scanning perceptual anticipatory requirements	Recall requirements	Interpreting requirements	Manipulative requirements	Likely human error
1									
1.1									
1.2									
2									
2.1									

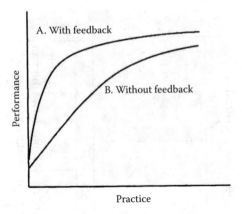

FIGURE 15.3 Training with feedback improves job performance significantly.

trained at the same time (Fitts and Posner, 1973; Holding, 1986). The transition between subtasks becomes difficult. Think of a pianist learning ten bars of music at a time—the interpretation and dynamics of the entire piece of music may suffer.

When should one use part-task and when should one use whole-task training? Researchers have not yet come to a conclusion. As often is the case in ergonomics, we can only give a general answer: it depends on the task.

15.5   USE OF JOB AIDS

Job aids are used for training as well as later on the job, when the operator has learned the task. Sometimes details are difficult to memorize, but still essential for good task performance. Most of us have used cheat-sheets, small pieces of paper which are posted to a computer terminal to help us remember log-on procedures and so forth. Such job aids can be critical, as the following example illustrates.

EXAMPLE: REMEMBERING ERROR CODES

The author was once called in to analyze the ergonomic design of several micro-scope workstations at IBM Corporation. The microscopes were used for quality control in the manufacturing of electronic boards. Management had anticipated recommendations to improve the seating posture, the illumination of the work-station, and the microscope itself. However, there was a much more important problem. The operators' primary duty was to report quality defects of compo-nents, and there were 24 different kinds of defect. For each defect the operators had to write an eight-number error code on a sheet of paper which accompanied the board. In case the error code had been forgotten, operators could find it in a 250-page manual. This was an awkward procedure, and operators frequently took a chance, hoping to report the correct error code.

> To help operators remember the error codes, we designed a job aid. We condensed the manual to a plastic cube with 8 cm sides. On each of the six sides, four quality defects were illustrated with a figure and the corresponding error code. This job aid simplified the task significantly.

There are many types of job aids. In discussing the use of computer manuals, Carroll (1993) concluded that manuals are rarely read. He advocated the use of a minimal manual, which consists of only a few pages. The point is that job aids must be easy to use. There are many possible formats that can be used for job aids (Kinkade and Wheaton, 1972).

- *Procedural instructions, flowcharts, tables, and codebooks.* The main concern here is to condense the information into a format that takes the minimum space, and yet is instructive.
- *Color coding* is sometimes used in a workstation to connect parts or procedures that belong together by using the same color.
- *Schematic diagrams and graphics* are often used in process control tasks. Sometimes they are painted on the control panel to suggest causal relationships.
- *Checklists* are used by an airplane pilot to follow the necessary procedure in taking off and landing an aircraft. Checklists are likewise used in industry to help people remember long procedural tasks.
- *Computer help systems* can be made available for any computerized task. Help functions must be easy to access and easy to understand. Most of us have bitter experience of computer help systems that are impossible to use. Ergonomics expertise is necessary in designing usable help systems.

15.6 THE POWER LAW OF PRACTICE

Practice makes perfect—but how much time will it take? The power law of practice can actually predict future performance times. This law is well known to industrial engineers who have knowledge of time-and-motion studies. It is well illustrated in a classic study reported by Crossman (1959). Crossman obtained performance records for a woman in Tampa, Florida, who was working in a tobacco factory rolling cigars. She would pick up tobacco leaves and form them into a cigar. She would then put the cigar into a cigar rolling machine that compressed the cigar and applied a wrapper. The improvement in performance over a time period of 7 years is illustrated in Figure 15.4.

During the first year the woman rolled approximately 1 million cigars and her performance time was getting close to the machine cycle time, but even after 7 years and 10 million cigars there were still improvements in performance. The results suggest that the performance of complex manual skills will continue to improve over time. For the woman in Tampa, the machine cycle time set a limit to how much she could improve.

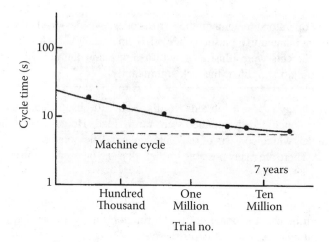

FIGURE 15.4 Improvement in performance time for rolling cigars.

The relationship between performance time (T_N) and the number of trials follows the power law of practice (Welford, 1968; Konz, 1990):

$$T_N = T_1 N^{-a}$$

where T_1 is the performance time on the first occasion, T_N is the performance time on the Nth occasion, and a is a constant. Rewritten in logarithmic form, we obtain a linear relationship (Figure 15.5):

$$\log T_N = \log T_1 - a \, \log N$$

The power law of practice is used to predict the performance times for industrial tasks; an example is given below.

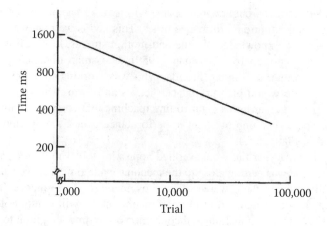

FIGURE 15.5 Performance time decreases with the number of trials. When plotted in a log-log diagram, the relationship is linear.

EXAMPLE: PREDICTION OF FUTURE ASSEMBLY TIME

A company is trying to estimate how the assembly time of a machine part will decrease in the future when the operator has more experience. They plan to use the predicted assembly time to estimate future manufacturing costs. The assembly time of the operator was measured. For the 1000th trial the assembly time was 14.5 seconds, and for the 2000th trial the assembly time was 11.5 seconds. What is the expected assembly time for the 5000th trial?

Solution: from the equation above, two expressions of T_1 are obtained:

$$T_{1000} \times 1000^a = T_{2000} \times 2000^a$$

$$a = 0.36$$

$$T_1 = T_{1000} \times 1000^{0.36} = 174s$$

$$T_{5000} = 174 \times 5000^{-0.36} = 8.3 \text{ s}$$

The improvements in assembly time are substantial and the manufacturing costs can be predicted to decrease accordingly.

LEARNING RATES

The slope of the learning curve (as in Figure 15.5) is used by industrial engineers to express the rate of improvement, or the learning rate. The learning rate is expressed as a percentage and is obtained by dividing the performance times for the $2N$th trial by the Nth trial. In our example, the learning rate can be calculated by dividing the assembly time of the 2000th trial by the assembly time for the 1000th trial, and we obtain:

$$\text{Learning rate} = 11.5/14.5 = 78\%$$

Konz (1990) reported learning rates for several different types of task, and part of this information is reproduced in Table 15.4. Two important issues are illustrated in this table.

1. Complex tasks improve more than simple tasks. For example, from Table 15.4 we note that truck-body assembly improves much more (68%) than a simpler task such as attending a punch press (95%). This principle is also illustrated in predetermined time systems that are used to predict assembly time. For example, in MTM-1 motions are broken down into ten categories: reach, move, turn, apply pressure, grasp, position, release, disengage, body motions, and eye motions. For the very simple types of motion such as "reach" and "move" there is very little improvement over time. These are primitive movements of the hand back and forth, and there is not much that can be improved. For more complex movements that

TABLE 15.4
Learning Rates for Several Types of Task

Learning Rate %	Type of Task
68	Truck body assembly
74	Machining and fitting small castings
80	Precision bench assembly
82	Grinding
83	Servicing automatic transfer machines
84	Cigar making
88	Welding (manual)
89	Punch press, milling
90	Punch press
92	Assembly with jig, welding
95	Punch press, screwdriver work
95	Word-class mail runner
98.5	Grinding, milling, assembly

require manual skills, such as positioning, there is more opportunity for improvement. Positioning typically involves intricate assembly movements for alignment and orientation of parts.

2. There is not one correct learning rate. For a task such as grinding, Table 15.4 gives two different learning rates (82% and 98.5%). This illustrates that grinding tasks can be very different; some of them involve complex movements for which there is a greater potential for improvement.

15.7 COGNITIVE TASK ANALYSIS AND WORK ANALYSIS

Work is becoming increasingly cognitive, and HFE needs tools to analyze the cognitive demands and the type of expertise that an operator needs to be successful at work. The recent developments in cognitive task analysis (CTA) and work analysis provide interesting avenues for measurement of cognitive demands. These developments are maybe the most important in HFE during the last ten years. They are, however, still under development. In this chapter we describe two approaches: the analytical approach and the situated approach.

1. The analytical approach is used to solve interface problems in complex systems (Vincente, 1999). This method is based on Rasmussen's (1986) approach to task description and task decomposition.
2. The situated approach is based on Klein's (1989) concept of recognition-primed decision making. This is more suitable for routine tasks, and is based on task analytical procedures.

15.8 RASMUSSEN'S ANALYTICAL APPROACH

This methodology was originally formulated for analysis of work in complex environments, such as nuclear power plants (Rasmussen, 1983, 1986). It has later been applied to maintenance work and analysis of emergency situations. To understand how an operator solves problems, the work task is analyzed in terms of abstraction levels and systems decomposition level. Table 15.5 shows how a power plant can be represented using five abstraction levels, from purpose to physical form. Rasmussen (1987) claimed that the use of five abstraction levels is sufficient to represent even the most complex analysis, such as that of a nuclear power plant operation.

The abstraction levels can be derived by asking the questions why, what, and how. Assume that we are working at abstraction level 3, analyzing connected pieces of equipment. If you ask the question why (Why is this useful?), you will get to abstraction level 2, and if you ask the question how (How can this be done?), you will get a detailed description at abstraction level 4. The questions why, what, and how are used to identify the functionality of any abstraction level. This analytical approach is hence a way of analyzing a system in terms of the work functions that it generates.

TABLE 15.5
Representation of a Nuclear Power Plant in Terms of Means–Ends and Whole–Part Decomposition

Abstraction Level	Means–Ends	Whole–Part	
1	Purpose	Functional purposes and goals in terms of energy, material flow, and distribution	Why
2	Abstract function	Mass, energy flow, and balance in terms of underlying generic functions	What, Why
3	Generalized function	Info on cooling, heat transfer, regulation in terms of connected pieces of equipment	How, What, Why
4	Physical function	Performance data on physical equipment in terms of information on components	How, What
5	Physical form	Installation and maintenance info on components: take-apart diagrams, illustrations	How

Adapted from Konz (1990). The lower the learning rate, the greater the improvement in performance time.

TABLE 15.6
Abstraction Levels and Means–Ends Decomposition
for Water Pump

	Abstraction Level	Means–Ends
1. Purpose	Supply water	Why
2. Abstract function	Pump water	What, Why
3. Generalized function	Use human–machine system	How, What, Why
4. Physical function	Use bellows	How, What
5. Physical form	Use mechanical movements	How

Another example is illustrated in Table 15.6 and Figure 15.6. This is a depiction of Vittorio Zonca's engraving of a bellows pump from 1607 (Helander, 1996). This system can be broken down into five abstraction levels, from the top purpose or goal of the system to the bottom, which illustrates design details. Let's say you are at the second level of abstraction, "Pump Water," and you ask yourself why. This question brings you to the first abstraction level, "Supply Water."

Rasmussen claimed that the analytical approach of a system is useful in order to analyze several usability questions, including the following:

- Determine the functionality of the device
- Tell what actions are possible
- Tell what state the system is in
- Determine mapping from intention to physical movement
- Determine mapping from system state to interpretation

One example is the analysis of the work of a maintenance technician that is illustrated in Figure 15.7 (Rasmussen, 1990). In this case there is a division according to abstraction level as well as systems level. The reason for the maintenance action is stated at number 1, and the solution to the problem is in the last two activities with numbers 14 and 15. From the figure we understand that in searching for the correct solution the maintenance engineer jumps between abstraction levels. As long as he knows how to identify a fault, he goes down in abstraction level, from the top goal ("No communication") to the physical form ("Looks burnt").

Most of the circles are located along the diagonal. This means that there is coupling between abstraction level and decomposition level. Obviously the purpose must be considered for the whole system and the physical form at the component level. Any excursion from the diagonal is therefore interesting to study. Sometimes the technician had to go up a level or two to check the purpose or the procedures. This is what happened for activities 5 (backtracked to purpose) and 9 (backtracked to abstract function). These are at a higher abstraction level than the previous

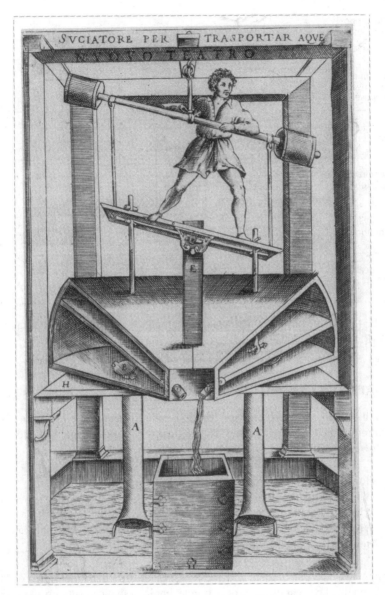

FIGURE 15.6 Vittorio Zonca's illustration of a bellows pump, 1607.

activities 4 and 8, and can therefore offer help of a more general nature—such as reminding the technician about the appropriate procedures.

We can learn from Figure 15.7 how an experienced technician approaches problem solving. The sequence of activities is informative for other technicians, who may want to understand how to troubleshoot this device.

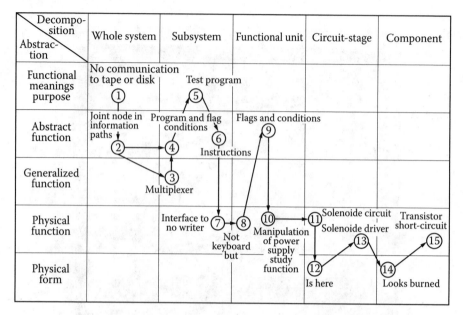

FIGURE 15.7 Analysis of a maintenance engineer's task to identify a fault in a circuit (Rasmussen, 1990).

15.9　APPLIED COGNITIVE TASK ANALYSIS

There are several methods, most of which are based on Klein's (1986) notion of recognition-primed decision making. The purpose of this type of analysis is to answer the following questions:

- How do people make decisions at work?
- What information do they use?
- What are the particular skills required?
- What are the differences between experienced and inexperienced operators?
- How do individuals collaborate in task performance?

Assuming that we can understand these issues, it would be possible to design information displays and other information technology that can support the operator in making decisions. One can also understand what types of skills an experienced worker has and, from this, what type of training an inexperienced worker would need. If the work is carried out in teams, it should be possible to understand how people collaborate in decision making.

Of particular interest is the study of decision making in emergency situations. Can the team collaborate in such situations of high stress, where the physical asset of the plant is at risk? Unfortunately it is not easy to study these aspects, since they rarely occur.

In this section we will review applied cognitive task analysis (ACTA), which was developed by Militello and Hutton (1998). An applied cognitive task analysis emphasizes the situational assessment that an operator has to do in order to make decisions. Let's assume that an operator recognizes a situation as familiar. This recognition will make it possible for her to consider important details of the task, including:

- Purpose of the task
- Information that is used for task performance
- Strategies that are used to predict task outcomes
- Action schemes to execute the task

Note that there is a similarity with Rasmussen's decision ladder, which was discussed in Chapter 5. The operator may recognize the situation, and since it is familiar, she would know immediately what to do. If the situation is unfamiliar, the operator may conduct a mental simulation of the most plausible course of action, and then decide what to do. The mental simulation is based on similar task scenarios, which will then support the operator in decision making. If there are no similar task scenarios the situation becomes much more difficult, and the operator will often be forced to act without information that can be used to predict the outcome.

ACTA is intended to help the HFE practitioner to extract meaningful information about cognitive task demands and skills required to perform a task. To obtain information about the cognitive elements, the HFE expert will ask subject matter experts (SMEs) using three main methods: task diagram, knowledge audit, and simulation interview. These methods are described below (Militello and Hutton, 1998).

TASK DIAGRAM

The purpose of the task diagram is as follows:

- Provide a broad overview of a task.
- Identify difficult cognitive elements.
- Provide a surface level evaluation of the cognitive elements. Which are the most difficult and relevant cognitive task elements? Which are the elements that take more time and resources than other task elements?

SMEs are first asked to break the task into three to six steps or subtasks. The goal is to get the SME to walk through the task in his or her mind, verbalizing the major steps. The limit of three to six steps is to ensure that time is not wasted on minute details during this surface-level interview. The details will be filled in at a later stage (Militello and Hutton, 1998).

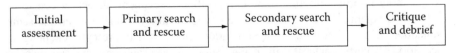

FIGURE 15.8 Task diagram for a fire commander (Militello and Hutton, 1998).

The SME is then asked to decompose and articulate the three to six tasks into steps or subtasks. The investigator prompts the SME:

- Think about what you do when you (task of interest).
- Identify which of the steps require cognitive skill. Cognitive skills refer to judgments, assessments, and problem solving and thinking skills.
- Of the steps you have just identified, which types of cognitive skills are necessary?

The results of the task analysis are then presented in a task diagram. This diagram is a for future interviews. Figure 15.8 presents an example of a task diagram for a fire commander.

KNOWLEDGE AUDIT

The purpose of the knowledge audit (KA) is to convey what types of expertise are required to perform a specific task. The KA is developed to capture the most important aspects of expertise. It is organized around knowledge categories that characterize expertise, such as diagnosing and predicting, compensating for equipment limitations, situation awareness, perceptual skills, applying tricks of the trade, and knowing how to improvise. Several questions are used in order to solicit detailed information (Militello and Hutton, 1998):

- Is there a time when you walked into the middle of a situation and knew exactly how things got there and where they were headed?
- Can you give me an example of what is important about the big picture for this task?
- Have you had experiences where details of a situation just popped out at you, where you noticed things going on that others didn't catch? What is an example?
- Are there ways of working smart, or accomplishing more with less, that you have found particularly useful?
- Can you think of an example when you have improvised in this task?
- Can you think of a time when you realized that you would need to change the way you were performing the task in order to get the job done?

SIMULATION INTERVIEW

Using information collected from the task diagram and the knowledge audit, the interviewer creates a challenging scenario on the job—an imagined problem. He then asks the SME how she would solve the problem. This allows the interviewer to get more information about how the situation assessment impacts a course of action and potential errors that a novice would be likely to make given the same situation. The main purpose of the simulation interview is to present a challenging scenario. All participants are given the same scenario, and answers can then be compared afterwards.

TABLE 15.7

An Example of One Difficult Cognitive Element in Firefighting

Difficult Cognitive Element	Why Is the Task Difficult?	Common Errors	Cues and Strategies
One example of a difficult cognitive element is to find victims in a burning building.	This is a difficult task because there are many distracting noises, and your own breathing makes it hard to hear anything else.	An error commonly committed by novices is that they mistake their own breathing sound as coming from victims.	One strategy is to hold your breath and listen. In particular listen for crying, and victims talking to themselves

This technique will produce different answers and thereby emphasize different aspects of the skills that are necessary for accomplishing the task. This method will therefore produce rich material that can be used for both training and systems design, such as design of information displays.

COGNITIVE DEMANDS TABLE

The data collected from the task diagram, knowledge audit, and simulation interview is then sorted, systematized, and analyzed (see Table 15.7). The table illustrates one particular problem; more complete information is given in Militello and Hutton (1998).

The table helps in finding common themes as well as conflicting information among the SMEs. The information may be used, for example, to develop a training course or design a new communication system.

VALIDITY OF ACTA

Militello and Huttton (1998) evaluated the validity of ACTA using two groups of experts: firefighters and operators working with electronic warfare. In average they thought that 93% of the information addressed cognitive issues. The content was important domain knowledge that novices could learn from. The results were hence important for training. On average 92% of the information was judged to be information that only highly experienced SMEs can produce. They were then asked if the generated information was accurate. The accuracy was rated higher for the fire fighting (92%) than for the electronic warfare (54%). Finally they were asked if the information was important for novices to learn. The ratings were 95% for firefighting and 93% for electronic warfare. This is convincing and impressive information. Since ACTA was published it has been extensively used to analyze cognitive difficulties as well as cognitive skills in task performance.

RECOMMENDED READING

For more information about the use of predetermined time systems, we refer the reader to Konz, S. and Johnson, S., 2004, *Work Design: Occupational Ergonomics*, Scottsdale, AZ: Holcomb Hathaway. The latest and most authoritative book on cognitive task analysis is Crandall, B., Klein, G., and Hoffman, R., 2006, in press, *Labors of the Mind: A Practitioner's Guide to Cognitive Task Analysis*, Cambridge, MA: MIT Press.

16 Shift Work

Work schedules differ in many ways, and more than 10,000 schedules are in use worldwide (Knauth, 1998).

16.1 INTRODUCTION

Shift work is not a new phenomenon. Scherrer (1981) reported that in ancient Rome, transportation of goods had to be performed at night in order to reduce traffic congestion. However, it is only during the last century, after Edison's invention of the lamp, that shift work has become widely adopted in industry. This is concomitant with several trends in industry and society:

1. *Process industries.* Many modern industries such as power plants and steel works cannot close at night.
2. *Economic pressures.* Companies often prefer to introduce a second and a third shift because production machinery is expensive and cannot be duplicated. In addition, shift work makes it possible for individuals to work overtime, which is less expensive and is often perceived as less risky than recruiting additional employees.
3. *Service sector demands.* In the service sector there are many types of job where people are needed around the clock (nurses, physicians, policemen, transportation workers, and restaurant employees).

In this chapter we take a broad definition of shift work as being anything outside the hours of 7:00 am and 6:00 pm (Monk and Folkard, 1992). With this broad definition, approximately 20–30% of the workforce participates in shift work. A study summarizing work patterns in the European Union found that 16.8% worked shift work and 18.2% worked at night (Costa et al., 2001). They noted that the definitions of shift work differ in the different European countries and it is therefore difficult to compare data. In the U.S., Beers (2000) estimated the percentage of shift workers in several job categories (see Table 16.1).

There are two types of operation: around-the-clock, usually involving three shifts; and operations involving fewer hours. The around-the-clock operation, and in particular the hours from 12:00 midnight to 4:00 am, cause severe problems in terms of health, fatigue, and lost productivity, and these problems are the major focus of this chapter.

Shifts are usually designated as morning shift, afternoon shift, and night shift. There are other common names: day shift, swing shift, and graveyard shift, or simply

TABLE 16.1
Percentage of Shift Workers in Various Occupations in the U.S.

Occupation	Shift Workers
Food service	42.0
Sales workers	28.5
Health service	30.1
Cleaning/building service	27.1
Machine operators	26.2
Protective services	55.1
Handlers, helpers, laborers	24.6
Teachers, college and university	13.9
Construction trades	4.4

TABLE 16.2
Typical Working Hours in Shift Work

Name of Shift	Typical Working Time
Morning shift	6:00 am–2:00 pm (600–1400 h)
Afternoon shift	2:00 pm–10:00 pm (1400–2200 h)
Night shift	10:00 pm–6:00 am (2200–0600 h)

shift 1, shift 2, and shift 3. In this chapter we use the first designation, as illustrated in Table 16.2.

EXAMPLE: POOR PRACTICES IN SHIFT WORK SCHEDULING ABOUND

The author once visited an underground metal mine in the southern U.S. During interviews with the workers it was obvious that many of them suffered fatigue from participating in shift work. It turned out that there were only two shifts, and the working hours had a beautiful symmetry:

- Shift 1: 7:00 am to 3:00 pm
- Shift 2: 7:00 pm to 3:00 am

We asked a manager why there were 4 hours of non-work starting at 3:00 pm. He gave the following explanation: the work procedures were identical for both crews. In the beginning of a shift miners first transport ore and rocks which had just been blasted by the previous shift. Then they would start drilling holes for blasting, and at the end of the shift they would blast. Many years ago it used to be that blasting agents produced lots of smoke, and it was necessary to ventilate the mine for 4 hours before the next crew could come in. However, with modern types of blasting agents ventilation is no longer necessary. We told the manager that the problems with shift work would be eliminated if the second crew could work from 3:00 pm to 11:00 pm. Our suggestion was not well received. The manager said: "We would hate to renegotiate the contract with the union!" Often the design of shift work schedules is contingent upon management policy and conflicts, rather than worker-centered.

16.2 CIRCADIAN RHYTHMS

The basic physiological problem with shift work is that the body establishes a 24-hour rhythm which is difficult to change. Figure 16.1 illustrates the so-called diurnal or circadian changes in oral temperature over 24 hours. The temperature is at a maximum at about 4:00 pm and at a minimum at about 4:00 am. Many other body mechanisms (heart rate, breathing rate, body temperature, excretion of many types of hormones, and urine production), follow the same sinusoidal pattern (Chapanis, 1971). Assume that a person starts working the night shift (10:00 pm to 6:00 am) instead of the morning shift (6:00 am to 2:00 pm). It would take about 1 week to flatten out the sinusoidal curve and about 3 weeks to reverse the waveform. As illustrated in Figure 16.1, the pattern is never quite reversed. The circadian changes are smaller for a person who works the night shift as compared to a person who works the morning shift. In other words, it seems to be impossible to adjust totally to night time work.

There are many reasons for this lack of adjustment. The most important component may be daylight. Daylight is a very forceful cue in indicating the time of day. In German, this phenomenon is referred to as a *Zeitgeber* (literally, time-giver). Recent research has shown that exposure to daylight levels (more than 2000 lux) of illumination increases alertness during night shifts, and suppresses the production of melatonin (a sleep-inducing hormone). But there are also many environmental and social *Zeitgebers* (Monk and Folkard, 1992). It is easier to sleep during the nighttime because there is less disturbing noise and there are no social activities. On the other hand, a night worker suffers more from daytime noise and daytime activities, and family and friends also disturb the sleeping pattern.

16.3 PROBLEMS WITH SHIFT WORK

There are many problems with working the night shift. Some of these problems have been well documented, whereas others have been suggested but not yet verified by research (Table 16.3). Some of the items listed in the Table 16.3 warrant comment. It is evident that shift workers have a much higher rate of stomach problems than daytime workers (Monk and Folkard, 1992). Part of the problem is that the sensation of appetite is tied to the circadian cycle. The appetite is suppressed while people are asleep and is greater during daytime. Individuals starting on the night shift will carry their daytime habits along until they have adjusted. Shift workers are hungry

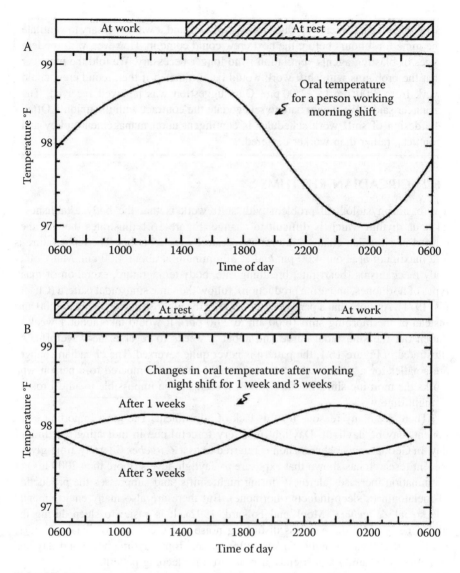

FIGURE 16.1 Diurnal (or circadian) rhythm of oral temperature. (A) The normal pattern for day work. (B) There is a flattening after 1 week and reversal of the curve after 3 weeks of working on the night shift.

at the wrong times, and go to the toilet at the wrong times. In addition, shift workers eat more junk food than do daytime workers. One reason for this may be that the company cafeteria is closed and there are no cooked meals available. A shift of the circadian cycle also disturbs the digestive functions.

One of the basic problems with research on shift work is that, while some individuals like it, about 20% of the population has severe difficulties and will never adjust. Perhaps their bodily constitutions are not robust enough to cope with shift

---

TABLE 16.3
Typical Problems Associated with Working the Night Shift

Fatigue. On average a night-shift worker sleeps 1.5 h less.

Health disorders. Stomach problems, digestive disorders, and possibly an increased rate of cardiovascular disease.

Disruption of social life. With family, friends, labor unions, meetings, and other gatherings.

Decreased productivity. More for knowledge-based tasks than skill- and rule-based tasks.

Safety. Accident rates may increase.

---

work, and perhaps those who remain in shift work are physically stronger and have better health. Therefore the population of study may be biased to start with, and it is difficult to arrive at a fair scientific comparison.

Some studies suggest that there is a greater possibility of an increased rate of cardiovascular disease among shift workers (Rick et al., 2002). A longitudinal study of 50 workers in a Swedish paper mill showed that after 10–15 years of exposure to shift work, the risk of heart disease was doubled compared with a population of workers on day shift (Knutsson et al., 1986). But there were many uncontrolled factors. In addition to shift work there might have been differences related to lifestyle, diet, and so forth, although some of these factors were taken into account in the study. A later study by Olsen and Kristensen (1992) indicated that shift workers are four times more likely to develop cardiovascular disease as compared to non-shift workers. Bøggild and Knutsson (2000) also indicated the increasing risk of cardio-vascular disorders among shift workers.

The disruption of social life is another important consequence of shift work. Night work can make it impossible to participate in gatherings of family and friends and other social functions. This is one of the major reasons why several countries in Europe propose a fast rotating shift-work schedule, with 2 or 3 days at most on each type of shift (Table 16.4).

16.4 EFFECTS ON PERFORMANCE AND PRODUCTIVITY

It has been difficult to establish in research whether productivity is reduced during the night shift. One of the problems is that the type of work tasks is often different, so there cannot be fair comparisons with daytime work. For example, some plants schedule maintenance work during the night shift, whereas in other plants maintenance work is performed during the day shift. There is also a lack of supervisors and managers during the night shift, which means that group morale can suffer.

The consensus from the research is that simple skill-based and rule-based tasks do not suffer as much during shift work as cognitive knowledge-based tasks which require deep thinking. Two interesting studies, however, show a detriment in skill-based task performance. A classic study by Bjerner and Swensson (1953) evaluated

TABLE 16.4
The Effect of Shift Work on Social Activities
and Leisure Activities

Not Enough Time For	Shift Workers %	Day Workers %
Social events	87	22
Cultural events	72	11
Friends	80	13
Family	72	11
Hobbies	67	17

Adapted from Knauth et al. (1983).

records of error frequency in reading meters at a gas company. The error frequency was greatest at 3:00 am. Browne (1949) evaluated the speed of switchboard operators. The slowest responses were obtained between 3:00 am and 7:00 am.

Several studies have pointed to the effect of circadian rhythms on accidents. Folkard et al. (1978) showed that the frequency of minor accidents is the greatest at 5:00 am. Harris and Mackie (1977) investigated accidents due to falling asleep that involved U.S. interstate truck drivers. They found that the accident rate was 20 times as high at 5:00 am as at 12:00 noon.

One of the most quoted events, although one of the least conclusive, is the fact that the Three Mile Island nuclear accident occurred during a night shift. The occurrence of this event was traced to human error and may not have occurred during daytime (Monk and Folkard, 1992).

16.5 IMPROVING SHIFT WORK

Guidelines that can be used for scheduling shift work and for selecting individuals to participate in shift work are listed below. The length of the shift should be related to the type of work. For light work, a 12-hour shift could be contemplated. In fact, most workers like 12-hour shifts (Miller, 1992). There is better job satisfaction, improved morale, and reduced absenteeism. But alertness and thus safety may decline, and workers may work at a slower pace. For heavy physical or complex mental (knowledge-based) work, shifts should be no more than 8 hours, and may be only 6 or 7 hours, during the night.

Visual inspection and visual monitoring is extremely difficult during the nighttime. This is a low vigilance task. The arousal level is low even during daylight hours and at night time many operators simply fall asleep. Rohmert and Luczak (1978) investigated operators sorting letters in the German Post Office. After working for only 2 hours on a night shift the fatigue became overwhelming. In addition, during the critical hours of 3:00 am to 5:00 am, the error rate in sorting letters increased significantly. Due to the problems with fatigue and because misrouted letters are extremely costly for the postal system, it was decided to abolish the night shift—a radical solution for any operation. For these reasons, visual inspection and quality control should not be scheduled for the early morning hours.

Miller (1992) suggested that the number of hours could be reduced for the night shift. It might be advisable to use a shift schedule of 8-hour morning, 9-hour afternoon, and 7-hour night (8M-9A-7N) or, alternatively, 8M-10A-6N or 9M-9A-6N. This may allow the worker to deal more appropriately with the greater amount of stress experienced during the night shift.

16.6 SHIFT WORK SCHEDULES

There is an infinite number of ways of arranging a shift work schedule. Here we restrict ourselves to the most difficult case: 7 days of operation using 4 shift crews. Knauth et al. (1979) pointed out that the 40-hour working week is cumbersome and limiting, and that a 42-hour week allows an even distribution of work time across workers on all shifts, because

7 days/week × 24 hours/day = 168 hours/week

42 hours/crew × 4 crews = 168 hours/week

Thus, the week is nicely divided into 4 segments of 42 hours each.

In Germany and the Scandinavian countries there is a clear preference for fast, forward-rotating shift schedules. The philosophy is that the number of consecutive night shifts should be as few as possible. Preferably, there should be only one consecutive night shift in a shift schedule. In the schedule in Table 16.5 several important principles have been incorporated (Knauth, 1997):

- It takes 4 weeks to go through the cycle. The shorter the cycle, the easier it is for the worker to keep track of it.
- After each night shift there are at least 24 hours of rest.
- The long weekend at the end of the first week is much appreciated.
- The shift assignments rotate forward, from morning to afternoon to night.

Forward rotation is advantageous because the true diurnal cycle is closer to 25 than 24 hours. That is, people have a tendency of wanting to go to bed 1 hour later every night. This has been proven in investigations where people live in isolation for a long period of time without any time cues (as if they are living in a dark, isolated cave).

The main philosophy behind this shift-work pattern is that workers are supposed to remain adjusted to the daytime schedule. Usually it is possible to work a single night shift without being overly tired. Of course the one disadvantage of this shift-work pattern is the sequence of 3 nights at the end of week 4 and the beginning of week 1. Labor unions in Germany and the Scandinavian countries have claimed that this type of schedule improves family life and social life (Rutenfranz et al., 1985). But the tradition elsewhere in the world is different. In the U.S. it is common to have a slowly rotating shift schedule with 1 week devoted to each shift. Monk and

---

TABLE 16.5
A Rapid Forward-Rotating 8-Hour Shift System with Four Crews and a 4-Week Cycle for a 42-Hour Week

Week	Mon.	Tues.	Wed.	Thur.	Fri.	Sat.	Sun.
1	N	—	M	A	N	—	—
2	—	M	A	N		M	M
3	M	A	N	—	M	A	A
4	A	N	—	M	A	N	N

Each crew will work 21 shifts of 8 hours each (total 168 hours).
M = morning shift; A = afternoon shift; N = night shift;— = rest.

Folkard (1992) suggested that this might be the worst possible policy, since there is insufficient time for the body to adjust to the new work patterns (see Figure 16.1). A much slower speed of rotation with 3 weeks or more in 1 shift would allow circadian adjustment. The main controversy has been discussed in detail by Monk (1986) and revolves around the loss of nocturnal orientation during free weekends, which break down the adjustment to the nighttime schedule.

Two alternative fast-rotating shift work schedules, the so-called metropolitan rota or 2-2-2 shift system, and the continental rota or 2-2-3 shift system, are displayed in Table 16.6 (Knauth, 1997). The numbers refer to the number of days on each shift. We provide these examples to illustrate the endless number of combinations that exist for shift work schedules. However, in the European tradition, the schedules illustrated in Tables 16.5 and 16.6 are among the better ones. There are social advantages in starting the morning shift either at 7:00 am or 8:00 am instead of 6:00 am; the family can have breakfast together. The preferred starting hours would then be 7-15-23 or 8-16-24.

Gauderer and Knauth (2002) may have found the ultimate schedule. They noted that drivers in transportation companies worked in rigid shift work schedules. New schedules were defined by incorporating driver's personal preferences. The preferred schedules were used as input to a computer program and personal work schedules were developed. The model was tested after a year in use. There were improvements in employee satisfaction as well as service quality.

16.7 SELECTING INDIVIDUALS FOR SHIFT WORK

Some individuals, although they volunteer to participate in shift work, may eventually have difficulties in coping. Usually they are at a disadvantage from the very beginning. There are several factors which can be used to predict if individuals can be expected to have difficulties with shift work (Tepas and Monk, 1986). Managers and workers should be informed about these factors, since they are linked to satisfaction and success on the job, as indicated in Table 16.7.

TABLE 16.6
The Metropolitan Rota (2-2-2) and the Continental Rota (2-2-3) Shift Systems

Week	Mon.	Tues.	Wed.	Thur.	Fri.	Sat.	Sun.
			Metropolitan Rota				
1	M	M	M	A	A	N	N
2	—	M	M	A	A	N	N
3	—	—	M	M	A	A	N
4	N	—	—	M	M	A	A
5	N	N	—	—	M	M	A
6	A	N	N	—		M	M
7	A	A	N	N	—		M
8		M	A	A	N	N	—
			Continental Rota				
1	M	M	A	A	N	N	N
2	—	—	M	M	A	A	A
3	N	N	—	—	M	M	M
4	A	A	N	N	—	—	—

Both systems assume 4 crews and a 42-hour work week. The metropolitan rota has an 8-week cycle and the continental rota a 4-week cycle. M = morning shift; A = afternoon shift; N = night shift;— = rest.

TABLE 16.7
Individual Factors that Are Likely to Cause Problems in Adapting to Shift Work

People living alone do not adjust as easily
More difficult for people with gastric or digestive disorders
People with inadequate sleeping facilities suffer more
Over 50 years of age
Morning-type individuals (larks)
Second job or heavy domestic duties
Epilepsy

Family members usually support a shift worker and make concessions. For example, a wife of a shift worker told me that she bought a white-noise generator for her husband to diminish the impact of noise during the daytime; the bedroom had special curtains to make it completely dark; and meals were served at special times to help her husband adjust to the shift-work schedule.

With increasing age it seems that individuals become more set in their circadian rhythms. There is also a change towards a pattern of "morningness," indicating that individuals tend to go to bed earlier and wake up earlier. Morningness is indeed one of the greater obstacles to shift work. Horne and Östberg (1976) published a questionnaire that can be used to distinguish between morning types and evening types. This questionnaire can be used to help select evening types who are more suitable for shift work.

Several medical conditions could disqualify an individual from shift work. People with gastrointestinal problems get worse. Epileptics have a higher rate of seizures during the night shift.

RECOMMENDED READINGS

An excellent text on shift work is Monk, T.M. and Folkard, S., 1992, *Making Shift Work Tolerable*, London: Taylor & Francis.

Recent research findings are summarized in Caruso, C.C., Hitchcock, E.M., Dick, R.B., Russo, J.M., and Schmit, J.M., 2004, *Overtime and Extended Work Shifts: Recent Findings On Illnesses, Injuries And Health Behaviors*, DHHS (NIOSH) Publ. 2004-143, Cincinnati, OH: National Institute of Occupational Safety and Health.

17 Design for Manufacture and Maintenance

"To be a machine, to feel, to think, to know how to distinguish good from bad, as well as blue from yellow, in a word, to be born with an intelligence and a sure moral instinct."—LaMettrie (1748). By considering the constraints in machine operation, one can also learn about human needs—as we will demonstrate below.

17.1 INTRODUCTION

Human factors professionals in manufacturing have mainly focused on two areas: design of industrial workstations and design of products to meet customer's needs and improve functionality and usability. The design of products has broader implications. It affects the types of jobs created in the assembly of the product and in the maintenance of the product. Products that are well designed will be easy to assemble. The manufacturing tasks may be distributed in an optimal fashion between manual labor and automated processes. The allocation of tasks to people or machines should be beneficial for the company, and should also create satisfying jobs for the employees. In the first part of this chapter we will address design for manufacturability. In the second part of the chapter, we will analyze how products should be designed in order simplify maintainability: design for maintenance.

It may seem that most of these issues are of engineering interest. Maybe so, but unfortunately engineers rarely deal with theses issues. HFE experts are required, and whether they come from psychology or engineering is not very important.

17.2 THE DESIRE TO AUTOMATE

During the last 20 years, manufacturing engineers vigorously pursued opportunities to automate, and sometimes the results were very disappointing. In the 1980s when robots suddenly became popular, General Motors invested U.S. $80 billion in automated manufacturing, but at least 20% of their spending on automation technology failed; the money was lost (*The Economist*, 10 August, 1991). Other major companies had similar experiences. In many cases the reason was surprising: manual labor, with its greater flexibility and adaptability, will outperform automation. The focus on automation, particularly by engineers, does not necessarily lead to increased productivity. We must understand how different work functions should be allocated between humans, machines, and computers, something that is not well understood by the engineering community. The issue of task allocation has since long interested

TABLE 17.1
Fitts' List: Humans versus Machines (Fitts, 1951)

Attribute	Machine	Operator
Speed	Much superior	Slow
Power	Superior	1500 W max, 150 W/h for a day
Consistency	Ideal for repetitions	Unreliable, subject to fatigue
Information capacity	Multi-channel, megabits/sec	Usually single channel, <10 bits/sec
Memory	Literal production; access restricted	Better for principles and strategies; innovative and versatile access
Reasoning	Deductive, tedious to program, fast, accurate, poor error recognition	Inductive, easy to program, slow, inaccurate, good error correction
Sensing	Specialized, narrow range, poor pattern correction	Wide energy ranges, multifunction, great pattern recognition
Perceiving	Poor interpretation of written/spoken material; suffers from noise	Good interpretation of variations in written/spoken material; suffers from noise

human factors engineers. The question is: what shall people do and what machines shall do (see Table 17.1) (Fitts, 1951)?

Fitts' list has been criticized for oversimplifying task allocation. Many tasks develop over time; a task that initially is ideal for a human operator may a second later be ideal for a machine. Such dynamic considerations make it impractical to allocate tasks.

Robots were first used in industry for fairly simple tasks such as welding and painting. At the beginning of the 1980s there was increasing interest in using robots for assembly. Early on there was a realization that robots can only be used for fairly simple assembly tasks, which are easy to describe and program. In order to enhance the utility of robots, assembly had to be simplified, and it became necessary to redesign products, so that they were easy to assemble by automation. In the last 20 years many design guidelines have been published, which prescribe the design of parts that are easy for a robot to assemble. This type of design is referred to as design for automation (DFA) or design for manufacturability (DFM) (Boothroyd and Dewhurst, 1983). However, the redesign of a product sometimes leads to a very surprising outcome, as illustrated in the following example.

EXAMPLE: THE ASSEMBLY OF A PAPER PICKING MECHANISM

In the early 1980s, IBM Corporation was manufacturing copier machines at the plant in Boca Raton, Florida. The paper picking mechanism in the copier machine had 27 parts, as shown in Figure 17.1 (Helander and Domas, 1986).

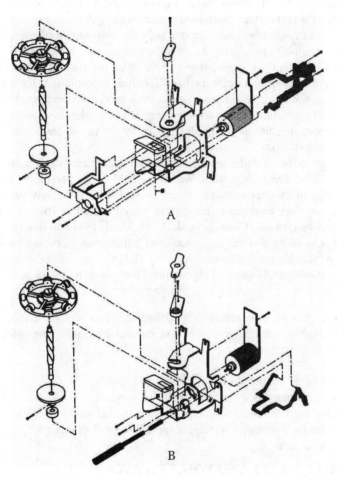

FIGURE 17.1 (A) The original design of the paper picking mechanism had 27 parts. (B) The redesigned mechanism had 14 parts.

One problem with the use of robot assembly is that individual parts often require individual feeding mechanisms, which present each part to the robot arm so that is easy to grip and easy to assemble. Even seven to eight part-feeding mechanisms would be too many, because the entire work envelope would be filled with part-feeding mechanisms. After the redesign the paper picking mechanism had 14 parts, 13 of which could be assembled by robots and automation. The fourteenth part required a complex insertion motion and had to be put in place using manual labor (thereby creating a highly repetitive task). The surprising outcome of the redesign was that manual assembly of the mechanism became so simple, that the use of automation and robots could no longer be cost-justified. This product was therefore assembled manually.

From our perspective it is ironic that only the introduction of automation has compelled engineers to investigate principles of manual assembly. Throughout the history of manufacturing, engineers have taken for granted that workers can adapt to any situation. Engineers have ignored opportunities for ergonomic improvements, which could increase productivity as well as operator comfort. Only in the last 20 years, through the advancement of automation have engineers been forced to consider alternative production methods, in this case manual labor instead of automation.

One may question under what circumstances automated devices are actually more productive than human labor. Within the IBM Corporation there are several similar cases of product redesign, where manual labor ultimately proved more cost-effective. One example is the printer manufactured at IBM in Charlotte, North Carolina (see also Genaidy et al., 1990; Mital, 1991). In this case manual assembly was faster and the introduction of automation could not be justified. Could it be that design for human assembly (DHA) is a viable method, and we do not need robots? As usual in ergonomics, it depends on the task.

Below we provide an overview of guidelines that may be used in product design to simplify both automated assembly and manual assembly. The information is presented in four sections:

1. What to do and what to avoid in product design.
2. Boothroyd's method for the redesign of products.
3. Use of predetermined time systems to diagnose product design.
4. Human factors design principles applied to product design.

17.3   WHAT TO DO AND WHAT TO AVOID IN PRODUCT DESIGN

In this section we provide examples of product design features that simplify assembly. Many of them are used for automation and have been published in guidelines for DFA. They apply equally well to manual assembly (Helander and Nagamachi, 1992).

USING A BASE PART AS A FOUNDATION AND FIXTURE

Design the product with a base part as the foundation and fixture for other parts. It should be possible to assemble the other parts from one direction, preferably from

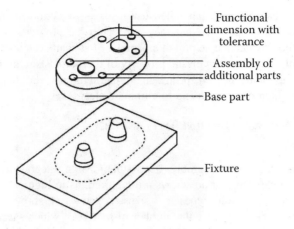

FIGURE 17.2 Provide a simple and reliable fixture for the base part. If possible the base part should also serve as a fixture.

above (Figure 17.2). It is also advantageous to use fasteners which are inserted from one direction, either from the front or from above. The base part should also serve as a fixture. If this arrangement is not feasible, pins can be used so that the base part can be easily positioned on a fixture, as shown in Figure 17.2. If this is not possible, a specially designed fixture is used. To make the product easy to transport, the fixture should have a flat bottom and a simple shape.

MINIMIZING THE NUMBER OF COMPONENTS AND PARTS

Integrate or combine parts, since they take less time to organize and less time to assemble. In some cases an entire subassembly can be replaced by a single part (compare with modular design in electronics). Integrated parts may be complex to handle, but they reduce the number of operations (Figure 17.3). Holdbrook and Sackett (1988) noted that it is difficult to combine parts if

- Parts move relative to each other.
- Parts are required to be of different materials.
- Parts must be separate for maintenance and service reasons.
- Parts are necessary to enable the assembly of remaining parts.

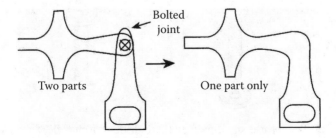

FIGURE 17.3 Integrate or combine parts.

Combined parts can often be fabricated using plastic injection molding. Another advantage with plastic parts is that they can easily be provided with chamfers, notches, and guides which are helpful in assembly. Metal parts can also be molded or mounted into plastic parts. The elastic property of thermoplasts (e.g., nylon) can be used to form snap joints, integral springs, and integral hinges. Thermoplasts can also be used to straighten other parts and to eliminate clearances.

ELIMINATE OR MINIMIZE DIFFERENT TYPES AND SIZES OF FASTENERS

- Use snap and insert assembly. If possible, design integral fasteners and clips into parts so that no screws are required, as in Figure 17.4.
- Minimize the various types and sizes of screws (Figure 17.5). Fewer number of parts decrease the number of part bins, which saves space. A smaller number of bins will also decrease the operator's choice-reaction time between bins. In addition, fewer parts will reduce the number of hand tools, which in turn decreases handling time and space requirements.

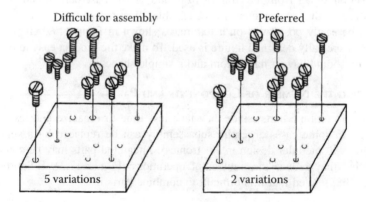

FIGURE 17.4 Minimize the various types and sizes of screws.

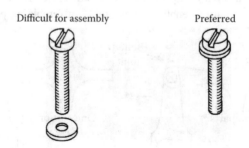

FIGURE 17.5 Do not use small parts that are difficult to handle, such as washers.

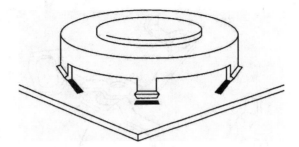

FIGURE 17.6 Use snap and insert assembly.

Do Not Use Small Parts Such as Washers

This requirement, which is mandatory for robotic assembly, also simplifies manual assembly (Figure 17.6). The use of washers increases the manual handling time. In most cases the operator would have to use pinch grips, which has been implicated as a cause of cumulative trauma disorder.

Facilitate Handling of Parts

- Improve parts handling by using parts that are easy to grip (Figure 17.7).
- Avoid using flexible parts, such as wires, cables, and belts, because they are difficult to handle. Sometimes components can be plugged together in order to eliminate the use of connecting wires.
- Avoid parts which nest or tangle. Close open ends and make part dimensions large enough to prevent tangling. For example, use springs with closed ends rather than open ends (Figure 17.8).

Facilitate Orientation of Parts

- Use symmetrical parts, because they are easy to orient (Figure 17.9). Symmetrical parts reduce the need for human information processing, since the operator does not have to decide whether to turn the part around. It also reduces manual handling time.
- If asymmetrical parts are used, provide visual aids for orienting parts (e.g., color coding or shape coding). If asymmetrical parts are used, it may be advantageous to exaggerate the asymmetry to improve visual cues

Difficult to grip   Preferred

FIGURE 17.7 Improve parts handling by making parts easy to grip.

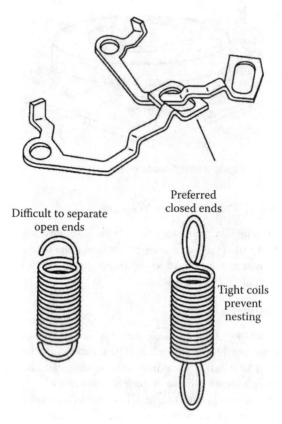

Difficult to separate open ends

Preferred closed ends

Tight coils prevent nesting

FIGURE 17.8 Avoid parts that nest or tangle.

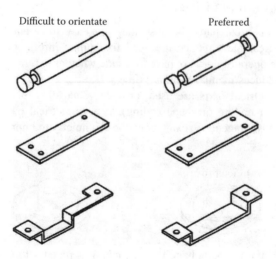

Difficult to orientate

Preferred

FIGURE 17.9 Use parts that are easy to orientate, such as symmetrical parts.

(Chhabra and Ahluwalia, 1990). Color coding of parts may be used to form families of parts; that is, parts which belong together in a subassembly. Color coding will enhance stimulus response compatibility in assembly. This results in reduced reaction time and better eye-hand coordination (Figure 17.10).

• Consider feeding parts. The use of vibratory bowl feeders or other types of electromechanical feeder simplify the presentation and grasping of parts (Figure 17.11). Alternatively, magazines for parts or trays of parts can be used by the operator. These devices were conceived for use in automated assembly. However, they are equally practical for manual assembly. In fact, some companies use automation to feed screws to human operators. The screws can be quite small and are difficult to grasp by hand.

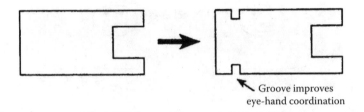

Groove improves
eye-hand coordination

FIGURE 17.10 Exaggerated asymmetry may enhance stimulus–response compatibility.

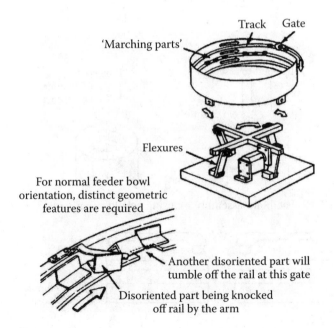

FIGURE 17.11 Use of a vibratory bowl feeder simplifies manual (and automatic) grasping of parts.

FACILITATE QUICK ASSEMBLY

- Use self-locating parts, such as parts with chamfers, notches, and guides for self-location that simplify assembly (Figure 17.12). The use of chamfers, for example, reduces the amount of manual precision required to insert the part. (The insertion time with and without chamfers can be modeled using Fitts' law (Fitts and Posner, 1973).
- Reduce tolerances in part mating. Figure 17.13 illustrates how a slotted hole may be used to simplify positioning and relax accuracy requirements.

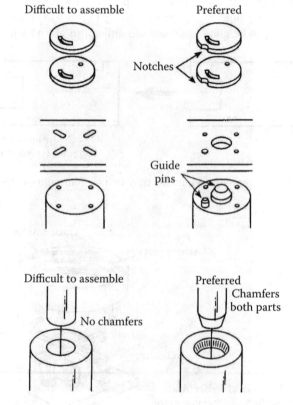

FIGURE 17.12 Facilitate assembly by using self-locating parts.

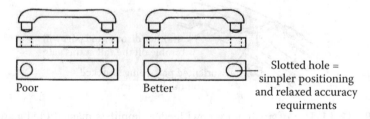

FIGURE 17.13 Reduce tolerances in part making.

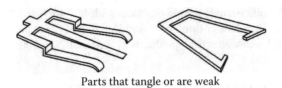

Parts that tangle or are weak

FIGURE 17.14 Avoid parts that are easily bent or parts that crack or chip. Microscope inspection tasks can be avoided if product designers choose materials that are less likely to chip or crack.

SELECT PARTS THAT ARE DURABLE AND STABLE

Parts that are weak or easily bent are difficult to assemble (Figure 17.14). These parts often cause extra work in quality control, visual inspection, and replacement. Grossmith (1992) noted that many microscope inspection tasks can be avoided if product designers select materials that are less likely to chip or crack.

17.4 DESIGNING AUTOMATION USING BOOTHROYD'S PRINCIPLES

The design principles formulated by Boothroyd and Dewhurst (1983; 1987) have been extremely influential in industry. Several companies, including Hitachi, Black & Decker, General Electric, General Motors, IBM, and Xerox, have used these principles to develop corporate DFM guidelines (Gager, 1986; Holbrook and Sackett, 1988).

In Boothroyd's technique, an existing product is disassembled. The necessity of each part is then analyzed. First, one must decide if a part is necessary for assembly or disassembly. If not, it may be possible to eliminate a part or integrate it with a mating part if

1. There is no relative motion between the two parts
2. The materials of which the two mating parts are composed do not have to be different

For each part the assembly time is measured. Boothroyd then makes the assumption that an ideal time for a part is 3 seconds. This is reasonable for a part that is easy to handle and insert. A measure of the manual assembly design efficiency (E_m) is then obtained using the equation:

$$E_m = 3N_m/T_m$$

where N_m is the minimum number of parts, and T_m is the total assembly time. If $E_m < 1$ then the design is inefficient, and if $E_m > 1$ the design is efficient. The value of E_m is, however, not always conclusive. Complex electromechanical products that require extensive wiring tend to have low design efficiencies, even when well designed. On the other hand, simple products with few parts can have high design efficiency. In their handbook, Boothroyd and Dewhurst (1987) provide many examples of successful redesign where productivity gains of 200–300% were obtained.

17.5 MTM ANALYSIS OF AN ASSEMBLY

Boothroyd's technique is useful for redesigning existing products, but it cannot be used in the design of new products at the conceptual stages of design. Predetermined time-and-motion studies (PTMS) can be used for this purpose. As a basis for our analysis, we use motion time measurement (MTM) (e.g. Konz, 1990). In MTM, an assembly is broken down into several constituent tasks, including reach, grasp (pick up and select), move, position part, and insert. MTM specifies the amount of time it takes for a trained worker to do each of these elemental tasks. However, the assembly time depends very much on how the product is designed. Table 17.2 illustrates time savings for a best design case as compared with less efficient design. For example, reaching to a fixed location is the best case and takes about 30% less time than reaching to a variable location or to small and jumbled parts. Grasping of easily picked up objects is 75% faster than for objects on a flat surface. Hence the design engineer should design parts that are easily reached and easily grasped. The parts should be presented at a fixed location. This can be accomplished by using part feeders (Figure 17.11). Much research has been performed to develop part feeders for robots (Boothroyd, 1982). These can also be used for manual assembly. A cost-benefit calculation can easily determine whether parts feeders for a manual assembly are cost-efficient. Simply calculate the time savings for assembly and compare to the cost for parts feeders.

Following the "pick-up" the part has to be transported and positioned for the final insertion step. Table 17.2 illustrates that moving a part against a stop (case A) requires about 15% less time than when a part is moved to a location without a stop (case B). In the latter case the absence of tactile feedback requires greater manual control. Ironically, most products are assembled as in case B. One objective of good design must therefore be to incorporate stops which provide tactile feedback (Furtado, 1990).

The time to position parts depends on whether the part is symmetrical (code S) or non-symmetrical (code NS). In the latter case the operator must turn the part, which takes 30% longer. To insert a 4-in part takes twice as long as to insert a part with no depth. If heavy pressure must be used to fit the part (code 3), it takes 5 times longer than if there is no pressure (code 1). To disengage two parts takes 5 times longer if the parts are tight-fitting, rather than loose fitting.

Luczak (1993) presented several methods which simplify the grasp of a part to be assembled (Figure 17.15). This is a complementary approach. Here, the parts are not redesigned; rather the process of grasping is redesigned. The process is redesigned by using various aids that simplify manual assembly. Process design is typically more abstract than product design, and hence more difficult to implement.

TABLE 17.2
Examples of Time Savings Obtained with the Best Case as Compared with Less Efficient Alternatives

	Best Case	Comparison	Approximate Time Saving for Best Case (%)
Reach	To fixed location (case A)	To variable location (case B)	30
		To small or jumbled objects (case C)	40
Grasp			
Pick up	Easily grasped (case 1A)	Object on flat surface (case 1 B)	75
		Small object, 1/2 in (case 1C2)	400
Select (for jumbled objects only)	Large jumbled objects (case 4A)	Object smaller than $1 \times 1 \times 1$ in (cases 4B, 3C)	50
Move	Against a stop or to other hand (case A)	To exact location without a stop or physical barrier (case C)	15
Position Part			
Symmetrical or Non-symmetrical	Symmetrical, e.g., round peg in round hole (code S)	Semi-symmetrical, 45° turn typical (code SS)	20
	Non-symmetrical, 75°	turn typical (code NS)	30
Insertion			
Depth of insertion	No depth	4 in insertion	100
Pressure to fit	Gravity, no pressure (code 1)	Light pressure (code 2)	210
		Heavy pressure (code 3)	500
Disengage (Two Parts)			
Class 1 fit	Loose	Tight	500
Ease of Handling	Easy	Difficult	40

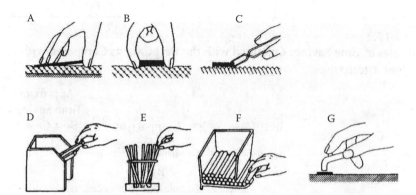

FIGURE 17.15 Gripping aids for an assembly workstation. From Luczak (1993). (A) and (B): gripping against a soft surface; (C) tweezers or tongs used against a rippled table surface; (D) container with inclined opening; (E) a ring holder with small bottom diameter; (F) automatic feeding of parts; (G) use of vacuum gripper.

17.6 HUMAN FACTORS PRINCIPLES IN DESIGN FOR ASSEMBLY

MTM and other PTMS methods are used to predict the time for manual assembly. These methods do not explicitly consider the time required for human information processing. Yet there are many design features that can affect the information processing time. Table 17.3 lists several human factors principles that are applicable to design for human assembly (DHA), including design features that reduce human information processing time.

To make it easy to manipulate parts, one can design for feedback. This will also reduce information processing time. An example of tactile feedback is the use of physical stop barriers. When a part is moved against a stop there is a sensation in the fingers, tactile feedback which indicates that the task has been completed. Auditory feedback is helpful not only with parts but also for hand tools and controls

---

TABLE 17.3
Human Factors Principles in DHA

Design for ease of manipulation
Design for tactile and auditory feedback
Design for visibility and visual feedback
Design for spatial compatibility
Design to enhance the formation of a
  mental model
Design for transfer of training
Design for job satisfaction

---

and for hand tools operating on parts. In this case a sound is produced that indicates task completion. For example, the clicking sound of a switch, or the ricketing noise of a hydraulic screwdriver, indicating that the task was completed.

Visibility and visual feedback play an important role in assembly. Everything that is used in the manufacturing task should be fully visible. Hidden or invisible parts cannot be pointed at. They become difficult to think of and are more abstract. When a task has been completed, there should be visual feedback—in other words, something should look different. Sometimes in automobile assembly a piece of tape is put on top of a part to indicate it is finished.

Spatial compatibility has to do with the spatial layout of a workstation and has been addressed previously (see Figure 6.6). Part bins can be located in sequential order so that the operator can pick parts from left to right in the same order as used in the assembly. Part bins can also be arranged so that their location mimics the product design. This could, for example, be used with components that are inserted in an electronic board. The best arrangement depends on the product design and the number of parts used. Obviously product design should consider spatial compatibility. One should also consider the locations of hand tools and controls. Typically items that belong together in task execution should be physically close. This is the proximity principle in Gestalt laws.

Workers develop mental models of the task they are performing; that is, they think of an assembly in a certain way. The concept of mental models has been used extensively in human–computer interaction. Software programmers have a different mental model than do users of the same software. Therefore, programmers fail to consider the needs of the user. Similarly, in manufacturing the product designer fails to consider mental models other than his or her own. There are, indeed, many different mental models (Baggett and Ehrenfeucht, 1991). A person assembling a product would have a different mental model than a person responsible for the quality control of the same product. They look for different things and they do different things, and the priorities are different. This observation is contrary to the notion that assembly operators should exercise their own quality control; it may be difficult to change a person's mindset (Shalin et al., 1994).

Transfer of training applies when a new product has only small modifications compared with the old product. A worker can then apply his skills to the new product. However, differences in product design and workstation layout may create confusion, and assembly times can increase drastically. Product designers have a responsibility here to make the assembly of new products similar to the assembly of previous products.

Design for job satisfaction is probably the most difficult aspect in planning for manufacturing. One problem is that people have different needs and are satisfied by different factors. We may understand better what factors lead to job dissatisfaction, and it could be easier to design to avoid job dissatisfaction. However, more research is required to understand the problem in depth.

Designers of manufacturing processes, facilities, and products must evaluate their design from the point of view of job satisfaction. There are several criteria (Locke, 1983). The design of a job should allow operators to

- Collaborate
- Talk to others.
- Receive performance feedback
- Have control over one's own work pace
- Use their own judgment and decision-making
- Be exposed to opportunities to learn new concepts and develop new skills

These factors are affected by engineering design and should be addressed in the design process.

EXERCISE: DESIGN FOR JOB SATISFACTION

Discuss the effects of product design and facilities design on job satisfaction. In particular, address the factors listed above. Provide examples of scenarios where these factors cannot considered and where they can be considered. You may think about how the manufacturing facility is laid out and how that affects these issues. For these scenarios, discuss what you think the effects will be on job satisfaction and job dissatisfaction.

Is it easier to predict when an individual will be satisfied or when an individual will be dissatisfied? Is there a difference in the types of issues that lead to satisfaction versus those that lead to dissatisfaction?

17.7 DESIGN FOR MAINTAINABILITY

With increased complexity in manufacturing and use of computers and automated devices, maintenance in a manufacturing plant has become more difficult. To maintain an automated piece of equipment or a robot, an operator needs knowledge of electronics, hydraulics, pneumatics, and programming. In a manufacturing plant there is also an increased use of specialized machines or one-of-a-kind machines. In this complex scenario, it is important that production equipment is designed from the very start with maintainability in mind. To avoid expensive downtime, production equipment must be easy to maintain and quick to service. The design of equipment then becomes extremely important, since machines that are designed with maintainability in mind can effectively reduce the amount of downtime.

We need only consider the military as a case to understand that increased complexity of machines has a severe outcome on machine availability. Bond (1986) claimed that, at any given moment, only about one half of the combat aircraft on a U.S. Navy carrier are able to fly off the ship with all the systems in "up" condition. Below we discuss four aspects of maintainability: fault identification, testability and troubleshooting, accessibility, and ease of manipulation (see Figure 17.16).

Ease of fault identification	Testability troubleshooting	Manual and visual accessibility	Ease of manipulation

FIGURE 17.16 The four steps of design for maintainability partly overlap.

EASE OF FAULT IDENTIFICATION

Equipment should be designed so that it is easy to identify faults. From our perspective there are two interesting aspects:

1. Increased use of diagnostic aids and software and automatic test equipment (ATE)
2. Reduced complexity of machine design to simplify human fault identification

ATE and other diagnostic tools are increasingly used in maintenance. Often, ATE is not helpful. Coppola (1984) reported that even in a well-seasoned system, such as the AWACS surveillance aircraft, the built-in test (BIT) capabilities were not of much use. Out of 12000 trouble indications, it turned out that 85% were false alarms. In the end only 8%, or about 1000, of the incidents were real problems.

ATE has become popular in civilian applications too. Advanced copier machines have many built-in sensors which may be monitored by a modem that is connected to a telephone line. Data are then transmitted at regular intervals from the copy machine to a central database, where the data are analysed remotely. Differences in data are used to determine malfunction of the equipment. However, there are often problems with false alarms.

A complementary and probably better approach is to design the new equipment with maintainability in mind. There could be ways of reducing the complexity of the equipment. In general, we want to design the equipment so that it is possible for the maintenance technician to "chunk" the different components. Due to the limitations in short-term memory, it is difficult to think of individual components. It is much easier to think of modularized or functional blocks.

The maintenance technician will use systems charts for fault identification. A good chart is typically more pictorial, hierarchical, and "chunkable" than the usual information given in schematic diagrams.

Once the critical functionalities and components have been identified, there is still the problem of finding one's way inside the equipment. Labeling and color coding of the various functional elements will help to identify components as well as in determining the functional relationships between components. Table 17.4 and Table 17.5 suggest some ways of using labeling and color coding.

Color coding can be used in functional diagrams of equipment. Diagrams can be attached to the inside of a cover door; the choice of different colors will affect the perception of the state of a function. Thus, red is typically associated with stop, danger, and hot. Yellow is typically associated with caution and near, whereas green is associated with go (equipment operating in a normal fashion) and on. Table 17.5 presents coding as used in military guidelines (Van Cott and Kinkade, 1972).

17.8 DESIGN FOR TESTABILITY

Most real troubleshooting activity seems to be opportunistic in nature; that is, the technician will typically not plan ahead of time or generate a list of possible faults. Instead, he or she will test what seems to be convenient to test. Testing points that

TABLE 17.4
Labels Can Be Used in Equipment to Simplify Maintenance

Label access ports with information about components that can be reached through them

Use labels to identify test points and present critical information; use short and clear messages

If any fasteners are not familiar or not common, label them to indicate how they should be used

Use labels to identify potential hazards; make the labels apparent to the casual operator

Place the labels where they will not be destroyed by dirt or wear

Adapted from VanCott and Kinkade (1972).

TABLE 17.5
Suggested Use of Color Coding in Industrial Military Environments

Color	Indicates	Application
Red	Emergency or danger condition	Mainline power on
	Warns of energized or unsafe condition	Main breaker on
	Malfunction requiring maintenance	
	Stop	
	(Flashing red is used when application requires a more compelling alert than a steady red light)	
Amber or yellow	Motors running, machine in cycle	Test
	Alerts to condition requiring response, but not necessarily maintenance	Attention; stand by
	(Flashing yellow is used when application requires a more compelling alert than a steady light)	Intervention required
Green	Safe condition	Start
	Cycle complete	Go, ready, proceed
	Go or start condition	Maintenance mode
White or clear	Major power not on	
	Normal conditions	
	Equipment operating conditions	AC on, AC off,
	Normal indications which do not have right, wrong, or alert significance	power off, and auto-select

are easily visible, and where the logic is clear, are likely to be tested first (Kieras, 1984). One of the best things an equipment designer can do for the maintenance person is to separate logically the different test points or units of a piece of equipment. If this is done, a relatively simple check sequence can be used to decide what part of the equipment is faulty.

17.9 DESIGN FOR ACCESSIBILITY

In design for accessibility we are interested in two aspects: visual accessibility and reach accessibility. To enhance the visibility of components, it may be desirable to install lamp fixtures inside machines. It is also important to make the openings large and prominent and to locate the service components in a prominent location where they are easy to reach.

Maintenance is often performed using access ports to the equipment. Several issues should be considered when designing such ports (Table 17.6). These issues clearly matter. In one of the few experiments ever performed on maintainability, Kama (1963) demonstrated the effect of accessibility on work time (Figure 17.17). The U.S. Department of Defense has taken much interest in accessibility, and many guidelines have been devised. Figure 17.18 shows the clearances for the hands necessary for equipment maintenance.

17.10 DESIGN FOR EASE OF MANIPULATION

To simplify maintenance one must consider the design of the components that are used in the equipment. Several guidelines for ease of manipulation of connectors and couplings are given in Table 17.7. The required clearance for the hand when using different hand tools is illustrated in Figure 17.18.

TABLE 17.6
Design of Access Ports for Maintenance

Consider the requirements of the maintenance task in terms of tool use, exertion of force, and depth of reach. Use this information to determine the dimensions of access ports.

Provide openings to components that need maintenance. Openings must be large enough to permit access by both hands. Openings must also offer visibility of components.

Locate access ports so that they do not expose maintenance operators to hot surfaces, electrical currents, or sharp edges.

Locate access ports so that the operator can monitor necessary display(s) while making adjustments.

Adapted from Van Cott and Kinkade (1972).

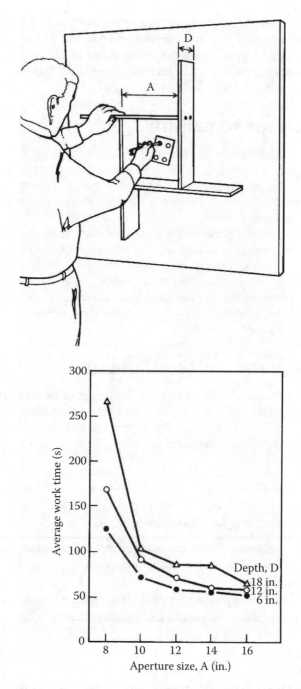

FIGURE 17.17 The width of the opening (A) and the depth of the opening (D) affect the average work time required for this maintenance task. Note that for aperture sizes less than 10 in there is a dramatic increase in work time, and even more so for a depth of 18 in than 6 in.

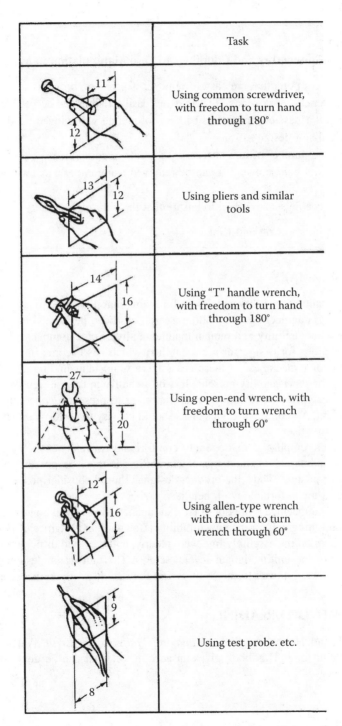

	Task
	Using common screwdriver, with freedom to turn hand through 180°
	Using pliers and similar tools
	Using "T" handle wrench, with freedom to turn hand through 180°
	Using open-end wrench, with freedom to turn wrench through 60°
	Using allen-type wrench with freedom to turn wrench through 60°
	Using test probe. etc.

FIGURE 17.18 Clearances (in cm) required for the hands. Additional information is available in U.S. Military Standard 1472F (U.S. Department of Defense, 2002).

TABLE 17.7
Design of Connectors and Couplings to Ease Manipulation in Maintenance

Provide access ports that are easy to remove; if possible, hinged

Design fasteners for covers so that they are easily visible and accessible

Fasteners on access covers should be easy to operate with gloved hand; e.g., tongue-and-slot design

Minimize the number of turns necessary to remove components

Use hexagonal bolt screws that can be removed using either a screwdriver or a wrench

Make replaceable seals visible, to ensure that they are replaced

Adapted from Van Cott and Kinkade (1972).

17.11   SUMMARY

Design for manufacturability and design for maintainability are extremely important for concurrent engineering. Very little research has been performed in the areas of human manufacturability and human maintainability. Some companies have established guidelines for design for serviceability, which has become important in the design of copier machines. In these cases there is no fault finding and diagnosis, and design for serviceability presents less of a challenge than design for maintainability. In design for serviceability it is important to present a coherent picture of the equipment. All common service tasks should be easy to perform and they should be obvious to a user.

We need to emphasize that operator comfort and convenience are paramount in design for maintainability and manufacturability; it must be possible to adopt a natural work posture. Thus, the product designer should consider positioning common items at a comfortable work height.

Mitsuo Nagmachi, a professor of HFE in Japan, told a story of how he redesigned an air-conditioning unit for manufacturability (Helander and Nagmachi, 1992). Most of the workers on the assembly line were elderly, and they had difficulties reaching to the back of the unit to tighten several screws. The unit was redesigned and the screws holes were moved to the front, where they could easily be accessed.

RECOMMENDED READING

Reason, J, and Hobbs, A., 2003, *Managing Maintenance Error*, Aldershot, UK: Ashgate Publishing. This book gives an account of what can be done to manage maintenance.

18 Accidents, Human Errors and Safety

18.1 INTRODUCTION

In order to understand safety problems we must first analyze accident statistics. It is important to know how common certain accidents are, in which context they occur, and how they can be prevented. Despite strong developments in data collection methods, accidents statistics are still difficult to document, especially those that relate to work accidents. According to Eurostat figures, accident rates for the European Union have not changed significantly since 1996. There are 4.5 million accidents resulting in 146 million working days lost (European Agency for Safety and Health at Work, 2002). Every year about 5500 people are killed in workplace accidents and about 3000 more are killed on their way to and from work.

Collecting and analyzing accident statistics is not easy. We discuss this issue first. There are many conceptual models of accidents. In engineering, the energy exchange models dominate, while in psychology, behavioral models take center stage. In human factors, systems safety has been given much emphasis (see Chapter 1). These models are explained below. We will then investigate models for risk taking and models for accident proneness. Finally we will give examples of ways to prevent accidents in the manufacturing industry.

18.2 INTERPRETING ACCIDENT STATISTICS

Statistics on fatalities, accidents, and injuries are commonly collected in the Western world. The purpose of these statistics is twofold: (1) to analyze how fatalities and injuries occur and (2) to understand how accidents can be prevented. Work activities that are undertaken in demanding environments are typically associated with high accident rates. Examples are construction work, mining, and agriculture, which top the statistics. But there are also many odd occupations with very high accident frequency, such as timber felling and ocean fishing.

There are many problems in collecting accident data for statistical purposes. Some of the problems are exemplified in Table 18.1, which shows the official accident statistics for one year from a country in Europe. We use this as an example to demonstrate how difficult it is to collect reliable and conclusive statistics.

Table 18.1 illustrates several well-known phenomena in accident statistics:

- Accidental deaths at home are slightly bit more common than road fatalities.
- The number of road fatalities is about ten times the number of fatalities in factories.

TABLE 18.1
Accident Statistics from a Country in Europe

Location	Fatalities	Injuries	Slight Injuries
Home	7561	120,000	1,500,000 (estimate)
Road	6810	88,563	253,835
Rail	216	920	11,570
Aircraft	147	?	?
Water transport	158	?	?
Factory	628	?	11,805 (3+ days away from work)
Farm	136	?	8,945 (3+ days away from work)

- Farming accidents, air traffic accidents, and water transportation accidents are about the same in this particular country.
- There are many question marks in the statistics, since the standards for reporting varies for different work activities (factory, farming, mining, rail transport, and water transport). This is because statistics are collected by different authorities with different legislations. This makes it difficult to compare the hazard of the different work activities.
- From the statistics one could conclude that the home environment is the most dangerous, but this is not true. In order to estimate the relative hazard of each environment one must divide the number of accidents by the number of hours that people spend in each environment per year. People spend about ten times as much time at home as they spend in cars. Driving is hence a more dangerous activity than staying at home.

At home, children and retirees have the most accidents. They spend almost their entire days at home, and they are therefore relatively overexposed to opportunities for having accidents. According to the statistics the most common type of home accident is tripping and falling, and the most frequent cause of fatality is falling down the stairs.

Some accident statistics are relatively unreliable, particularly in the case of developing countries. A colleague of mine who worked for the World Health Organization (WHO) found out that statistics on work fatalities are rarely recorded in China. However, information on widow's pensions could be used, since it mentioned the reason for the pension. In the Western world it can also be difficult to get reliable statistics. Most safety experts in the U.S. agree that only about 50% of the fatalities at work are reported. The main reason seems to be that fatalities at work often lead to lawsuits, and employers therefore try to protect themselves by not reporting them. Nonetheless, fatality statistics are more accurate than injury statistics.

18.3 SOCIAL AND DEVELOPMENTAL FACTORS IN ACCIDENT STATISTICS

There have been interesting trends over the years in accident statistics. Because of advances in technology society has become more hazardous. In 1870, 8% of the accidents in France were traffic accidents; today the figure is about 40%. Power hand tools, nuclear power plants, and machinery today generate accidents that did not occur 50 years ago. Yet society matures with time, and there is a general trend downwards for all types of accidents—for fatalities as well as injuries. For traffic accidents, Smeed (1952, 1968, and 1974) and Smith (1999) proved an interesting trend, which is referred to as Smeed's Law (see Figure 18.1).

In the lower right corner, the number of vehicles per capita is very high. This part of the figure refers to countries such as Australia, France, Sweden, and the U.S. They also have a very low fatality rate. In the upper left corner are developing countries, including many African and Asian countries. Their fatality rate is 20 times as high as for developed countries. A very high correlation coefficient was obtained for the data in Figure 18.1 with $R^2 = 0.97$.

Smeed explained the results in following way: with time and with greater motorization, procedures and traffic measures in countries improve. There are improvements in legislation, roads, and vehicle inspections. Developing countries lack much of the infrastructure to support traffic. The legislation, driver training,

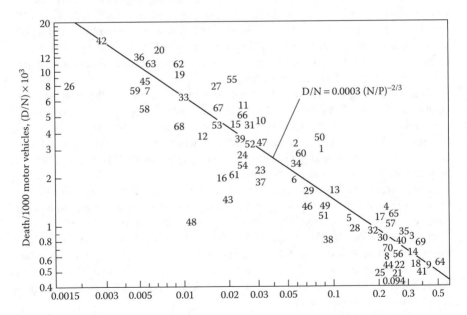

FIGURE 18.1 Numbers of fatalities D per 1000 motor vehicles (vertical axis) plotted as a function N- of the number of motor vehicles per capita (horizontal axis). Each number corresponds to a country.

and traffic rules and laws have not been fully developed. But over time developing countries will develop; driver training and traffic legislation will improve. Also, with an increasing number of vehicles the infrastructure improves with better roads, regulations, and training.

As the number of vehicles increases in a country, it moves down the regression line in Figure 18.1, and as a result there is on average a 3% reduction of traffic fatalities per year. We can imagine a government official formulating a goal for traffic safety: "Over the next 5 years we will improve traffic safety by 15%." This is a safe bet—and it does neither more nor less than what is predicted by the automaticity of Smeed's law.

Clearly there must be similar mechanisms for work fatalities and accidents. National safety and health programs generally report a decreasing trend over the years. Behind the reduction in accidents and fatalities lies the concerted action of employees and managers in industries as well as government officers. There is an increasing awareness of safety and a constant fine tuning of programs. There are improved legislation, testing programs, training facilities, better reporting in news media, and sometimes even a social protest regarding high death rates. From what I know, Smeed's approach has not been investigated in occupational safety, but I feel that it could explain why occupational safety generally improves over time.

The reporting of injuries and accidents is also affected by sociological and legal factors. A study by Hadler (1989) compared disabling back injuries in France, Switzerland, and the Netherlands. He observed that the legislative programs for workers compensation was different in the three countries, and the patterns of reported injuries were also different. The conclusion was that individuals will report certain injuries and accidents because they are recognized by the country's legislation or by society, and therefore the affected worker will get economic compensation. Different countries allow different injuries.

One interesting difference is between computer operators in the Scandinavian countries and the U.S. In the U.S. there is a prevalence of injuries due to cumulative trauma disorder (CTD) and tenosynovitis of the hand and of the wrist. These types of injuries are rare in Scandinavian countries, where operators complain about pain in the neck and the shoulder. Certainly the injury patterns must be similar, but the prevalent ethic of reporting is different.

There are many other social factors in operation that affect safety records over time. Some of them are surprising. When Sweden shifted from driving on the left hand side to driving to the right hand side in 1964, the number of road fatalities dropped from about 1000 per year to 500 per year. The following year the fatality rate was back to 1000 per year. Can it have been that the safety propaganda sensitized drivers, so that they became much more careful? Why did the fatality rate return back to the original rate after a year? Does it take a year to adapt and relax one's fears? Many questions remain.

Starr (1959) claimed that people accept much higher hazard levels if an activity is voluntary, than if the activity is offered or arranged by society. People who fly their own aircraft take a much greater risk than people who fly with commercial aviation. It is worth the extra risk—because it is fun. The safety of public facilities, such as electrical power and public transportation, are taken for granted. Users do

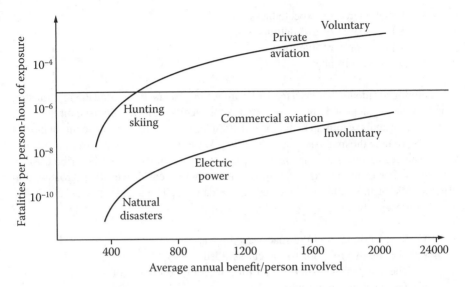

FIGURE 18.2 Starr's (1969) risk-taking model. The horizontal line at 10^{-6} represents the natural death rate due to old age.

not accept a high accident rate with city buses or with subways. There would be a public outcry if anything serious happened. Hence we can explain the two curves in Figure 18.2. Involuntary, public fatality rates are much lower than voluntary fatality rates. This is because the voluntary activities offer additional benefits; people enjoy hunting, skiing, and flying their own aircraft. The greater the personal benefits from an activity, the greater the risk users can accept.

EXAMPLE

A railway accident happened in Turkey in August, 2004, and the Turkish government faced new accusations over the country's poor railway system after 8 people were killed and 88 injured in the train collision, the second major rail tragedy in the country in 3 weeks. Experts were quick to blame the accident on a possible error in the decade-old signalization system, which authorities allegedly neglected, and to maintain. The government had already come under fire after a newly inaugurated express train derailed in the same region in the northwest, killing 37 people and leaving some 80 others injured, on July 22, 2004 (Channel News Asia, 2004).

18.4 MODELS AND DEFINITIONS OF ACCIDENTS

Over the last 50 years researchers have tried to define what an accident is. A common definition is that an accident is something without apparent cause; an unexpected, unintentional act, mishap, chance occurrence, or act of God. To qualify as an accident, Suchman (1961) noted that there should be

- A low degree of expectedness
- A low degree of avoidability
- A low degree of intention
- Quick occurrence

In defining accidents there is a tendency among engineers to focus on energy transfer, and on the physical and chemical aspects of an accident. Psychologists, on the other hand, often analyze human behavior and attitudes surrounding the accident. In human factors we take the analysis one step further. We look at the accident from a systems perspective, recognizing that many elements and various causes contribute to an accident. For example, consider an accident where a car drove off the roadway in the middle of the night and incurred severe damage. The accident could be caused by a variety of factors:

- It was raining and the road was slippery
- The tires were worn so the car skidded
- The accident took place on a rural road in the middle of a curve
- The road was quite narrow
- There was an oncoming car and the headlights caused glare for the driver
- The windshield wipers did not work well, which reduced visibility
- The driver collided with a tree
- The driver was distracted since he was worried about a recent incident at work
- The driver had a beer before he took off and this made him a little drowsy

In analyzing this accident, an engineering approach would focus on the energy exchange or on the worn tires and the windshield wipers. A psychologist may be interested in investigating the effects of driver fatigue and the driver's psychological state. In human factors engineering, we take a systems approach, where we try to understand how all the contributing factors interacted: environmental, mechanical, and human factors. Below we will explain sevral accident models, including systems safety.

Energy Exchange Model

Schutzinger (1984) claimed that accidents result from the integration of a constellation of forces. Likewise, Haddon (1964) thought of accidents as an occurrence of an "unexpected physical or chemical damage" to living or non-living structures. This takes an engineering approach. Injuries are produced by energy transfer and exchange of mechanical, thermal, or electrical energy. When the car hits a tree along the roadway, it stops suddenly, and the driver may continue through the windscreen, unless he wears a seat belt.

The energy exchange model does not attribute causes to accidents, but the model can be useful to suggest effective barriers to accidents, such as soft road sides, seat belts, airbags, and removal of roadside trees. Through these measures the energy is distributed over a greater area and absorbed over a longer time. The deceleration distance of the body and body parts until they come to rest is increased, and forces are thereby reduced. Serious injuries may be prevented. We may still not understand the cause of the accident, but that is not the issue. Prevention is good enough according to this model.

CHAIN OF EVENTS

Arbous and Kerrick (1981) considered an accident an "unplanned event in a chain of planned or controlled events." This definition implies that accidents and errors, although not planned, are in some way similar to purposeful behavior, since they develop in a sequential fashion. According to this model each accident is the result of a series or chain of events. No singular cause exists; many factors influence the accident. Our driver was fatigued, and he decided to drive his car in a poor condition; it starts raining; and the driver enters a poor quality road. He is worried about work, and therefore does not focus on the driving. Note that all these states can be modeled as a "chain" of conditions. When he enters a narrow curve and there is glare from an opposing car he loses control of the vehicle. If one can break the chain of events at any of its links, the accident will be avoided. For example, improve the road quality: take away the narrow curve, and the accident would never have happened.

Authorities are well aware of the influence of road quality on accident rates. About seven times as many accidents happen per vehicle kilometer on narrow rural roads as compared to freeway driving. Rebuilding the road is a matter of cost–benefit ratio. There are many alternative road projects available to the authorities and the projects compete for money. There are benefits in building a better road, such as reduction in accident rate, savings of traveler's time, and savings in car expenses including fuel. This must be compared to the expense of building the new road, so that a cost–benefit ratio can be calculated. Chances are that the winding rural road will never be rebuilt. There are not many cars on it, and therefore the economic benefits of savings in travel time, and so forth, are not substantial enough.

SYSTEMS SAFETY

The systems safety approach demonstrated in Figure 18.3 is similar to the systems approach presented in Chapter 1. We learn from this that accidents have multiple causations, and must be analyzed as such. There are no simple answers to accident causation.

An accident does not happen as a result of one factor alone. In the case of the car that drove off the road we listed nine different reasons. First there are predisposing factors (see Figure 18.3). The worn tires and windshield wipers belong to the equipment factors. The winding and slippery road and the rain are environmental factors. Driving through a curve and meeting other cars are task factors. If these factors lead to a failure of the system, such as a blown tire or a skidding vehicle, then there is a precipitating factor. The operator can save the situation if he responds quickly and appropriately to the situation, for example by counteracting the skidding by steering. However, the driver was fatigued and distracted by events that had happened at work, so he did not respond in time, and the accident could not be avoided.

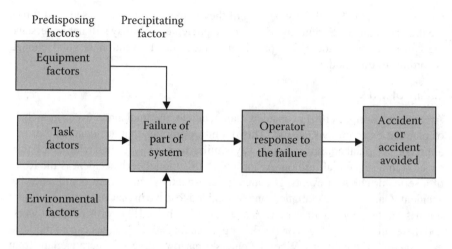

FIGURE 18.3 Example of a systems approach to accident analysis. There are predisposing factors such as worn tires, wet road, and glare. These can lead to precipitating factors and eventually an accident.

Ramsey (1985) took a systems approach that emphasized human information processing. First there is exposure to a hazardous situation. The victim must then be able to perceive the hazard, understand the gravity of the hazard, make a decision to avoid, and finally take the correct action. But there is still a chance that the action that is taken will not be enough (see Figure 18.4). Here is an example:

- Old lady sees water puddle when crossing the road
- She recognizes the slipping hazard
- She decides to avoid the puddle
- But she does not step to the side quickly enough
- She slips and falls!

INTERACTIVE MODEL OF ACCIDENTS

This model represents the current thinking in accident causation. As we noted above, human error is often attributed as a major cause of accidents. This is misleading. No doubt one can claim that without a human operator an accident would not have taken place. But this does not allow us to understand the cause of an accident. To illustrate, let's return to the road accident that we discussed above.

The accident could have been avoided if

- The road had been straight, and not curving
- The tires had been in good condition, not worn
- There was no other car on the road
- The road surface was dry, not wet from rain

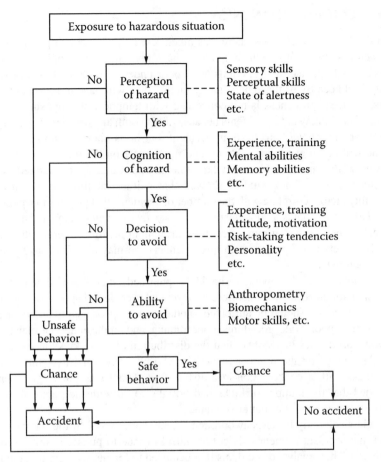

FIGURE 18.4 Ramsey's information processing model.

The state of the accident scene can be described using several different factors. None of these factors has precedence over the others. The accident was caused *by interactivity among the factors—not by a single cause.*

The constant blaming of the operator represents an oversimplification of causality. There is always an operator present, and therefore an accident investigator may find it convenient to blame the operator. It does not always rain, so rainy weather is not a convenient cause. A problem with interactive accidents is that the number of factors and the number of interactions and potential causes becomes very large, and therefore accidents become difficult to analyze and describe. This will frustrate the accident investigator who is looking for quick fixes.

We should not overemphasize the role of the human in accident causation. The environment and the task must be equally blamed for an accident. Accidents occur as interactions between many systems elements. Unfortunately most investigators like to blame the operator—the blame and shame attitude is common in our society. Someone has to take responsibility! This attitude is clearly wrong.

18.5 ACCIDENT PRONENESS

From the 1920s to the 1960s, there was much research on accident proneness. This approach assumes that certain individuals have permanent personality characteristics, such as high risk taking, clumsiness, poor understanding, and lack of responsibility, and because of these personality factors they will have more accidents than others. Accident proneness is not concerned with temporary factors such as youth. A young person may have accidents because of his youth and inexperience in driving. He is hence more *liable* to have an accident, but he is not more *prone*. Ten years later he will be an experienced driver and much safer on the road.

A very large amount of research has gone into this field, as summarized by Shaw and Sichel (1971). Their studies showed that neither intelligence nor any other personality factors correlate with the number of accidents that persons in a population incur. However, poor social adjustment is one factor which may make a person accident prone, because it may affect him throughout his life. A person drives as he lives; if he gets into fights with family members, he may also get into fights with fellow motorists.

The notion that the operator should be punished or a person should be made responsible for an accident is therefore unwarranted. Accidents happen by chance! The blame for accidents should be attributed to other factors: poor quality and poor design of equipment, poor work procedures, poor training, and, in the end, poor management.

Von Bortkiewics in 1880 studied the distribution of horse kicks in the Prussian army. He compared the number of people who had zero, one, two, three, or four horse kicks with a chance (Poisson) distribution. There was good agreement. He concluded that the number of kicks had nothing to do with the personality of the recipient. It was totally a matter of chance.

A classical study by Greenwood and Wood (1919) gives further support to the notion that accident proneness is not a fruitful avenue to pursue in prevention and in research. They analyzed accidents that occurred to female workers in a munitions plant that produced bullets and explosives. The results of studies are shown in Table 18.2.

TABLE 18.2
Greenwood and Wood (1919) Study of Female Munitions Workers

Number of Accidents	Number of Women with N Accidents	Expected with Poisson Chance Distribution	Expected with Single Biased	Expected with Unequal Liabilities
0	441	406	452	442
1	132	189	117	140
2	42	45	56	45
3	21	7	18	14
4	3	1	4	5
5	2	0.1	1	2
3 + 4 + 5	26	8.1	23	21

As can be seen from Table 18.2, 441 women had zero accidents, 132 had 1 accident, 42 had 2 accidents, 21 had 3 accidents, 3 had 4 accidents, and 2 had 5 accidents. The next column in the table shows the expected number of accidents, assuming a Poisson (chance) distribution. This gives a fairly good estimation of the number of accidents. The third column shows how the Poisson distribution can be manipulated. Single bias implies that a person who already had an accident has an increased chance of a second accident. But for a person without an accident, the chances of an accident decrease.

Finally, the last column shows the effect of unequal liabilities. In this case the assumption is that people learn from the first accident, so they have fewer accidents in the future. In conclusion the support for accident proneness is minimal, and it is not a viable idea to lay off certain workers based on accident proneness.

The idea of accident proneness lingers on in society. A recent newspaper quoted that 12–15% of workers have close to 100% of the accidents, implying that these people were accident prone. Another newspaper clipping mentioned that 5% of the factory workers accounted for 42% of the accidents.

Greenwood's and Wood's results can be critiqued, because different workers have different risk exposure. It would be better to investigate differences in risk exposure to accidents. Obviously some of the women in the munitions plant must have worked with processes that were more dangerous than others and therefore had more accidents. However, it is extremely difficult to assess the differences in risk exposure objectively. A study by Hakkenen (1972) compared the accident records of bus drivers in Helsinki, the capital of Finland. His conclusions were that since every bus driver drives a different route, each is exposed to a different level of hazard. Therefore the accident data could not be used to investigate accident proneness.

18.6 HUMAN ERROR

A review of the literature suggests that human error is a primary cause of 60–90% major accidents. Accidents happen because of the operator's misinterpretation, wrong decision, lack of knowledge, or silly mistake. This does not mean that humans must be held responsible for the errors they commit. Human errors are very frequent in daily life and they appear in all walks of life. Doctors and nurses make an average of 1.7 errors per patient. In most cases they are not consequential, since the doctor can compensate for the error and recover from it—if not, the patient will have to recover. A study from the beginning of the personal computer age showed that there was a 30% error rate in command selection in word processing (Card et al., 1980). Most of these errors are the results of poor system designs and poor organization, rather than irresponsible actions. There are many reasons for errors, including poor discriminability, memory lapses, and communication breakdown. These issues are illustrated below, using the example of aircraft accidents.

POOR VISUAL DISCRIMINABILITY

The flight disaster with Singapore Airlines flight SQ 006 in Taipei is an example of poor visual discriminability. The pilot had been cleared for taking off with a Boeing

747 to Los Angeles (Lim, 2004). It was night time, and there were strong winds and heavy rain as it was typhoon season. This made it difficult to see the runway, which was also very poorly illuminated. Several light bulbs that delineated the taxiway leading to the runway were out. The pilot could not see the taxiway. The only runway that the pilot could see was an old runway that was being rebuilt into a taxiway. For reasons we don't understand, it was lit up. The pilot selected the only visible option and took off on the taxiway. Halfway down the taxiway he saw the construction vehicles that blocked the runway. The pilot could not see them when he started taking off. The aircraft crashed into the construction vehicles. Eighty-one of 179 passengers were killed. Neither the correct runway nor the construction activities that led to the accident could be seen.

You wonder if the blame should not have placed on the airport authorities rather than the pilot. International standards require that old runways under repair should be blocked off with visible signs, and that all runways must be clearly illuminated. The pilot and the copilot were declared guilty of causing the accident. A human factors systems approach to the accident would take into consideration all contributing factors before blaming the pilots alone.

Memory Lapses

A system with mode control can be difficult to control, and sometimes a user selects the wrong mode. The system operates so that one action is appropriate in one mode of operation, and another action in another mode; in Mode A the operator must take action X, in mode B he must do Y, and so forth. The problem with mode controls is that people are easily distracted and forget what mode they are in. This has happened several times with aircraft.

In the 1990 Bangalore air disaster, Air India Airlines flight with Airbus A320 crashed on the final approach to Bangalore airport (Sarter and Woods, 1995). All 90 persons on board were killed. This was a clear automation mode error. The pilots accidentally selected a control mode called "Open Descent." In this mode the aircraft cuts back engine power and maintains its speed by progressively losing height. As a result, the rate of descent was too great and the aircraft landed half a mile before the runway. The pilots discovered their error only 10 seconds before impact—too late to correct it.

This and other aircraft accidents have highlighted pilot's vulnerability in using mode controls. Airbus has since redesigned the aircraft. There are no mode controls, because pilots may be distracted and forget what mode they are in.

Communication Breakdown

There are frequently communication breakdowns in the cockpit (Wiener and Nagel, 1988; Wiener, Kanki, and Helmreich, 1993). The main problem is that the captain and the first officer do not communicate appropriately. This was a contributing reason for the Air Florida flight that crashed into the 14th Street Bridge over the Potomac River in Washington D.C. in January 1982. Seventy-eight persons died. It was discovered that the cause of the crash was that the plane's wings were covered with frozen ice just prior to the crash. Despite the freezing weather and snowy conditions, the crew failed to activate the anti-ice systems. Some blame was put on the captain and first officer, since they did not communicate about the icing problems.

Although there may be several reasons for this behavior, one common reason is the difference in status. Sometimes the captain is dominating and the first officer is submissive (common in Asia). To break down this barrier, flight crews around the world are today given annual training in crew resource management (CRM). For example, when a captain first meets the flight crew he must immediately try to develop a friendly and trusting atmosphere. This can be done by talking to the other crew members and inviting a discussion regarding flight related matters.

JAMES REASON'S TERMINOLOGY

James Reason's (1990) model can help in consolidating many of the issues that we have discussed. According to Reason there are three different types of human errors: mistakes, slips, and lapses or mode errors. Figure 18.5 shows a flow diagram of human information processing. An operator will perceive a stimulus which must be interpreted, and she then makes an assessment of the situation. The operator then formulates a plan for action, and finally executes an action. There are two types of mistakes: knowledge-based mistakes and rule-based mistakes. These result in failure to formulate the right intention or plan for action. They depend on shortcomings in perception, memory, and information processing.

James Reason used Jens Rasmussen's (1986) distinction between knowledge-based mistakes and rule-based mistakes. Knowledge-based mistakes occur when an operator lacks knowledge and therefore cannot interpret what is going on. As a result the operator is overwhelmed by the complexity of evidence and cannot interpret the information correctly. Take, for example, a scenario in a nuclear power plant when

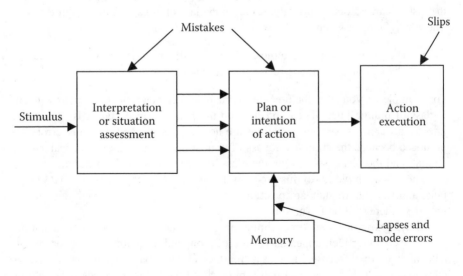

FIGURE 18.5 James Reason's error model (1990).

something goes wrong. There are hundreds of alarms, both auditory and visual, and the operator has great difficulty in understanding what is going on and what actions to take. An inexperienced operator may simply not have enough knowledge about the functionality of the plant. An operator is well aware of his lack of knowledge, and after he comitted an error, he will testify that the accident was his fault, because he did not have enough knowledge.

RULE-BASED MISTAKES

These mistakes are different in nature. They don't require deep thinking, as with a knowledge-based situation. In this case the operator uses a set of rules which can be formulated as if–then conditions. If A, then do X; if B, then do Y; if C, then do Z. If dialing to the U.S., use country code 1; for Sweden, use 46; for Singapore, use 65. Sometimes I mix them up and I have no idea that I made a mistake.

Another example comes from driving in wintertime. If the road turns to the right, I steer to the right. If the road turns to the right and there is ice on the road and I skid, then I first steer to the left. This will compensate for the skid and the car will be appropriately repositioned on the road. Then I steer to the right to take the curve.

Rule-based mistakes tend to be done with much confidence: "strong but wrong." A person can mention all the decision rules he uses, and he may say that it is inconceivable that he could have made a mistake. In fact, he just simply forgot what he did. Rule-based decisions are more or less automatic, so people don't attend to their own behavior—hence it is easy to forget what one just did.

SLIPS

Slips are different from mistakes. In this case the person has the right intention, but the task is carried out incorrectly. A common class of slips is called capture errors. These may happen when

- The intended action is almost the same as the routine action
- The action sequence is relatively automatic

For example, as you sit at the breakfast table reading a newspaper you pour yourself a glass of orange juice. Then you pick up the glass to drink, but it is not orange juice; it's water. These routine actions are not attended to, and therefore errors are produced because the stimuli—a glass of orange juice versus a glass of water—are similar, and the response—lifting the glass to the mouth—is also similar.

There is a simple fix to this problem. Design the breakfast table so that orange juice and water have standard positions, water to the right and orange juice to the left. If so there will be fewer slips.

In flying there are sometimes similar problems. The controls for flaps and landing gears have similar feel, appearance, and location, and both are relevant for takeoff and landing. In this case, it is easy for the pilot to make a slip—with serious consequences. It has happened that a pilot raised the landing wheels although the flight was not yet off the ground. People are well aware of their slips, and they will admit that the cause of an accident was that they shipped.

LAPSES

These are failures to carry out an action due to forgetfulness; the flow of information from the memory is disrupted and people forget what they did. Imagine, for example, that you are sitting in your office and talking to a colleague and the telephone rings. You pick up the telephone and engage in a conversation for 2 minutes. When you return to the discussion, you may have to ask, "What was I saying"? The telephone call interrupted your train of thoughts.

MODE ERRORS

Mode errors are different. Depending on the mode that a system is in, different control actions will be necessary. One action is appropriate in one mode of operation, and another action in another mode. The reason why users have problems is that they forget what mode they are in. Very careful design is required to avoid mode errors. There have been disastrous accidents with airplanes; for example, the India Airlines A320 crash that was mentioned above.

IMPLICATIONS OF REASON'S MODEL FOR DESIGN
AND TRAINING

There are interesting implications of Reason's model. To avoid knowledge-based mistakes, the operator should have more knowledge—he must be trained. To avoid rule-based mistakes, there must be a combination of training and redesign of the task environment. If there are slips, then the task and the environment should be redesigned; and if there are lapses, the task should be redesigned (see Table 18.3).

This thinking fits well with the human factors design philosophy, but not necessarily with common practice. In industry it is common to train operators, but not to redesign work equipment to get rid of awkward procedures or taxing work postures. In human factors, we design workplaces, tools and machines, and we also write training programs and training procedures. Often the equipment is difficult and awkward to use, and it takes much training to compensate for the poor design and teach operators how to handle the equipment. It is a money-saving proposition if some training can be replaced by good design. For example, design the controls of a fighter aircraft so that the aircraft is easier to use, and pilot will not have to train so much.

TABLE 18.3
Implications of Reason's Error Model for Training and Redesign

Knowledge-based mistakes	Train the operator
Rule-based mistakes	Training and redesign
Slips	Redesign the task/environment
Lapses and mode errors	Redesign the task

Errors in an Organizational Context

Human errors in organization may be blamed on organizational deficiencies (Reason, 1997). Accidents, because they are so visible, are often analyzed, but less visible organization errors are not analyzed. Accidents represent the tip of the iceberg. Underneath the surface there are many organizational deficiencies that limit productivity as well as quality. Such deficiencies may be due to poor communication policies and poor operating procedures. The same organizational characteristics that cause unsafe acts and accidents will also decrease productivity and quality. We may note the analogy with poor usability of software, which goes hand in hand with poor quality and poor software definition (see Chapter 7).

Table 18.4 shows the results of analysis of errors committed by operators of mining equipment (Conway et al., 1981). In addition to slips and mistakes, there was also violation of company procedures. The implications of these are noted in terms of redesign of the operator's compartment, training, organizational changes, and management enforcement. Note that depending on the error type there are different implications for design, training, and disciplinary measures.

Human errors are commonplace. Many of the errors that people commit in operating equipment systems are the results of poor design and poor organizational structure, rather than irresponsible action (Reason 1997; Woods and Cook, 1999). Although human errors are statistically identified as a contributing cause to most accidents, we must consider that the human errors are accompanied by organizational errors. Before the human error occurred there were several mechanical and organizational errors. But the organizational errors are more difficult to identify. They are abstract and not so visible. They affect the system and weaken its defenses (Perrow, 1984). The notion that the operator should be punished or personally made responsible is unwarranted, unless of course there is a clear violation of regulations. The

TABLE 18.4
Analysis of Errors Committed by the Operators of Mining Equipment

Potential Error	Error Type	Action
Hauler not returned for service according to schedule	Violation	Management
Setting off with parking brake on	Slip	Design
Operator driving machine while sitting on rear bumper	Violation	Design/Training
Misreading of displays	Slip	Design
Operator leaning out of cabs while traveling	Violation	Design/Training
Insufficient warning to people behind haulers while backing	Slip	Design
Instability prevents effective use of fire extinguishers	Mistake	Design
Incorrect operation of machine controls	Mistake	Training/Design
Tires not maintained according to accepted practice	Mistake	Training

Examples from study by Conway et al. (1981).

blame for accidents and poor quality falls on poor design, procedures, and training, and, in the end, poor management. Environments must be designed for human beings, taking into account all their vulnerabilities and competencies—if not, the system is not appropriately designed.

18.7 ERGONOMICS FOR PRODUCTIVITY, SAFETY, HEALTH, AND COMFORT

In many industries ergonomics is implemented as a means for reducing high injury rates and high insurance premiums. In the U.S., a construction worker's compensation insurance premiums can amount to 15% of the salary. This is because there are many back injuries due to materials handling and injuries to the joints in the arms, shoulders, and neck. During the past 5 years many injuries due to cumulative trauma disorders, carpal tunnel syndrome, and tenosynovitis have been reported. At the same time, the number of back injuries remains high, and is still the main cause of industrial injury. It is estimated that the annual cost of musculoskeletal disease in the U.S. exceeds U.S. $50 billion per year (National Research Council, 2001).

While the reduction of injuries and improved health of workers are important reasons for implementing ergonomics, to the management it may be a negative issue. Management is forced to implement safety legislation, and many mangers view this just as another chore imposed by government. I am concerned that the negative message dominates, and industry leaders ignore a more important driving factor for ergonomics—namely, increases in productivity, and satisfaction. Ergonomic improvements in workstations, industrial processes, and product design can be undertaken for the sake of enhancing productivity, and there can be tremendous benefits. Management often do not understand how poor working conditions can decrease productivity. Workers in plants and in offices usually adapt and don't complain, but the cost is increased production time, lower quality, and of course, increased injury rate. The case study in Chapter 2 clearly illustrates the potential of ergonomics to improve productivity.

Human factors and ergonomics are highly related to industrial safety. If workers can perceive hazards, if there are relevant warning signs, if controls are easy to use, if work postures are acceptable, if noise and other environmental stressors are reduced, if there is collaboration between workers and management based on mutual understandings, and if there is good housekeeping, then safety will improve. Ergonomics safety measures focus on the operator and are different from the conventional approach taken in industrial safety. Ergonomics can improve safety through worker's attitudes, perception, decision-making, and risk-taking behavior.

EFFECTIVE SAFETY PROGRAMS

Formulation of effective safety programs is very important, particularly in industries where safety is a recognized component of everyday business. The chemical industry and the nuclear industry are examples of safety critical industries, where there have been very large investments, and many excellent safety programs are now in place. The National Safety Council in the U.S. hands out annual awards for excellence in safety, and these are typically won by the chemical industries. In their case safety is a necessity. An accident such as Bhopal in India would be an environmental and social disaster (Meshkati, 1989). But a large number of accidents occur in manufacturing and construction, and it is equally important for them to implement efficient safety programs.

The perennial problem in safety concerns motivation. Most people believe that they are responsible and highly skilled individuals: "I am an excellent driver, and accidents happen to others, not to me." And even if there are risks they are perceived as very light or manageable: "I can handle it!"

Employees rationalize the fact that they work in an unsafe environment. For example, construction work is one of the most dangerous jobs, and the most common type of accident is due to falling. Painters fall the most, because they stand on ladders (Helander, 1991). Zimolong (1985) asked construction employees to estimate the risk of common objects and associated tasks that are often involved in accidents. The objects and task were ladders, scaffolds, roofs, steps and stairs, and openings in walls, ceilings, and floors. There was significant agreement among the occupational groups about which situations were the most hazardous. However, workers underestimated the risk of situations that they were familiar with. For example, painters underestimated the risk of ladders, and scaffold assemblers underestimated the risk of scaffolds. Perhaps this is a psychological necessity—you have got to rationalize the fact that you are working in a high-risk environment. What would your family members say if you told them that you are employed in a very hazardous occupation?

Bring some painters into a safety briefing and they will say, "Don't tell me about the risk of ladders—I use them all the time! What do you know about ladders?" Safety managers in industry have a tough job.

In order to get the collaboration of employees we must find safety methods that break down the motivational problems and lead to real improvements in workplace safety. So what kind of safety program works? Guastello (1993) reviewed 52 accident prevention programs (see Table 18.5).

TABLE 18.5
Summary of Effect Sizes for Ten Types of Accident Prevention Programs

Type of Program	Number of Studies	Effect (%)
1. Personnel selection	26	3.7
2. Technological interventions	4	29.0
3. Behavior modifications	6	38.6
4. Poster campaign	2	14.0
5. Quality circle	1	20.0
6. Exercise and stress management	2	15.0
7. Near-miss accident reporting	2	0.0
8. International safety rating	4	17.0
9. Comprehensive ergonomics	3	51.6
10. Finnish national program	2	18.3

Half of the studies related to personnel selection select safe individuals. This shows how the accident proneness concept still dominates industry. As demonstrated here, an effect size of 3.7% is not convincing. We can write off this approach as inadequate. Unfortunately the number of studies for the other programs was small, so one can only draw limited conclusions. We will comment below on the programs that had three or more studies.

TECHNOLOGICAL INTERVENTIONS

These studies had an average effect size of 29.0 %. They focused on accident prevention for robots and automation. Automation and robots bring about new types of workplace injuries. For example, a robot can catch an operator in its work envelope and press him against a structure. This was in fact the scenario for the first robot fatality. The robot kept pressing the operator against a pillar, and the operator could not breathe and he suffocated. Thus, robots may require specific safeguards, such as sensors in the floor and on the robot itself. We will return to these issues in the following section.

BEHAVIOR MODIFICATIONS

The effect size was 38.6%. A typical program consisted of basic safety information and training in safe behavior. This was followed by a period of observation and feedback. Feedback could be provided by supervisors to employees. One can also display accident statistics and graphs in the workplace. Some of these programs included goal setting and/or incentives to encourage the observation and feedback process among employees.

COMPREHENSIVE ERGONOMICS

The effect size was 51.6%. Most of the studies involved a redesign of the work place or equipment to improve the work conditions. Employees actively participated in the program. One study emphasized the safety climate or safety culture. The following were also used:

- Monthly tracking of safety statistics
- Safety seminars
- Ergonomics expert advice

INTERNATIONAL SAFETY RATING SYSTEM (ISRS)

The effect size was 17.0%. This is a safety audit program that addresses many topics, including management and employee training, planned inspections, task analyses, group meetings, personal communication, and accident analyses.

For this approach, operations personnel met voluntarily to discuss safety issues and problems, and to develop action plans for safety improvement. This approach is analogous to quality circles where employees who perform similar types of work meet regularly to solve problems of product quality, productivity, and cost.

BEHAVIORAL SAFETY

During the last 10 years industry has started using a new approach to safety called behavioral safety (Geller, 2002). This builds on studies such as Gustallo's. There is a realization that effective safety programs can be formed if there is a consensus among workers and management. Behavioral safety has several building blocks:

- Participation by workers and management
- Work in groups—similar to quality circles
- Use members in the group to establish a list of safety critical behaviors (SCBs)
- Collect statistics on SCBs
- Provide feedback to employees on their SCBs
- Evaluate and validate the list of SCBs; remove old items and add new items

In summary, behavioral safety is a participatory approach to safety that is based on group consensus and feedback.

It may be difficult to establish a valid list of SCBs. Employees have a tendency to select behaviors that can be easily observed, while many safety critical behaviors are actually dynamic and not so easy to observe. Nonetheless I believe that from the group consensus emerges a true interest and motivation to fight accidents and injuries. Through the frequent interaction with peers there is not only consensus building, but also a conditioning effect which is very helpful in establishing a good safety standards and safety morale.

18.8 MACHINE SAFETY

Below we explain how different safety devices can be used to protect workers. In the last 15 years safety problems with robots have received much attention. Many different types of safety devices have been developed to protect workers. Our interest is, however, broader, and includes all types of machines and automation. It turns out that the robotic safety devices are also applicable to other machines.

There are many different types of safety device that can be used to shut down a machine, in case the operator gets too close. Table 18.6 gives an overview of the most commonly used devices. The devices are divided into two categories: work area intrusion, and inside machine movement zone.

PHYSICAL BARRIERS

Physical barriers such as fences, guard rails, chains, and curtains are used to prevent access to the working area. Fences are also used to protect people from flying objects that may be accidentally thrown by a robot. Depending upon the installation, fences can be designed to permit a flow of parts in and out of the working envelope. Access doors must be interlocked.

TABLE 18.6
Current Safety Devices

Safeguarding Devices	Description	Typical Application and Restrictions
Fencing	Guards with interlocked gates	Heavy material handling
Guard rails	Awareness barriers with interlocked gates	Light material handling
Chains and posts	Passive guard	Small assembly; light-duty applications
Curtain	Flexible screen	Protection from welding and heat in addition to safety
Photoelectric beams	Photocell/optical	Often used in combination with other devices
Pressure-sensitive mats and surfaces	On floor or attached to machine; can sense walking or touching	Often used in combination with other devices
Floor markings	Painted floor warnings	Indicates robot/machine work envelope

CHAINS

Chains are passive guards that work as awareness barriers. They can be easily overcome by intruders. However, they are appropriate for small robots and machines used for light duty work, and the main function is to remind the operator.

WELDING CURTAIN

The main purpose of a welding curtain is to protect operators from radiation and ultraviolet light from welding. They also have a secondary purpose, since they are in effect physical barriers.

PHOTOELECTRIC BEAMS

Photo cells can be used for outlining the work barrier. Alternative devices such as light curtains and magnetic curtains are also sometimes used to detect personnel intruding into the working envelope.

PRESSURE-SENSITIVE MATS

These are used to detect a person walking towards the work envelope. Their use is impractical if parts have to be rolled in and out of the work envelope. Pressure-sensitive surfaces or skins may also be attached to a critical work surface or a machine, so if an operator presses against or touches this area, the robot or machine is stopped.

INFRARED SENSORS

All objects in the environment emit infrared radiation. Infrared sensors have been developed that are tuned to the well-defined spectrum range of human infrared energy. Rahimi and Hancock (1988) have pointed out that many of these types of sensor can be used in hybrid workstations where operators and robots are cooperating.

18.9 CASE STUDY: ROBOT SAFETY AT IBM CORPORATION/LEXMARK

The extensive safety precautions that surround robotic workplaces have actually been a deterrent to the development and use of automation, as this case study shows. The IBM Corporation manufacturing plant in Lexington, Kentucky, specialized in printers and typewriters. During the 1980s an extensive system for automation was introduced. This was an engineer's dream. In the beginning of the 1990s this plant was spun off to form an independent company, Lexmark, which manufactures printers. Most of the automation was then surprisingly discarded. One important reason for doing away with the robotic installation was the additional cost of robotic safety. Table 18.7 lists problems encountered at the IBM/Lexmark plant.

Although the automation for manufacturing was successful from a technological point of view, it was determined too costly. It was in fact cheaper and faster for some of the products to be assembled manually. Design for automation was considered impractical because of the short product life. For automation to be successful it must be possible to manufacture large volumes.

TABLE 18.7
Problems of Automation and Robotic Safety at IBM Lexington

- Robot safeguards increase the cost of the automated manufacturing system.
- Little space was left over after robot installation, causing robots to bump elbows. Entrapment became a concern.
- Safety regulations did not consider maintenance of robots, where the needs were the greatest. They were designed for normal operation.
- Robot workstations were difficult to change. Additional engineering updates usually involved major modifications of the robot system.
- The line was down much of the time, necessitating an employee to hand assemble the product. No provisions had been made for manual intervention stations and thus there were many ergonomic problems with bad work postures.
- The constant movement of conveyors, robots, and other automation produced much noise.
- The manufacturing lines backed up when the operator had to turn off a robot to enter the work envelope.
- Operators began losing production and circumventing safety systems. Near-miss accident reports increased.

Additional information about robotic safety is given in robotics safety standards issued by the Robotic Industries Association (1987, 1989). An excellent collection of papers on robotics safety has been edited by Rahimi and Karwowski (1993).

EXERCISE

When a person learns to drive, many different types of errors can happen. Make a list of things that can go wrong. This should include vehicle-related errors as well as traffic-related errors. Once you have finalized the list classify the errors using Reason's framework. Make a table to present your results, such as Table 18.4.

Discuss how you can use the list to

1. Propose better design of the vehicle/road environment
2. Propose further training of the driver
3. In particular, discuss how you can classify traffic violations

---

RECOMMENDED READINGS

Harms-Ringdahl, L., 2002, *Safety Analysis: Principles and Practices in Occupational Safety*, London: Taylor & Francis.
Geller, E. S., 2001, *Working Safe: How to Help People Actively Care for Health and Safety*, Boca Raton, FL: CRC Press.

Harms-Ringdahl's book gives a good overview of methods in occupational safety, and Geller's book gives a good introduction to behavior-based safety.

References

Andersson, G.B.J. and Örtengren, R., 1974, Lumbar disc pressure and myoelectric back muscle activity during sitting, *Scandinavian Journal of Rehabilitation Medicine*, 3, 115–121.

ANSI, 1986, *Octave-Band and Fractional Octave-Band Analog and Digital Filters*, ANSI S1.11, New York: American National Standards Institute.

ANSI, 1991, *Specifications for Personal Noise Dosimeters*, ANSI S1.25, New York: American National Standards Institute.

Armstrong, T.J. and Chaffin, D.B., 1979, Carpal tunnel syndrome and selected personal attributes, *Journal of Occupational Medicine*, 21, 481–486.

ASHRAE, 1997, Thermal comfort, in *ASHRAE Handbook: Fundamentals*, Atlanta: American Society for Heating, Refrigeration, and Air-Conditioning Engineers.

Åstrand, I., 1969, Aerobic work capacity in men and women with special reference to age, *Acta Physiologica Scandinavica*, 49 (Suppl. 169).

Åstrand, P.-O., Rodahl, K., Dahl, H.A., and Stromme, S.B., 2003, *Textbook of Work Physiology*, 4th ed., Champaign, IL: Human Kinetics Publisher.

Ayoub, M.M., 1973, Workplace design and posture, *Human Factors*, 15, 265–268.

Baggett, P. and Ehrenfeucht, A., 1991, Building physical and mental models in assembly tasks, *International Journal of Industrial Ergonomics*, 7, 217–228.

Bailey, R.W., 1996, *Human Performance Engineering: A Guide for Systems Designers*, Englewood Cliffs, NJ: Prentice-Hall.

Baldur, R. and Baron, L., 1988, Sensors for safety, in Dorf, N. (Ed.), *International Encyclopedia of Robotics*, New York: Wiley.

Beers, T.M., 2000, Flexible schedules and shift work: replacing the 9-5 workday? *Monthly Labor Review*, June, 33–34.

Bennett, C., Chitlangia, A. and Pangrekar, A., 1977, Illumination levels and performance of practical visual tasks, *Proceedings of the Human Factors Society 21st Annual Meeting*, Santa Monica, CA: The Human Factors and Ergonomics Society, 322–325.

Beranek, L. and Newman, R., 1950, Speech interference levels as criteria for rating background noise in offices, *Journal of the Acoustical Society of America*, 22, 671.

Beranek, L., Balzier, W., and Figwer, J., 1971, Preferred noise criteria (PNC) curves and their application to rooms, *Journal of the Acoustical Society of America*, 50, 1223–1228.

Berger, E.H. and Casali, J.G., 1997, Hearing protection devices, in Crocker, M.J. (Ed.), *Encyclopedia of Acoustics*, New York: Wiley.

Berger, E.H, Ward, W.D., Morrill, J.C., and Royster, L.H., 2000, *Noise and Hearing Conservation Manual*, American Industrial Hygiene Association.

Bergqvist, U., 1986, Bildskärmsarbete och hälsa, in *Arbete och Hälsa*, Vol. 9, Stockholm: Arbetarskyddsverket.

Bergum, B.O. and Bergum, J.E., 1981, Population stereotypes: an attempt to measure and define, *Proceedings of the Human Factors Society 25th Annual Meeting*, Santa Monica, CA: The Human Factors and Ergonomics Society, 662–665.

Bernard, P.J., Grudin, J., 1988, Command names, in Helander, M.G. (Ed.), *Handbook of Human Computer Interaction*, Amsterdam: North Holland.

Bernard, T.E., 2002, Thermal stress, in Plog, B. (Ed.), *Fundamentals in Industrial Hygiene*, 5th ed., Chicago: National Safety Council.

Beyer, H. and Holtzblatt, K., 1998, *Contextual Design*, San Francisco: Morgan Kaufmann.

Bjerner, B. and Swensson, A., 1953, Shiftwork and rhythm, *Acta Medica Scandinavica*, 278, 102–107.

Blackwell, H.R., 1964, Further validation studies of visual task evaluation, *Illuminating Engineering*, 59, 627–641.

Blackwell, H.R., 1967, The evaluation of interior lighting on the basis of visual criteria, *Applied Optics*, 6, 1443–1467.

Blackwell, O. and Blackwell, H., 1971, IERI report: visual performance data for 156 normal observers of various ages, *Journal of Illuminating Engineering Society*, 1, 2–13.

Bøggild, H, and Knutsson, A., 2000, Meta analyses of the epidemiological literature on shift work and heart disease, *Zeitschrift für Arbeitswissenschaft*, 5.

Bond, N.A., Jr., 1986, Maintainability, in Salvendy, G. (Ed.), *Handbook of Human Factors*, New York: Wiley, 1329–1355.

Boothroyd, G. and Dewhurst, P., 1983, *Design for Assembly*, Amherst, MA: Department of Mechanical Engineering, University of Massachusetts.

Boothroyd, G. and Dewhurst, P., 1987, *Product Design for Assembly*, Wakefield, RI: Boothroyd Dewhurst.

Boothroyd, G, and Knight, W., 1993, Design for assembly, *Spectrum IEEE*, September, 53–55.

Boyce, P. R., 1981a, Lighting and visual performance, *International Review of Ergonomics*, 2, London: Taylor & Francis.

Boyce, P.R., 1981b, *Human Factors in Lighting*, New York: Macmillan.

Braidwood, R., 1951, *Prehistoric Men*, Chicago, IL: Natural History Museum.

Branton, P., 1984, Back shapes of seated persons—how close can the interface be designed? *Applied Ergonomics*, 15, 105–107.

Bratko, I., Tancig, P., and Tancig, S., 1986, Detection of positional patterns in chess, *Advances in Computer Chess*, 4, 31–56.

Broadbent, D., 1977, Language and ergonomics, *Applied Ergonomics*, 8, 15–18.

Broadbent, D., 1978, The current state of noise research: reply to Poulton, *Psychological Bulletin*, 85, 1052–1067.

Brogmus, G.E. and Marko, R., 1990, Cumulative trauma disorders of the upper extremities, *Proceedings of IEA Conference on Human Factors in Design for Manufacturability and Process Planning*, Santa Monica, CA: The Human Factors and Ergonomics Society, 49–59.

Brooke, J., 1996, SUS: A quick and dirty usability scale, in Jordan, P., Thomas, B. Weerdmeester, B,. and McClelland. I. (Eds.), *Usability Evaluation in Industry*, London: Taylor & Francis, 189–194.

Brown, A.C. and Brengelmann, G., 1965, Energy metabolism, in Ruch, T.C. and Patton, H.D. (Eds), *Physiology and Biophysics*, Philadelphia, PA: Saunders.

Brown, C.R. and Schaum, D.L., 1980, User-adjusted VDU Parameters, in Grandjean, E. and Vigliani, E. (Eds), *Ergonomic Aspect of Visual Display Terminals*, London: Taylor & Francis.

Browne, R.C., 1949, The day and night performance of the tele-printer switchboard operators, *Occupational Psychology*, 23, 121–126.

Boyce, P.R., 2003, *Human Factors in Lighting*, 2nd ed. London: Taylor & Francis.

Burri, G. and Helander, M.G., 1991a, A field study of productivity improvements in the manufacturing of circuit boards, *International Journal of Industrial Ergonomics*, 7, 207–216.

Burri, G. and Helander, M.G., 1991b, Implementation of human factors principles in the design of manufacturing process, in Pulat, B.M. and Alexander, D.C. (Eds.), *Industrial Ergonomics Case Studies*, Norcross, GA: Industrial Engineering and Management Press.

Cakir, A., Reuter, H.-J., von Schmude, L., and Armbruster, A., 1978, *Anpassung von Bildschirmarbeitsplätzen an die physische Funktionsweise des Menchen*, Bonn: Der Bundesminister für Arbeit und Sozialordnung.

Card, S.K. and Moran, T.P., 1986, History of personal workstations, *Proceedings of the ACM Conference on the History of Personal Workstations*, New York: ACM, 183–198.

Card, S.K., Moran, T.P., and Newell, 1983, *The Psychology of Human-Computer Interaction*, Mahwah, N.J.: Lawrence Erlbaum.

Carlsson, L., 1979, *Ljus- och belysningskrav vid arbete med bildskarmar pa tidningsforetag*, Stockholm: Tidningarnas Arbetsmiljo-Kommittee.

Carroll, J.M., 1993, Techniques for minimalist documentation and user interface design, in Jansen, C., Vander Poort, P., Steehouder, M. and Verhejen, R. (Eds), *Quality of Technical Documentation: Utrecht Studies in Language and Communication*, Vol. 2, Amsterdam: Rodopi.

Carroll, J.M., 1985, *What Is in a Name?* New York: Freeman.

Caruso, C.C., Hitchcock, E.M., Dick, R.B., Russo, J.M., and Schmit, J.M., 2004, *Overtime and Extended Work Shifts: Recent Findings on Illnesses, Injuries and Health Behaviors*, DHHS (NIOSH) Publ. 2004-143, Cincinnati, OH: National Institute of Occupational Safety and Health.

Casali, J.G. and Park, M.Y., 1990, Attenuation performance of four hearing protectors under dynamic movement and different user fitting conditions, *Human Factors*, 32, 9–25.

Chaffin, D.B., 1969, A computerized biomechanical model: development of and use in studying gross body actions, *Journal of Biomechanics*, 2, 429–441.

Chaffin, D.B., Andersson, G.B.J., and Martin, B.J., 1999, *Occupational Biomechanics*, 3rd ed., New York: Wiley.

Channel News Asia, 2004, <www.channelnewsasia.com/stories/afp_world/view/100523/1/.html>.

Chapanis, A., 1971, The search for relevance in applied research, in Singleton, W.T., Fox, J.G., and Whitfield, D. (Eds.), *Measurement of Man at Work*, London: Taylor & Francis, 1–14.

Chapanis, A., 1974, National and cultural variables in ergonomics, *Ergonomics*, 17, 153–176.

Chapanis, A., 1995, Ergonomics in product development: a personal view, *Ergonomics*, 38, 1625–1638.

Chapanis, A., 1990, The International Ergonomics Association: its first 30 years, *Ergonomics*, 33, 275–282.

Chapanis, A., 1996, *Human Factors in Systems Engineering*, New York: Wiley.

Chapanis, A. and Kinkade, R.G., 1972, Design of controls, in Van Cott, H.P. and Kinkade, R.G. (Eds.), *Human Engineering Guide to Equipment Design*, Washington, DC: U.S. Government Printing Office.

Chapanis, A. and Lindenbaum, L., 1959, A reaction time study of four control-display linkages, *Human Factors*, 1, 1–7.

Chapanis, A. and Moulden, J.V., 1990, Short-term memory for numbers, *Human Factors*, 32, 123–137.

Chase, W.G. and Simon, H.A., 1973. Perception in chess, *Cognitive Psychology*, 4, 55–81.

Chabra, S.L. and Ahluwalia, R.S., 1990, Rules and guidelines for ease of assembly, in *Proceedings of the International Ergonomics Association Conference on Human Factors in Design for Manufacturability and Process Planning*, Santa Monica, CA: Human Factors and Ergonomics Society, 93–99.

Chengalur, N., Rodgers, S.H., and Bernard, T.E., 2004, *Kodak's Ergonomic Design for People at Work*, 2nd ed. New York: Wiley.

Childe, G., 1944, *The Story of Tools*, London: Cobbet.

Cohen, H., 1979, *Conveyor Safety*, Washington, DC: U.S. Department of HEW, NIOSH.

Collins, A.M., 1973, Decrements in tracking and visual performance during vibration, *Human Factors*, 15, 379–393.

Collins, B. and Lerner, N., 1983, *An Evaluation of Exit Symbol Visibility*, NBSIR Report 82-2685, Washington, DC: National Bureau of Standards.

Conrad, R. and Hull, A.J., 1968, The preferred layout for numerical data-entry keysets, *Ergonomics*, 11, 1–22.

Conway, E.J., Helander, M.G., Sanders, M., Krohn, G., and Schurick, J., 1981, *Human Factors Problem Identification in Surface Mines*, Tech. rep. 10-3071, Westlake Village, CA: Canyon Research Group, Inc., 1981.

Corlett, E.N., 1995, The evaluation of posture and its effect, in Wilson, J.R. and Corlett, E.N. (Eds.), *Evaluation of Human Work*, London: Taylor & Francis, 662–713.

Corlett, E.N. and Bishop, R.P., 1976, A technique for assessing postural discomfort, *Ergonomics*, 19, 175–182.

Coppola, A., 1984, Artificial intelligence applications to maintenance, in *Artificial Intelligence in Maintenance*, Brooks AFB, TX: Air Force Human Resources Laboratory, 23–44.

Costa, G., et al., 2001. Flexible Working Hours, Health and Well-Being in Europe: Some Considerations from a SALISA Project, *Chromobiology Int.*, 21, 831–844.

Courtney, A.J., 1986, Chinese population stereotypes: color associations, *Human Factors*, 28, 97–99.

Crandall, B., Klein, G., and Hoffman, R., 2005, *Labors of the Mind: A Practitioner's Guide to Cognitive Task Analysis*, Cambridge, MA; MIT Press.

Crossman, E.R.F.W., 1959, A theory of the acquisition of speed-skill, *Ergonomics*, 2, 153–166.

Czerwinski, M., 2000, Personal communication.

Das, B. and Sengupta, A.K., 1996, Industrial workstation design: a systematic ergonomics approach, *Applied Ergonomics*, 27, 157–163.

Dedobbeleer, N. and Beland, F., 1989, Safety climate in construction sites, *Proceedings of International Conference on Strategies for Occupational Accident Prevention*, Stockholm: The National Institute for Occupational Health.

Della Rocco, P.S., and Cruz, C.E., 1995, *Shift Work, Age, and Performance: Investigation of the 2-2-1 Shift Schedule Used in Air Traffic Control Facilities I. The Sleep/Wake Cycle*, N95-29261. Washington, D.C.: FAA Office of Aerospace Medicine.

Deutsches Institut für Normung, 1981, DIN Standard 66234, Parts 1 to 9, Berlin: DIN.

Dickinson, C.E., Campion, K., Foster, A.F., Newman, S.J., O'Rourke, A.M.T., and Thomas, P.G., 1992, Questionnaire development: an examination of the Nordic Musculoskeletal questionnaire, *Applied Ergonomics*, 23, 197–201.

Drillis, R.J., 1963, Folk norms and biomechanics, *Human Factors*, 5, 427–441.

Drury, C.G., 1975, Application of Fitts' law to foot pedal design, *Human Factors*, 17, 368–373.

Ducharme, R.E., 1973, Problem tools for women, *Industrial Engineering*, September, 46–50.

Dumas, J. and Redish, J., 1999, *A Practical Guide to Usability Testing*. U.K.: Intellect.

Dumas, J., 2003, User-based evaluations, in Jack, J.O. and Sears, A. (Eds.), *The Human-Computer Interaction Handbook*, Mahwah, New Jersey: Lawrence Erlbaum, 1093–1117.

Eastman Kodak Co., 2004, *Kodak's Ergonomic Design for People at Work*, 2nd ed., New York: Wiley.

EC Council Directive L156, *Official Journal of the European Communities*, 1990, 21 June, 9–13.

Eckstrand, G.A., 1964, *Current Status of the Technology of Training*, Report AMRL-TDR-64-86, Ohio: Aerospace Medical Laboratories, Wright-Patterson Air Force Base.

Ekstrom, E.B., French, J.W., Harman, H.H., and Dermen, D., 1976, *Manual for Kit of Factor-Referenced Cognitive Tests*, Princeton, N. J.: Educational Testing Service.

Elliott, T.K. and Joyce, R.D., 1971, An experimental evaluation of a method for simplifying electronic maintenance, Human Factors, 13, 217–227.

Endsley, M.R., 1995, Towards a theory of situation awareness, *Human Factors*, 37, 65–84.

Engelbart, D., English, W.K., and Berman, M.L., 1967, Display-selection techniques for text manipulation, *IEEE Transactions on Human Factors in Electronics,* Vol. HFE-8, No. 1, Mar. 1967.

English, W.K., Engelbart, D.C., and Berman, M.A., 1976, Display-selection techniques for text manipulation, *IEEE Transactions on Human Factors in Electronics*, HFE-8, 5–15.

Ericsson, K.A., Chase, W.G., and Faloon, S., 1980, Acquisition of a memory skill, *Science*, 208, 1181–1182.

Ericsson, K.A. and Simon, H.A., 1993, *Think Aloud Protocol*, Cambridge, MA: MIT Press.

Eriksson, R., 1976, Personal communication, International Labor Organization, Geneva, Switzerland.

Fu, L., Salvendy, G., and Turley, L. (1998), Who finds what in a usability evaluation, *Proceedings of the Human Factors and Ergonomics Society, 42nd Annual Meeting*, Santa Monica, CA: Human Factors and Ergonomics Society, 1341–1345.

Etherton, J., 1986, The use of safety devices and safety controls at industrial machine workstations, in Salvendy, G. (Ed.), *Handbook of Human Factors*, New York: Wiley.

Etherton, J.R., 1988, *Safe Maintenance Guidelines for Robotic Workstations*, Morgantown, WV: National Institute for Occupational Safety and Health.

Fanger, P.O., 1970, *Thermal Comfort*, New York: McGraw-Hill.

Farrell, R.J. and Booth, J.M., 1984, *Design Handbook for Imagery Interpretation Equipment*, Seattle, WA: Boeing Aerospace Company.

Faulkner, T.W. and Murphy, T.J., 1973, Lighting for difficult visual tasks, *Human Factors*, 15, 149–159.

Fellows, G.L. and Freivalds, A., 1991, Ergonomics evaluation of a foam rubber grip for tool handles, *Applied Ergonomics*, 22, 225–230.

Fitts, P.M. and Posner, M., 1973, *Human Performance*, Englewood Cliffs, NJ: Prentice-Hall.

Fitts, P.M. and Seeger, C.M., 1953, S-R compatibility: spatial characteristics of stimulus and response codes, *Journal of Experimental Psychology*, 46, 199–210.

Fitts, P.M. (Ed.)., 1951, *Human Engineering for an Effective Air Navigation and Traffic Control System*, Washington, D.C.: National Research Council.

Floderus, B., Stenlund, C., and Tornqvist, S., 1993, An update of recent epidemiological and experimental studies, *Proceedings of Mini-Symposium on Electromagnetic Fields and Cancer*, Solna, Sweden: National Institute for Occupational Health.

Folkard, S., Monk, T.H., and Lobban, M.C., 1978, Short and long-term adjustment of circadian rhythms in "permanent" night nurses, *Ergonomics*, 21, 785–799.

Fothergill, L.C. and Griffin, M.J., 1977, The evaluation of discomfort produced by multiple frequency whole-body vibration, *Ergonomics*, 20, 263–270.

Franco, G. and Fusetti, L., 2004, Berrnardino Ramazzini's early observation of the link between musculoskeletal disorders and ergonomic factors, *Applied Ergonomics*, 35, 67–70.

Furtado, D., 1990, Principles of design for assembly, *Proceedings of the International Ergonomics Association Conference on Human Factors in Design for Manufacturability and Process Planning*, Santa Monica, CA: Human Factors Society, 147–52.

Gager, R., 1986, Design for productivity saves millions, *Appliance Manufacturer*, January, 46–51.

Garg, A. and Herrin, G.D., 1979, Stoop or squat? A biomedical and metabolic evaluation, *Transactions of American Institute of Industrial Engineers*, 11, 293–302.

Garg, A., Chaffin, D.B. and Freivalds, A., 1982, Biomechanical stresses from manual load lifting: a static vs. dynamic evaluation, *Transactions of the American Institute of Industrial Engineers*, 14, 272–281.

Gauderer, P.C. and Knaut, P, 2000. Participatory design of work schedules in public local transport (PLT) through ìpersonalizedî duty rotas. *Zeitschrift für Arbeitswissenschaft*, 5.

Gawron, V., 1982, Performance effects of noise intensity, psychological set, and task type and complexity, *Human Factors*, 24, 225–243.

Gigerenzer G. and Selten, R., 1999, *Rethinking Rationality: Report of the 84th Dahlem Workshop on Bounded Rationality*, Berlin: The Adaptive Toolbox.

Geller, E.S., 2001, *Working Safe: How to Help People Actively Care for Health and Safety*, Boca Raton, FL: Lewis Publishers.

Gemme, G., Lundstrom, R., and Hansson, J.-E., 1993, Disorders induced by work with hand-held vibrating tools: a review of current knowledge for criteria documentation, *Arbete och Hälsa*, 6.

Genaidy, A.M., Duggai, J.S. and Mital, A., 1990, A comparison of robot and human performances for simple assembly tasks, *International Journal of Industrial Ergonomics*, 5, 73–81.

Gibbs, W., 1997, Taking computers to task, *Scientific American*, July, 82–89.

Gilbert, T.F., 1974, On the relevance of laboratory investigation of learning to self-instructional programming, in Lamsdaine, A.A. and Glaser, R. (Eds.), *Teaching Machines and Programmed Instructions*, Washington, DC: National Educational Association.

Gilovich, T., Griffin, D., and Kahneman, D., 2002, *Heuristics and Biases: The Psychology of Intuitive Judgment*, Cambridge, U.K.: Cambridge University Press.

Goldstein, I.L., 1980, Training in work organizations, *Annual Review of Psychology*, 31, 229–272.

Goldstein, I.L., 1986, The relationship of training goals and training systems, in Salvendy, G. (Ed.), *Handbook of Human Factors*, New York: Wiley.

Gould, J.D., 1997, How to design usable systems, in Helander, M.G., Landauer, T.K., and Prabhu, P.V. (Eds.), *Handbook of Human-Computer Interaction*, Amsterdam: Elsevier Science.

Gould, J.D. and Grischkowsky, N., 1984, Doing the same work with hard copy and cathode-ray tube (CRT) computer terminals, *Human Factors*, 26, 323–338.

Grandjean, E. (Ed.), 1984, *Ergonomics and Health in Modem Offices*, London: Taylor & Francis.

Grandjean, E., 1988, *Fitting the Task to the Man*, London: Taylor & Francis.

Grandjean, E., 1986, Design of VDT workstations, in Salvendy, G. (Ed.), *Handbook of Human Factors*, New York: Wiley, 1359–1398.

Greenburg, L. and Chaffin, D., 1977, *Workers and Their Tools*, Midland, MI: Pendall.

Greenstein, J.S., 1997, Pointing devices, in Helander, M., Landauer, T., and Prabhu, P. (Eds.), *Handbook of Human-Computer Interaction*, Amsterdam: North Holland.

Gregersen, N.P., Brehmer, B., and Morén, B., 1996, Road safety improvement in large companies: an experimental comparison of different measures, *Accident Analysis and Prevention*, 28, 297–306.

Grether, W.F., 1971, Vibration and human performance, *Human Factors*, 13, 203–216.

Griffin, M.J., 1996, *Handbook of Human Vibration*, Elsevier, Amsterdam, 1996.

Griffin, M.J., 1997, Vibration and motion, in Salvendy, G. (Ed.), *Handbook of Human Factors and Ergonomics*, Wiley, 1997.

Grössinger, C., 1996, *The Misericords of Manchester Cathedral*, Manchester, UK: Manchester Cathedral.

Grossmith, E.J., 1992, Product design considerations for the reduction of ergonomically related manufacturing costs, in Helander, M. and Nagamachi, M. (Eds.), *Design for Manufacturability: A Systems Approach to Concurrent Engineering and Ergonomics*, London: Taylor & Francis.

Gruber, G.J., 1976, *Relationships between Whole-Body Vibration and Morbidity Patterns among Interstate Truck Drivers*, San Antonio, TX: Southwest Research Institute.

Guastello, S.J., 1993, Do we really know how well our occupational accident prevention programs work? *Safety Science*, 16, 445–463.

Hadler, N.M., 1986, Industrial rheumatology, *The Medical Journal of Australia*, 144, 191–195.

Hadler, N., 1989, Personal communication, Denver, CO, U.S.

Hagberg, M. and Sundelin, G., 1986, Discomfort and load on the upper trapezius muscle when operating a word processor, *Ergonomics*, 29, 1637–1645.

Hale, A.R. and Glendon, A.L., 1987, *Individual Behavior in the Control of Danger*, Amsterdam: Elsevier.

Handbuch fuer Beleuchtung, 1975, Essen: Verlag Girardet.

Hansson, J.E., Klussel, L., Svensson, G., and Winström, B.O., 1976, Working environment for truck drivers—an ergonomic and hygienic study, *Arbete och Hilsa*, Vol. 6, Stockholm: Arbetarskyddsstyrelsen.

Harms-Ringdahl, L., 2001, *Safety Analysis, Principles and Practice in Occupational Safety*, London: Taylor & Francis.

Harris, W. and Mackie, R.R., 1972, *A Study of the Relationships among Fatigue, Hours of Service and Safety of Operations of Truck and Bus Drivers*, Technical Report 1727-2, Goleta, CA: Human Factors Research, Inc.

Hasselquist, R.J., 1981, Increasing manufacturing productivity using human factors principles, in *Proceedings of the 28th Annual Meeting of the Human Factors Society*, Santa Monica, CA: The Human Factors and Ergonomics Society, 204–206.

Health and Safety Commission, 1991, *Handling Loads at Work Proposals for Regulation and Guidance*, London: Health and Safety Executive.

Health and Safety Commission, 2003, *Health and Safety Highlights 2002/03: National Statistics*. London: Health and Safety Executive.

Heckel, P., 1984, *The Elements of Friendly Software Design*, New York: Warner Books.

Helander, E.A.S., 2002, Personal communication, United Nations Development Program, Geneva, Switzerland.

Helander, M.G. 1997, Forty years of IEA: some reflections on the evolution of ergonomics, *Ergonomics*, 40, 952–961.

Helander, M.G., 1986, Design of visual displays, in Salvendy, G. (Ed.), *Handbook of Human Factors and Ergonomics*, New York: Wiley.

Helander, M.G., 1988, *Handbook of Human-Computer Interaction*, Amsterdam: North Holland.

Helander, M.G., 1990, Ergonomics and safety considerations in the design of robotics workplaces: a review and some priorities for research, *International Journal of Industrial Ergonomics*, 6, 127–149.

Helander, M.G., 1991, Safety hazards and motivation for safe work in the construction industry, *International Journal of Industrial Ergonomics*, 8, 205–224.

Helander, M.G., 1996, Means-ends analysis in ergonomics design, Unpublished keynote address, *Fifth International Conference on Human Factors of Advanced Manufacturing and Hybrid Automation*, Maui, Hawaii, U.S.

Helander, M.G, 1997, *A Guide to the Ergonomics of Manufacturing*, London: Taylor & Francis.

Helander, M.G. and Burri, G., 1995, Cost effectiveness of ergonomics and quality improvements in the electronics industry, *International Journal of Industrial Ergonomics*, 15, 137–151.

Helander, M., Landauer, T.K., and Prahbu, P.V., 1997, *Handbook of Human Computer Interaction*, 2nd ed., Amsterdam: North Holland.

Helander, M.G. and Domas, K., 1986, Task allocation between humans and robots in manufacturing, *Material Flow*, 3, 175–85.

Helander, M.G., Little, S.E., and Drury, C.G., 2000, Adaptation and sensitivity to postural change in sitting, *Human Factors*, 43, 617–629.

Helander, M.G. and Nagamachi, M., 1992, *Design for Manufacturability: A Systems Approach to Concurrent, Engineering and Ergonomics*, London: Taylor & Francis.

Helander, M.G. and Palanivel, T., 1990, *Anthropometric Survey of Employees at IBM Corporation*, San Jose, Buffalo, NY: Ergonomics Research, Inc.

Helander, M.G. and Rupp, B., 1984, An overview of standards and guidelines for visual display terminals, *Applied Ergonomics*, 15, 185–195.

Helander, M.G. and Schurick, J.M., 1982, Evaluation of symbols for construction machines, in *Proceedings of the 26th Annual Meeting of the Human Factors Society*, Santa Monica, CA: The Human Factors and Ergonomics Society.

Helander, M.G. and Skinnars, Ö., 2000, Use of cognitive walkthrough for evaluation of cockpit design, *Proceedings of the XIVth Triennial Congress of the International Ergonomics Association and the 44th Annual Meeting of the Human Factors and Ergonomics Society*, Santa Monica, CA: Human Factors and Ergonomics Society, 1-616–1-619.

Helander, M.G. and Waris, J.D., 1993, Effect of spatial compatibility in manual assembly of performance, in Marras, W.S., Karwowski, W., Smith, J.L., and Pacholski, L. (Eds.), *The Ergonomics of Manual Work*, London: Taylor & Francis.

Helander, M.G., Billingsley, P.A., and Schurick, J.M., 1984, An evaluation of human factors research on visual display terminals in the workplace, *Human Factors Review*, 1, 55–129.

Hickling, E.M., 1985, *An Investigation of Construction Sites as Factors Affecting the Acceptability and Wear of Safety Helmets*, Loughborough: University of Technology, Institute for Consumer Ergonomics.

Hill, S.G. and Kroemer, K.H.E., 1989, Preferred declination and the line of sight, *Human Factors*, 28, 127–134.

Hocking, B., 1987, Epidemiological aspects of 'repetition strain injury' in Telecom, Australia, *The Medical Journal of Australia*, 147, 218–222.

Holbrook, A.E.H. and Sackett, P.J., 1988, Design for assembly guidelines for product design, *Assembly Automation*, 8, 202–211.

Holding, D.H., 1986, Concepts of training, in Salvendy, G. (Ed.), *Handbook of Human Factors*, New York: Wiley.

Hollnagel, E. and Woods, D.D., 1983, Cognitive systems engineering: new wine in new bottles, *International Journal of Man-Machine Studies*, 18, 583–600.

Hollnagel, E., 2003, *Handbook of Cognitive Task Design*, Mahwah, N.J.: Lawrence Erlbaum.

Hopkinson, R.G. and Collins, J., 1970, *The Ergonomics of Lighting*, London: Macdonald.

Horne, J.A. and Östberg, 0., 1976, A self-assessment questionnaire to determine morningness-eveningness in human circadian rhythms, *International Journal of Chronobiology*, 4, 97–110.

Hornick, R., 1973, Vibration, in *Bioastronautics Data Book*, 2nd ed., (NASA SP-3006), Washington, DC: National Aeronautics and Space Administration.

Human Factors and Ergonomics Society, 2003, *BSR/HFES100—Human Factors Engineering of Computer Workstations, Draft Standard for Trial Use*, Santa Monica, CA: The Human Factors and Ergonomics Society.

The International Ergonomics Association, 2000, *The Discipline of Ergonomics*, <www.iea.cc>.

Illuminating Engineering Society, 1982, *Office Lighting*, New York: ANSI/IES.

Illuminating Engineering Society of North America, 2000, *IE Lighting Handbook*, 9th ed., New York, NY: Illuminating Engineering Society of North America

International Labor Office, 1987, *Checklist for Workplace Inspection for Improving Safety, Health and Working Conditions*, Geneva: ILO.

International Labour Organization, 1972, *Kinetic Methods of Manual Handling in Industry*, Occupational Health Series No. 10, Geneva, ILO.

International Standards Organization, 2004, *ISO Series 9241*, Geneva: ISO.

International Standards Organization, 1997, *ISO 2631—Human Exposure to Wholebody Vibration*, Geneva: ISO.

International Standards Organization, 1994, *ISO 7730—Moderate Thermal Environments: Determination of the PMV and PPD Indices and Specification of the Conditions for Thermal Contrast*, Geneva: ISO.

International Standards Organization, 1989a, *ISO 7933—Hot Environments: Analytical Determination of Thermal Stress Using Calculation of Required Sweat Rate*, Geneva: ISO.

International Standards Organization, 1989b, *ISO 7243—Hot Environments: Estimation of the Heat Stress on Working Man, Based on the WBGT-Index (Wet Bulb Globe Temperature)*, Geneva: ISO.

International Standards Organization, 1998, *ISO 9241-5—Ergonomic Requirements for Office Work with Visual Display Terminals (VDTs)*, Part 5. Workstation layout and postural requirements, Geneva: ISO.

International Standards Organization, 1998, *ISO 9241-7—Ergonomic Requirements for Office Work with Visual Display Terminals (VDTs)*, Part 7. Requirements for displays with reflections, Geneva: ISO.

Jensen, S., 2002, *The Simplicity Shift*, Cambridge, U.K.: Cambridge University Press.

Johansson, G. and Backlund, F., 1970, Drivers and road signs, *Ergonomics*, 13, 749–759.

Jordan, P., 1998, *An Introduction to Usability*, London: Taylor & Francis.

Jorgensen, C., Hamel, W., and Weisbin, C., 1986, Autonomous robot navigation, *BYTE*, 4, 223–235.

Kahneman, D., 1973, *Attention and Effort*, Englewood Cliffs, NJ: Prentice-Hall.

Kahneman, D., Slovic, P., and Tversky, A., 1982, *Judgment under Uncertainty: Heuristics and Biases*, Cambridge, UK: Cambridge University Press.

Karlsson, K., 1989, *Bullerskador*, Stockholm: Arbetsmiljók ommissionen.

Karhu, O., Härkönen, R., Sorvali, P., and Vepsäläinen, P., 1981, Observing working postures in industry: examples of OWAS applicati*Applied Ergonomics*, 12, 13–17.

Karwowski, W., 1991, Complexity, fuzziness, and ergonomic incompatibility issues in the control of dynamic work environments, *Ergonomics*, 34, 671–686.

Kaufman, J. and Christensen, J. (Eds.), 1984, *IES Lighting Handbook*, New York: Illuminating Engineering Society of North America.

Keegan, J.J., 1953, Alterations of the lumbar curve related to posture and seating, *The Journal of Bone and Joint Surgery*, 35A, 589–603.

Keyserling, W.M. and Chaffin, D.B., 1986, Occupational ergonomics: methods to evaluate physical stress on the job, *American Review of Public Health*, 7, 77–104.

Khalid, H.M., 2000, Human factors of virtual collaboration in product design, in K.Y. Lim (Ed.), *Proceeding of APCHI—Asean Ergonomics 2000 Conference*, Amsterdam: Elsevier Science, 25–38.

Khalid, H.M., 2004, Usability evaluation of G2B/licencing portal, Unpublished report, Kuala Lumpur: Damai Sciences.

Kemmlert, K. and Kihlbom, 1986, *Questionnaire for Identifying Musculoskeletal Stress Factors*, Solna, Sweden: National Board of Occupational Health, Work Physiology Unit.

Kieras, D.E., 1984, The psychology of technical devices and technical discourse, in *Artificial Intelligence in Maintenance*, Brooks AFB, TX: Air Force Human Resources Laboratory, 227–254.

Kinkade, R.G. and Wheaton, G.R., 1972, Training device design, in Van Cott, H.P. and Kinkade, R.G. (Eds.), *Human Engineering Guide to Equipment Design*, Washington, DC: U.S. Government Printing Office.

Kirakowski, J., 1996, The software usability measurement inventory (SUMI): background and usage, in Jordan, P., Thomas, B., Weerdmeester, B., and McClelland, I. (Eds.), *Usability Evaluation in Industry*, London: Taylor & Francis, 169–177.

Klein, G.A., 1986, Recognition-primed decisions, in Rouse, W. (Ed.), *Advances in Man-Machine Systems Research*, Greenwich, CT: JAI Press, 47–92.

Klein, G.A., Orasanu, J., Calderwood, R., and Zsambok, E. (Eds.), 1993, *Decision Making in Action*, Norwood, NJ: Ablex.

Knauth, P., 1997, Changing schedules: shiftwork, *Cronobiology International*, 14, 159–171.

Knauth, P., 1998, Innovative worktime arrangements, *Scandinavian Journal of Work, Envirnoment and Health*, 24 (Suppl. 3), 13–17.

Knauth, P., Rohmert, W., and Rutenfranz, J., 1979, Systematic selection of shift plans for continuous production with the aid of workphysiological criteria, *Applied Ergonomics*, 10, 9–15.

Knauth, P., Kiesswetter, E., Ottman, W., Karvonen, M.J., and Rutenfranz, J., 1983, Time-budget studies of policemen in weekly or swiftly rotating shift systems, *Applied Ergonomics*, 14, 247–252.

Kokoschka, S. and Haubner, P., 1985, Luminance ratios at visual display workstations and visual performance, *Lighting Research and Technology*, 17, 138–145.

Konz, S. and Johnson, S., 2004, *Work Design: Occupational Ergonomics*, Scottsdale, AZ: Holcomb Hathaway.

Konz, S., 1992a, Macro-ergonomic guidelines for production planning, in Helander, M. and Nagamachi, M. (Eds.), *Design for Manufacturability*, London: Taylor & Francis, 281–300.

Konz, S., 1992b, Vision of the workplace, part 2, *International Journal of Industrial Ergonomics*, 10, 139–160.

Kroemer, K.H.E., 1989, Engineering anthropometry, *Ergonomics*, 32, 767–784.

Kroemer, K., Kroemer, H., and Kroemer-Elbert, K., 2001, *Ergonomics: How to Design for Ease and Efficiency*, 2nd Edition, Englewood Cliffs, NJ: Prentice-Hall.

Krohn, G.S., Sanders, M.S., and Peay, J., 1984, Work vests (life vests) used for dredge mining, *Proceedings of Human Factors Society Annual Meeting*, Santa Monica, CA: The Human Factors and Ergonomics Society, 478–482.

Krohn, R. and Konz, S., 1992, Best hammer handles, *Proceedings of the Human Factors Society*, Santa Monica, CA: The Human Factors and Ergonomics Society, 413–417.

Kroemer, K.H.E. and Kroemer, A.D., 2002, *Office Ergonomics*, Boca Raton, FL: CRC Press.

Kryter, K.D., 1985, *The Effects of Noise on Man*, 2nd ed., New York, Academic Press.

Knutsson, A., Akerstedt, T., and Orth-Gorn6r, K., 1986, Increased risk of ischemic heart disease in shift workers, *Lancet*, 12, 89–92.

Kuorinka I., Jonsson B., and Kilbom A., 1987, Standardized Nordic questionnaire for the analysis of musculoskeletal symptoms, *Applied Ergonomics*, 18, 233–237.

La Mettrie, J.O., 1748, *L'homme Machine*, Leyden, The Netherlands: Elie Luzac. Translated into English: <www.cscs.umich.edu/crshalizi/LaMattrie/Machine/>.

Landa, K. (Ed.), 2000, *Ergonomic Software Tools in Product and Workplace Design: A Review of Recent Developments in Human Modelling,* Stuttgart: Verlag ERGON GmBH.

Landauer, T.K., 1995, *The Trouble with Computers: Usefulness, Usability, and Productivity,* Cambridge, MA: MIT Press.

Lawrence, J.S., 1955, Rheumatism in coal miners, part 11, *British Journal of Industrial Medicine,* 12, 249–261.

Lehto, M.R., 1992, Design warning signs and warning labels, part II: Scientific basis for initial guidelines, *International Journal of Industrial Ergonomics,* 10, 115–138.

Lehto, M.R. and Miller, S.M., 1986, *Warnings, Vol. 1: Fundamentals, Design, and Evaluation Methodologies,* Ann Arbor, MI: Fuller Technical.

Lewis, C. and Wharton, C.,1997, Cognitive walkthrough, in Helander, M.G., Landauer, T.K., and Prabhu, P. (Eds.), *Handbook of Human-Computer Interaction,* Amsterdam: Elsevier Science, 717–731.

Lim, K.Y., 2004, A simulation band analysis of the field of view afforded to pilots, *Proceedings of the 7th International Conference on Work with Computing Systems,* Kuala Lumpur: Damai Sciences.

Liberty Mutual, 2004, *Workplace Safety Index,* <http://www.libertymutual.com/>.

Lindh, M., 1980, Biomechanics of the lumbar spine, in Frankel, V.H. and Nordin, M. (Eds.), *Basic Biomechanics of the Skeletal System,* Philadelphia, PA: Lea & Febiger.

Little, Steven, Personal communication, Maya Group, Inc., 2002.

Locke, E.A., 1983, The nature and causes of job satisfaction, in Dunnette, M.D. (Ed.), *Handbook of Industrial and Organizational Psychology,* New York: Wiley.

Loeb, M., 1986, *Noise and Human Efficiency,* Chichester: Wiley.

Loewenthal, A. and Riley, M.W., 1980, The effectiveness of warning labels, in *Proceedings of the 24th Annual Meeting of the Human Factors Society,* Santa Monica, CA: The Human Factors and Ergonomics Society, 389–391.

Luczak, H., 1997, Task ananylsis, in Salvendy, G. (Ed.), *Handbook of Human Factors and Ergonomics,* 2nd ed., New York: Wiley, 340–416.

Luczak, H., 1993, *Arbeitswissenschaft,* Berlin: Springer-Verlag.

Mackie, R.R. (Ed.), 1979, *Vigilance: Theory, Operational Performance, and Physiological Correlates,* New York: Plenum Press.

Mackie, R.R., O'Hanlon, J.F., and McCauley, M.E., 1974, *A Study of Heat, Noise, and Vibration in Relation to Driver Performance and Physiological States,* Technical Report 1735, Goleta, CA: Human Factors Research, Inc.

Magora, A., 1974, Investigation of the relation between low back pain and occupation, *Scandinavian Journal of Rehabilitation Medicine,* 6, 81–88.

Mann, C.C., 2002, Why software is so bad, *Technology Review,* July-August, 33–38.

Margolis, W. and Kraus, S.F., 1987, The prevalence of carpal tunnel syndrome symptoms in female supermarket checkers, *Journal of Occupational Medicine,* 29, 953–959.

Maxion, R., 1984, Artificial intelligence approaches to monitoring systems integrity, in *Artificial Intelligence in Maintenance,* Brooks AFB, TX: Air Force Human Resources Laboratory, 257–273.

McConville, J.T., Robinette, K.M. and Churchill, T., 1981, *An Anthropometric Data Base for Commercial Design Applications,* Yellow Springs, OH: Anthropology Research Project.

McGill, S.M. and Norman, R.W., 1986, Dynamically and statistically determined low back movement during lifting, *Journal of Biomechanics,* 18, 877–885.

McGrath, J.J., 1976, *Driver Expectancy and Performance in Locating Automotive Controls,* SAE Report SP-407, Santa Barbara, CA: Anacapa Sciences, Inc.

Meister, D., 1971, *Human Factors: Theory and Practice*, New York: Wiley.

Mellor, E.F., 1986, Shift work and flextime: how prevalent are they? *Monthly Labor Review*, 109, 14–21.

Meshkati, N., 1989, An etiological investigation of micro- and macroergonomic factors in the Bhopal disaster: lessons for industries of both industrialized and developing countries, *International Journal of Industrial Ergonomics*, 4, 161–175.

Michael, E.D., Hutton, K.E., and Horvath, S.M., 1961, Cardiorespiratory responses during prolonged exercise, *Journal of Applied Physiology*, 16, 997–1000.

Michel, D.P. and Helander, M.G., 1994, Effect of two types of chairs on stature change and comfort for individuals with healthy and herniated discs, *Ergonomics*, 37, 1231–1244.

Michie, S., 2002, Causes and management of stress at work, *Occupational and Environmental Medicine*, 59, 67–72.

Militello, L.G. and Hutton, R.J.B., 1998, Applied Cognitive Task Analysis (ACTA): a practitioner's toolkit for understanding cognitive task demands, *Ergonomics*, 41, 1618–1641.

Miller, G.A., 1956, The magical number seven plus or minus two: some limits on our capacity for processing information, *Psychological Review*, 63, 81–97.

Miller, J.C., 1992, *Fundamentals of Shift Work Scheduling*, Lakeside, CA: Evaluation Systems, Inc.

Miller, J.C. and Horvath, S.M., 1981, Work physiology, in Helander, M.G. (Ed.), *Human Factors/Ergonomics for Building and Construction*, New York: Wiley.

Mital, A. (Eds), 1991, Economics of flexible assembly automation: influence of production and market factors, in Parsaei, H.R. and Mital, A. (Eds.), *Economic Aspects of Advanced Production and Manufacturing Systems*, London: Chapman & Hall.

Mital, A., Nicholson, A.S., and Ayoub, M.M., 1993, *A Guide to Manual Materials Handling*, London: Taylor & Francis.

Monk, T.H., 1986, Advantages and disadvantages of rapidly rotating shift schedules—a circadian viewpoint, *Human Factors*, 28, 553–557.

Monk, T.M. and Folkard, S., 1992, *Making Shift Work Tolerable*, London: Taylor & Francis.

Moray, N., 1986, Monitoring behavior and supervisory control, in Boff, K.R., Kaufman, L., and Thomas, J.P. (Eds.), *Handbook of Perception and Human Performance*, New York: Wiley.

Montemerlo, M.D. and Eddower, E., 1978, The judgmental nature of task analysis, in *Proceedings of the 22nd Annual Meeting of the Human Factors Society*, Santa Monica, CA: The Human Factors and Ergonomics Society, 247–250.

Morris, N.M. and Rouse, W.B., 1984, *Review and Evaluation of Empirical Research in Troubleshooting*, Report 8402-1, Norcross, GA: Search Technology, Inc.

Muller, M.J., Haslwanter, J.H., and Dayton, T., 1997, Participating Practices in the Software Lifecycle, in: Helander, M.G., Landauer, T.K., and Prabhu, P. (Eds.), *Handbook of Human-Computer Interaction*, Amsterdam: Elsevier Science.

Nagamachi, M. and Yamada, Y., 1992, Design for manufacturability through participatory ergonomics, in Helander, M. and Nagamachi, M. (Eds.), *Design for Manufacturability*, London: Taylor & Francis, 219–229.

National Aeronautics and Space Administration, 1978, *Anthropometry Source Book, Volume II: A Handbook of Anthropometric Data*, Houston, TX: NASA.

National Institute for Occupational Safety and Health, 1989, *NIOSH Criteria for a Recommended Standard: Occupational Exposure to Hand-Arm Vibration*, Report DHHS-NIOSH Publ. 89-106, Cincinnati, OH: NIOSH.

National Institute for Occupational Safety and Health, 1992, *Health Hazard Evaluation Report*, HETA 89-299-2230, Cincinnati, OH: NIOSH.

National Academy of Sciences, 2003, *Musculoskeletal Disorders and the Workplace*, Washington, D.C.: National Academy Press.

National Research Council, 1983, *Video Displays, Work, and Vision*, Washington, DC: National Academy Press.

Neal, A.S., 1977, Time intervals between keystrokes, records, and fields in data entry with skilled operators, *Proceedings of the Human Factors Society*, 163–168, Santa Monica, CA: The Human Factors Society.

Nichols, D.L., 1976, Mishap analysis, in Ferry, T.S. and Weaver, D.A. (Eds.), *Directions in Safety*, Springfield, IL: Charles C. Thomas.

Nielsen, J., 1994a, Heuristic evaluation, in Nielsen, J. and Mack, R.L. (Eds.), *Usability Inspection Methods*, New York: Wiley.

Nielsen, J., 1994b, *Usability Engineering*, San Francisco: Morgan Kaufmann.

Nielsen, J. and Molich, R., 1990, Heuristic evaluation of user interfaces, *Proceedings of ACM CHI'90 Conference on Human Factors in Computing Systems*, New York: ACM, 249–256.

Norman, D.A., 2004, *Emotional Design*, New York: Basic Books.

Nylen, P., 2002, Comparison of stationary LCD and CRT screens—some visual and musculoskeletal aspects, in Luczak, H., Cakir, A.E., and Cakir, G. (Eds.), *Proceedings of WWDU 2002*, Berlin: Ergonomic Institute fur Arbeits- und Socialforschung Forschungsgesellschaft mbH, 682–684.

Olsen, O. and Kristensen, T.S., 1992, *Impact of Work Environment on Cardiovascular stress in Denmark*, Journal Epidemiology and Community Health, 45, 4–10.

Östberg, O., 1976, *Design of VDT Workplaces* (in Swedish), Stockholm: Liber Förlag.

Östberg, 0., 1980, Accommodation and visual fatigue in display work, in Grandjean, E. and Vigliani, E. (Eds.), *Ergonomic Aspects of Visual Display Terminals*, London: Taylor & Francis.

Parasuraman, R., 1985, Detection and identification of abnormalities in x-rays: effects of reader skills, disease prevalence, and reporting standard, in Eberts, R.E. and Eberts, C.G. (Eds.), *Trends in Ergonomics/Human Factors II*, Amsterdam: North Holland, 59–66.

Parrish, R.N., Gates, J.L., Munzer, S.J., Grimma, P.R., and Smith, L.T., 1982, *Development of Design Guidelines and Criteria for User-Operator Transactions with Battlefield Automated Systems, Final Report*, Alexandria, VA: U.S. Army Research Institute for the Behavioral and Social Sciences.

Parsons, K.C., 1993, *Human Thermal Environments*, 2nd ed., London: Taylor & Francis.

Patrick, J., 1992, *Training: Research and Practice*, San Diego, CA: Academic Press.

Pheasant, S.T., 1998, *Bodyspace, Anthropometry, Ergonomics and Design*, London: Taylor & Francis.

Pheasant, S.T. and Stubbs, D., 1992, *Lifting and Handling: An Ergonomic Approach*, London: National Back Pain Association.

Poulton, E., 1978, A new look at the effects of noise: a rejoinder, *Psychological Bulletin*, 85, 1068–1079.

Poulton, E.C., 1979, *The Environment at Work*, Springfield, IL: Charles C. Thomas.

Prabhu, G.V., Helander, M.G., and Shalin, V., 1992, Cognitive implications product structure on manual assembly performance, in Brödner, P. and Karwowski, W. (Eds.), *Ergonomics of Hybrid Automation*, Amsterdam: Elsevier, 259–272.

Purswell, J.L., Krenek, R.F., and Dorris, A., 1987, Warning effectiveness: what we need to know, in *Proceedings of the Human Factors Society 31st Annual Meeting*, Santa Monica, CA: The Human Factors and Ergonomics Society, 1116–1120.

Putz-Anderson, V. (Ed.), 1988, *Cumulative Trauma Disorders: A Manual for Musculoskeletal Diseases of the Upper Limbs*, London: Taylor & Francis.

Putz-Anderson, V. (Ed.), 2005 (in press), *Cumulative Trauma Disorders: A Manual for Musculoskeletal Diseases of the Upper Limbs*, London: Taylor & Francis.

Putz-Anderson, V. and Waters, T.R., 1991, Revision in NIOSH guide to manual lifting, presentation at the Conference on a National Strategy for Occupational Musculoskeletal Injury Prevention Implementation Issues and Research Needs, Cincinnati, OH: National Institute of Occupational Safety and Health.

Rahimi, M. and Hancock, P.A., 1988, Sensor integration, in Dorf, N. (Ed.), *International Encyclopedia of Robotics*, New York: Wiley.

Rahimi, M. and Karwowski, W. (Eds.), 1993, *Human-Robot Interaction*, London: Taylor & Francis.

Ramazzini, B., 1940 (1717), Wright, W. (Trans.), *The Disease of Workers*, Chicago, IL: University of Chicago Press.

Rasmussen, J., 1983, Skills, rules, knowledge: signals, signs and and symbols and other distinctions in human performance models, *IEEE Transactions on Systems, Man and Cybernetics*, SMC-13(3), 257–267.

Rasmussen, J., 1986, *Information Processing and Human Machine Interaction: An Approach to Cognitive Engineering*, Amsterdam: North Holland.

Rasmussen, J, 1990, A model for the design of computer integrated manufacturing systems: identification of information requirements of decision makers, *International Journal of Industrial Ergonomics*, 5, 5–16.

Ray, R.D. and Ray, W.D., 1979, An analysis of domestic cooker control design, *Ergonomics*, 22, 1243–1248.

Rea, M.S. (Ed.), 2000, *Lighting Handbook: Reference and Application*, 9th ed. New York, NY: Illuminating Engineering Society of North America.

Reason, J., 1990, *Human Error*, Cambridge University Press, Cambridge, UK.

Reason, J. 1997, *Managing the Risks of Organizational Accidents*, Brookfield, VT, Ashgate Publishing Limited.

Redmill, F. and Rajan, J. (Eds.), 1997, *Human Factors in Safety-Critical Systems*, Butterworth Heinemann, London.

Rhea, J.T., Potsdaid, M.S., and DeLuca, S.A., 1979, Errors of interpretation as elicited by a quality audit of an emergency facility, *Radiology*, 132.

Rick, J., Thomson, L., Briner, R.B., O'Regan, S., and Daniels, K., 2002, *Review of Existing Supporting Knowledge to Underpin Standards for Key Work-Relates Stressors, Phase 1*, London: Health and Safety Executive.

Robinette, K., Blackwell, S., Daanen, H., Fleming, S., Boehmer, M., Brill, T., Hoeferlin, D., and Burnsides, D., 2002, *Civilian American and European Surface Anthropometry Resource (CAESAR), Final Report, Volume I: Summary*, AFRL-HE-WP-TR-2002-0169, Wright-Patterson AFB, OH: Human Effectiveness Directorate, Crew System Interface Division.

Robotic Industries Association, 1987, *American National Standard R15.06 for Industrial Robots and Robot Systems Safety Requirements*, Ann Arbor, MI: Robotic Industries Association.

Robotic Industries Association, 1989, *Proposed American National Standard of Human Engineering Design Criteria for Hand Held Control Pendants*, Ann Arbor, MI: Robotic Industries Association.

Roebuck, J.A., 1995, *Anthropometric Methods: Design to Fit the Human Body*, Santa Monica, CA: The Human Factors and Ergonomics Society.

Roebuck, J.A., Kroemer, K.H.E., and Thomson, W.G., 1975, *Engineering Anthropometry Methods*, New York: Wiley.

Rosson, M-B., and Carroll, J., 2000, *Usability Engineering*, San Francisco, CA: Morgan Kaufmann Publishers.

Rohmert, W. and Luczak, H., 1978, Ergonomics in the design and evaluation of a system for postal video letter coding, *Applied Ergonomics*, 9, 85–95.

Rosenbrock, H.N., 1983, Seeking an appropriate technology, in *Proceedings of IFAC Symposium on Systems Approach to Appropriate Technology Transfer*, Vienna: IFAC.

Rubin, J.,1994, *Handbook of Usability Testing*, New York: Wiley.

Rutenfranz, J., Haider, M., and Koller, M., 1985, Occupational health measures for night-workers and shiftworkers, in Folkard, S. and Monk, T.H. (Eds.), *Hours of Work: Temporal Factors in Work Scheduling*, New York, Wiley, 199–210.

Safir, A. (Ed.), 1980, *Refraction and Clinical Optics*, Hagerstown, PA: Harper & Row.

Salmoni, A.W., Schmidt, R.A., and Waller, C.B., 1984, Knowledge of results and motor learning: a review and critical appraisal, *Psychological Bulletin*, 95, 355–386.

Sanders, M.A., 1980, Personal communication, Canyon Research Group, Westlake Village, CA, U.S.

Sanders, M.S. and McCormick, E.J., 1993, *Human Factors in Engineering and Design*, New York: McGraw-Hill.

Sarter, N.B. and Woods, D.D., 1995. How in the world did we ever get in that mode? Mode error and awareness in supervisory control, *Human Factors*, 37, 5–19.

Sauter, S.L., Chapman, L.S., and Knutson, S.J., 1985, *Improving VDT Work*, Madison, WI: Department of Preventive Medicine, University of Wisconsin.

Shannon, C.E. and Weaver, W., 1949, *The Mathematical Theory of Communication*, Urbana, IL: University of Illinois Press.

Scherrer, J., 1981, Man's work and circadian rhythm through the ages, in Reinberg, A., Vieux, N., and Andlauer, P. (Eds.), *Night and Shift Work: Biological and Social Aspects*, Oxford: Pergamon Press, 1 –10.

Schoenmarklin, R. and Marras, W., 1989, Effect of hand angle and work orientation on hammering: II. Muscle fatigue and subjective ratings of body discomfort, *Human Factors*, 31, 413–420.

Scholey, M. and Hair, M., 1989, Back pain in physiotherapists involved in back care education, *Ergonomics*, 32, 179–190.

Sen, R.S., 1989, Personal communication, SUNY Buffalo, Buffalo, NY, U.S.

Shalin, L., Prabhu, G.V., and Helander, M.G., 1995, *A cognitive perspective of manual assembly, Ergonomics*, 38, 2007–2029.

Shaw, L. and Sichel, H., 1971, *Accident Proneness: Research Occurrence, Causation and Prevention of Road Accidents*, Pergamon Press, London.

Sheridan, T.B., 2002, *Humans and Automation: System Design and Research Issues*, New York: Wiley.

Shute, S.J. and Starr, S.J., 1984, Effects of adjustable furniture on VDT users, *Human Factors*, 26, 157–170.

Silverstein, B.A., Fine, L.J. and Armstrong, T.J., 1987, Occupational factors and carpal tunnel syndrome, *American Journal of Industrial Medicine*, 11, 343–358.

Simon, H.A., 1996, *The Science of the Artificial*, 3rd ed., Cambridge, MA: MIT Press.

Simon, H.A., 1974, How big is a chunk? *Science*, 183, 482–488.

Singleton, W.T., 1962, *Ergonomics for Industry*, London: Department of Scientific and Industrial Research.

Slovic, P. and Lichtenstein, S., 1988, Decision making, *Steven's Handbook of Experimental Psychology*, 2, 673–728.

Smeed, R. J., 1953, The international comparison of accident rates, *International Road Safety and Traffic Review*, 1 (1), 43–52.

Smeed, R.J., 1968, Variations in the pattern of accident rates in different countries and their causes, *Traffic Engineering and Control*, 10 (7), pp 364–371.

Smeed, R.J., 1974, The frequency of road accidents, *Zeitschrift für Verkehrssicherheit*, 20 (2), 151–159.

Smith, M., and Cohen, W.J., 1997, Design of computer terminal workstations, in Salvendy, G. (Ed.), *Handbook of Human Factors and Ergonomics*, 2nd ed., 1637–1688, Wiley, New York.

Smith, I., 1999, Road fatalities, model split and Smeed's Law, *Applied Economic Letters*, 6, 215–217.

Snyder, H.L., 1985, Class notes from Virginia Polytechnic University, Department of Industrial Engineering, Blacksburg, VA.

Snyder, H.L., 1988, Image quality, in Helander, M.G. (Ed.), *Handbook of Human-Computer Interaction*, Amsterdam: North Holland.

Sperry, W., 1978, Aircraft and airport noise, in Lipscomb, D. and Taylor, A. (Eds.), *Noise Control: Handbook of Principles and Practices*, New York: Van Nostrand Reinhold.

Swain, A.D. and Guttman, H.E., 1980, *Handbook of Human Reliability Analysis with Emphasis on Nuclear Power Plant Applications*, NUREGICR-1278, Washington, DC: U.S. Nuclear Regulatory Commission.

Swedish Work Environment Fund, 1985, *Making the Job Easier: An Idea Book*, Stockholm: SWEF.

Swets, J.A., Pickett, R.M., Whitehead, S.F., Getty, D.J., Schnur, J.A., Swets, J.B., and Freeman, B.A., 1979, Assessment of diagnostic technologies, *Science*, 205, 753–759.

TechSmith, 2004, *Camtasia*, <http://www.techsmith.com/>.

Tepas, D.I. and Monk, T.H., 1986, Work schedules, in Salvendy, G. (Ed.), *Handbook of Human Factors*, New York: Wiley.

T.G. and R.L., 1975, Conveyor belt sickness, *National Safety News*, 117, 37.

Tichauer, E.R., 1966, Some aspects of stress on forearm and hand in industry, *Journal of Occupational Medicine*, 8, 63–71.

<http://www.asktog.com/basics/firstPrinciples.html>

Tullis, T.S., 1997, Screen design, in Helander, M., Landauer, T.K., and Prahbu, P.V. (Eds.), *Handbook of Human Computer Interaction*, 2nd ed., Amsterdam: North Holland.

U.S. Department of Defense, 2002, *Military Standard 1472F*, Washington, DC: U.S. Department of Defense.

U.S. Department of Energy, 1995, *Energy Efficient Lighting*, GOE/GO-10095-056, Washington, D.C.: U.S. Department of Energy, National Renewable Energy Laboratory.

U.S. Department of Labor, 1980, *Noise Control: A Guide for Workers and Employers*, Washington, DC: USDOL/OSHA.

U.S. Department of Labor, 1982, *Back Injuries Associated with Lifting*, Bulletin No. 2144, Washington, DC: Bureau of Labor Statistics.

Van Cott, H.P. and Kinkade, R.G., 1972, *Human Engineering Guide to Equipment Design*, Washington, DC: U.S. Government Printing Office.

Van Wely, P., 1970, Design and disease, *Applied Ergonomics*, 1, 262–269.

Van Welie, M., van der Veer, G.C. and Eliëns, A., 1999, Breaking down usability, in *Proceedings of Interact 99, 30th August–3rd September 1999, Edinburgh, Scotland*, <http://www.cs.vu.nl/~martijn/gta/docs/Interact99.pdf>.

Verplank, W.L., 1988, Graphic challenges in designing object-oriented user interfaces, in Helander, M.G. (Ed.), *Handbook of Human Computer Interaction*, Amsterdam: North Holland.

Vincente, K.J., 1999, *Cognitive Work Analysis*, Mahwah, NJ: Lawrence Erlbaum.

Virzi, R.A., Source, J.E., and Herbert, L.B., 1993, A comparison of three usability evaluation methods: heuristic, think-aloud, and performance testing, *Proceedings of the Human Factors and Ergonomics Society, 37th Annual Meeting*, Santa Monica, CA: Human Factors and Ergonomic Society, 309–313.

Vora, P.R., Helander, M.G., and Shalin, V.L., 1994, Evaluating the influence of interface styles and multiple access paths in hypertext, *Proceedings of CHI 94*, New York: ACM, 323–329.

Wang, M.J., Wang, E.M., and Lin, Y., 2001, *Anthropometric Data Book of the Chinese People in Taiwan*, Hsinchu: National Ergonomics Society of Taiwan, National Tsinghua University.

Ward, W.D., 1976, Transient changes in hearing, in *Proceedings of the International Congress on Man and Noise*, Turin: Edizioni Minerva Medica, 111–122.

Waters, T.R., Putz-Anderson, V., Garg, A., and Fine, L.J., 1993, Revised NIOSH regulation for the design and evaluation of manual tasks, *Ergonomics*, 36, 749–776.

Webb, R.D.G., 1982, *Industrial Ergonomics*, Toronto: Industrial Accident Prevention Association.

Webb, L.H. and Parsons, K.C., 1998, Case studies of thermal comfort for people with physical disabilities, *ASHRAE Transactions*, 104 (1), 1–12.

Webster, J., 1969, *Effects of Noise on Speech Intelligibility*, ASHA Report 4, Washington, DC: American Speech and Hearing Association.

Weinstein, N., 1977, Noise and intellectual performance: a confirmation and extension, *Journal of Applied Psychology*, 62, 104–107.

Welford, A.T., 1968, *Fundamentals of Skills*, London: University Paperback.

White, B. and Samuelson, P., 1990, Repetitive motion trauma in automotive parts manufacturing, in Karwowski, W. and Rahimi, M. (Eds.), *Ergonomics of Hybrid Automated Systems*, Vol. 2, Amsterdam: Elsevier.

Whitefield, A., 1986, Human factors aspects of pointing as an input technique in interactive computer systems, *Applied Ergonomics*, 17, 97–104.

Whitestone, J. and Robinnette, K., 1997, Fitting to maximize Performance of HMD Systems, in Melzer, E. and Moffit, K. (Eds.), *Head Mounted Displays: Designing for the User*, McGraw-Hill, New York.

Wickens, C.D. and Hollands, J.G., 2000, *Engineering Psychology and Human Performance*, 3rd ed. New York: Prentice Hall.

Wiener, E.L. and Nagel, D.C. (Eds.), 1988, *Human Factors in Aviation*, San Diego, CA: Academic Press.

Wilson, J.R. and Corlett, E.N. (Eds.), 2002, *Evaluation of Human Work*, 2nd ed., London: Taylor & Francis.

Wiener, E., Kanki, B., and Helmreich, R., 1993 (Eds.), *Cockpit Resource Management*. San Diego, CA: Academic Press.

Winkel, J., 1990, Personal communication, Buffalo, NY, U.S.

Wogalter, M.S., Desaulniers, D.R., and Godfrey, S.S., 1985, Perceived effectiveness of environmental warnings, in *Proceedings of the 29th Annual Meeting of the Human Factors Society*, Santa Monica, CA: The Human Factors and Ergonomics Society, 664–669.

Wogalter, M.S., Laughery, K.R., and Young, S.L., 2002, *Human Factors Perspectives on Warnings: Selections from Human Factors and Ergonomics Society Annual Meetings, 1994–2000*, Santa Monica, CA: Human Factors and Ergonomics Society.

Woodson, W.E., 1981, *Human Factors Design Handbook*, New York: McGraw-Hill.

Woodson, W.E. and Conover, D.W., 1964, *Human Engineering Guide to Equipment Design*, Berkeley, CA: University of California Press.

Wotton, E., 1986, Lighting the electronic office, in Lueder, R. (Ed.), *The Ergonomics Pay Off: Designing the Electronic Office*, Toronto: Holt, Rhinehart & Winston.

Wright, G. and Rea, M., 1984, Age, a human factor in lighting, in *Proceedings of the 1984 International Conference on Occupational Ergonomics*, Rexdale, Ontario: Human Factors Association of Canada, 508–512.

Yerkes, R.M. and Dodson, J.D., 1908, The relation of strength of stimulus to rapidity of habit-formation, *Journal of Comparative Neurology of Psychology*, 18, 459–482.

Zandin, K.B., 1990, *MOST Work Measurement System*, New York: Marcel Dekker.

Zenz, C., 1981, Physical health hazards in construction, in Helander, M.G. (Ed.), *Human Factors/Ergonomics for Building and Construction*, New York: Wiley.

Zhang, G., and Simon, H.A., 1985, STM capacity for Chinese words and idioms: chunking and acoustical loop hypotheses, *Memory and Cognition*, 13, 193–201.

Ziegler, J.E. and Fähnrich, K.-P., 1988, Direct manipulation, in Helander, M. (Ed.), *Handbook of Human Computer Interaction*, Amsterdam: North Holland, 123–133.

Zimolong, B., 1985, Hazard perception and risk estimation in accident causation, in Eberts, R.E. and Eberts, C.G. (Eds.), *Trends in Ergonomics/Human Factors*, Vol. 2, Amsterdam: North Holland, 463–470.

Ziefle, M., 2001, Aging, visual performance and eyestrain in different screen technologies, *Proceedings of the Human Factors and Ergonomics Society 45th Annual Meeting*, Santa Monica, CA: The Human Factors and Ergonomics Society, 262–266.

Zipp, von P., Haider, E., Halpern, N., Mainzer, J., and Rohmert, W., 1981, Untersuchung zur ergonomischen Gestaltung von Tastaturen, *Zentralblatt für Arbeitsmedizin, Arbeitsschutz, Prophylaxe und Ergonomi*, 31, 326–330.

Zwaga, H. and Easterby, R., 1982, Developing effective symbols for public information: the ISO testing procedure, in *Proceedings of the 1982 Congress of the International Ergonomics Association*, Santa Monica, CA: The Human Factors and Ergonomics Society, 512–513.

APPENDIX
The Use of Human Factors/ Ergonomics Checklists

There are two main applications of checklists:

1. As a memory aid during inspection of workplaces
2. As a tool for systematic data collection

CHECKLIST AS A MEMORY AID

In the case of a memory aid, we assume that during a workplace inspection it may be difficult to think of and remember important design details. A checklist makes it possible to cover all the important ergonomic issues systematically. However, there is a danger in using checklists. Just as in the case of task analysis, there is no fixed method.

The checklist presented below is problem oriented and intended to inspire design improvements. It can be used as a basis for discussions at work, in groups or between individuals. This can lead to many innovations and improvements of both the task and the workplace.

CHECKLIST FOR SYSTEMATIC INVENTORY OF ERGONOMICS PROBLEMS

The other application is to use a checklist as a tool for the systematic investigation of workplaces. Imagine, for example, that there are 200 computer workstations in an office. It may be of interest to collect statistics on the ergonomic design features associated with such a workplace. Issues of adjustability, work posture, and illumination may be particularly important. In this case, the statistics gathered can serve as a management tool for evaluating how many workstations need refurbishing and how many are in good condition. The checklist can also be used to produce a list of priorities for upgrading workstations. It would then be possible to predict the cost of certain types of upgrade and to compare this with the expected benefits.

DEVELOPING A CHECKLIST

The types of items included in the checklist depend on the application. One would devise a very different checklist for automated manufacturing than for manually oriented manufacturing. It is therefore necessary to develop a checklist that is suitable for the purpose. One should not accept uncritically the checklist supplied below. It needs to be complemented to cover the items that are of particular importance for the environment being investigated. It may be a matter of negotiation between management, workers, and labor unions to decide which items should be included. It is also possible that a checklist could be developed in a quality circle or by an ergonomics task force. Furthermore, checklists can have an emphasis on or a bias toward different issues, such as environmental hygiene, safety and injuries, mental workload, productivity, or operator comfort and convenience. Thus, the items on the checklist will depend on the criteria under evaluation.

Checklists can be made more detailed. Instead of a checkmark there could be an evaluation on a scale of, say, one to five, or the checklist could be complemented with a questionnaire so that workers themselves do the checking.

A somewhat less ambitious approach is the use of a survey checklist (Chengalur, Rodgers, and Bernard, 2004). This type of a checklist is not as complete as many others that have been developed for ergonomic surveys. Rather, a survey checklist is more generic and problem oriented. The intention is that the list of items could lead to further discussion and a more detailed analysis.

SOME PRECAUTIONS

Improvements in HFE come at a price. It is therefore important to discuss the costs and benefits of proposed improvements. A work environment is rarely 100% perfect; it would be too costly to make it perfect. A cost–benefit analysis would be helpful in identifying what projects should be performed first, and which ones should be delayed or scrapped. This methodology is commonly used for prevention of road accidents. How many accidents will be prevented if, say, a section of a curvy road is straightened out, or if an intersection is made safer by using traffic lights? In the road safety scenario, there are often accident statistics available that can help to answer such questions. But similar statistics of accidents and injuries are rarely available for a manufacturing plant or an office. Nonetheless, one can propose a list of priorities based on assumptions of improvements—or based on ideas that emerge from a group discussion.

There is a final precaution. Checklists are superficial instruments. They do not analyze the problem at hand, which one must do in order to solve design problems. A checklist is intended to act as a reminder or to inspire a discussion. Once a problem has been noted, the work starts: analysis, alternative solutions, benefits, costs, and proposals.

ERGONOMIC CHECKLIST TO ENHANCE PERFORMANCE, SAFETY, AND COMFORT

   A. Physical Demands
      Are the hands at a convenient working height for the task?
      Are the joints mostly in a convenient neutral position?

Are the wrists mostly in a straight, neutral posture? Can the operator assume several different postures while working?

Is this a dynamic rather than a static task?

Can the task be performed with the torso and the head facing forward?

Are primary items located within easy reach?

Is frequent lifting below 20 kg (45 lb)?

Is occasional heavy lifting less then 25 kg (55 1b)?

Are items to be lifted positioned between knuckle and shoulder height?

Are there convenient aids for manual materials handling?

Are there handles on items which are otherwise difficult to lift?

Are hand tools appropriate for the task?

Are hand tools comfortable and safe to use?

For sitting tasks:

    Are the feet firmly supported on the floor or with a footrest?

    Can the backrest be utilized while performing the task?

    Are the elbow joints mostly at an intermediate angle?

    Are primary items located within easy (5th percentile) reach (about 40 cm)?

    Is the head bent slightly forward, rather than backward?

B. Task Visibility

Are displays and dials easy to see from the normal work position?

Is printed or displayed text large enough for reading? The characters should have a size of about 18–25 min of arc.

Are eyeglasses appropriate for the task viewing distance?

Is the illumination level uniform throughout working area?

Are illumination levels appropriate (about 500 lux for VDT work; about 1000 lux for coarse assembly; about 2000 lux for fine assembly)?

Is direct glare from illumination sources and windows avoided?

Is indirect (reflected) glare avoided?

Is the luminance contrast ratio in the immediate task area less than 20:1?

C. Mental Demands

Does the task involve moderate rather than high short-term memory load?

Does the task involve few simultaneous factors, rather than several?

Is operator performance unpaced rather than paced by the task?

Is the task varying, rather than repetitive and monotonous?

Can operator errors and slips be corrected easily?

Are special memory aids used?

Do displays and controls follow population stereotypes?

Is the task easy to learn, rather than difficult?

D. Machine Design

Are tasks appropriately allocated between operators and machines?

Are manual controls easy to reach?

Are manual controls easy to distinguish from each other?

Are all machine functions and displays visible to the operator?
Can machine functions be handled through one command/control?
Are all controls on the machine necessary for the job?
Are the locations of controls and tools the same for similar machines?
Are memory aids used for difficult task information?
Is it possible to operate the machine without bending, twisting, and excessive reaching?
Is there adequate body clearance for handling and maintenance tasks?
Are machine symbols and icons readily understood?
Are labels used to inform and remind operators of task information?
Are labels/symbols used to designate locations for frequently used items?

E. Tasks at Computer Workstations
Are screens positioned perpendicular to windows?
Can reflected glare on the screen be avoided?
Is the display located below a horizontal plane through the eyes?
Do the locations of display, documents, and keyboard make it possible to sit straight without twisting the body?
Is a QWERTY keyboard used?
Are software functions understood and easy to use?
Are software functions and computertask routines easy to access?

F. Safety
Are there appropriate warning signs as a reminder of task hazards?
Is the wording on warning signs relevant and informative?
Are warning signs positioned where operators look?
Is the workplace organized and clean with excellent housekeeping?
Are the floors even without drains or pit marks?
Is it possible to perform the task without safety glasses or protective clothing?
Has the company established safety procedures and rules?
Are safety rules and procedures prioritized by management and enforced?
Does the company analyze each reported accident or injury to improve safety?
Do newly hired workers receive safety training?
Do safety training programs present relevant, task-specific information?
Are potential hazards clearly visible from the operator's position?
Have machine safety devices been installed (e.g., lockouts and guards)?

G. Ambient Environment
Is the ambient noise level below 85 dBA to protect against hearing damage?
Is the ambient noise level below 55 dBA to facilitate verbal communication?
Is there a program to reduce noise pollution by redesigning machines and the work environment?
Are vibration levels and frequencies low enough so as not to affect job performance?
Are the temperature and humidity within a comfortable range?

Is it possible to perform work tasks without protective equipment?
Can all work tasks be performed without risk of electric shock?

H. Product and Process Design

Has the product design been modified to improve productivity?
Has the product design been modified to create better jobs?
Have machines been selected that maximize productivity?
Have machines been selected that maximize operator convenience?
Have processes been located so as to improve productivity?
Have processes been located to improve operator convenience?
Have machines and processes been selected to optimize task allocation between operators and machines?

Index

A

Abbreviation strategies, 122
Abstraction levels, 285–286, *286*
Accessibility, design, 321, *321–323*
Access ports, 321, *321*
Accidents, *see also* Errors; Safety
    basics, 325
    chain of events, 331
    developmental factors, *327,* 327–329, *329*
    energy exchange model, 330–331
    interactive model, 332–333
    models and definitions, 329–333
    prevention programs, *342*
    proneness, 8, *334,* 334–335
    social factors, *327,* 327–329, *329*
    statistical interpretation, 325–326, *326*
    systems safety, 331–332, *332–333*
Acclimation and acclimatization, 231
Accommodation of eye, 42–43, *56*
Acromion height, 153
ACTA, *see* Applied cognitive task analysis
    (ACTA)
Active processing and attention, 116
Actuation force, controls, 94, *95*
ADA, *see* Americans with Disabilities Act (ADA)
Adenosine diphosphate (ADP), 225
Adenosine triphosphate (ATP), 225–226
ADP, *see* Adenosine diphosphate (ADP)
Adrenaline, 36
Aerobic process, 226
Aesthetics, user experience, 128
Affirmative statements, 114–115
Africa, 9, 327
Aging eye, 44, 55–58, *56–59*
Agreement and comprehension, 116–117
Aids, material handling, 203–204, *206*
Air Florida flight, 336–337
Air India Airlines flight, 336, 339
Airplanes and pilots
    anthropometric measurements, 153, 156
    cockpit, *132*
    cognitive walkthrough, 130, *132*
    communication breakdown, 336–337
    control-response compatibility, 103
    design, 5–6
    memory lapses, 336

reference points, 156, *157*
situation awareness, 86
slips, 338
training, 9, 339
visual discriminability, 335–336
Alternative actions, decisions, 80
Ambient environment, 13, 370–371
Americans, 104–105, *105, see also* United States
Americans with Disabilities Act (ADA), 162
Anaerobic process, 226
Analysis
    data, 38, *38–39*
    noise, 242, *243*
Annoyance, noise, 248
Anthropometric workstation design
    basics, 147, *148*
    developments, 160–162, *161*
    disabled employees, 162, *163–164*
    human dimensions measurements, 147–150,
        149, 151–152, 152–153
    measures defined, 153–154, *154–155,*
        156–158, *157*
    procedures, *154, 158–160*
    standards, 162–163, 165, *165*
    three-dimensional models, 160–162, *161*
Apple computers, 119, 126
Applied cognitive task analysis (ACTA), 288–291
Appropriateness, manual controls, *68,* 94–95, *95*
Arithmetic calculations, ignoring, 77
Arm movements, energy inefficiency, 226, *227*
Arm reach, 161
Arm rests, *261,* 263–264
ASHRAE, 232
Asia, 9, 327
As if heuristic, 76–77
Assembly work
    design, *316,* 316–318
    posture, 174
    quick, 312, *312*
Astigmatism, 44
ATE, *see* Automatic test equipment (ATE)
ATP, *see* Adenosine triphosphate (ATP)
AT&T, 96
Attention, sustained, 89–92, *90, 92*
Attention and active processing, 116
Auditory image store, 75
Auditory nerve damage, 241

Australia, 9, 327
Australian Telecom, 214
Automatic test equipment (ATE), 319
Automation, design, 303–306, *304*
Automobiles, *see* Vehicles
Availability heuristic, 77

B

Back
    computer workstations, 262–263, *263*
    injuries, manual materials handling, 188–191,
        188–191
    working at conveyors, 178
Bangladesh, 225
Base part, foundation/fixture, 306–307, *307*
Basic metabolic rate (BMR), 227
Bathroom taps, 100
Behavior, safety, 343–344
Bell Laboratories, 96
Bennett, John, 221
Bhopal, India, 342
Bible bump, *210*
Bin-assembly, car brakes, *103*
Biomechanics, 190–192, *191–193*
BIT, *see* Built-in test (BIT)
Black & Decker, 313
Blame, 333
Blink rate, 31
BMR, *see* Basic metabolic rate (BMR)
Body dimensions, *154–155, see also*
    Anthropometric workstation design
Brakes and braking, 32–33, 103, *103, see also*
    Vehicles
Broadbent theories, 246–247, *246–247*
Built-in test (BIT), 319
Burners (stove-top), 101–102, *102*
Buttock breadth, 162
Buttock-knee depth, *155*
Buttock-popliteal depth, *155*

C

CAESAR, *see* Civilian American and European
    Surface Anthropometry Resource
    (CAESAR) project
Calories, 227–228
Capture errors, 338
Car brakes, bin-assembly, *103*
Card assembly, cost-benefit analysis, 17–18,
    18–19
Car interior, *151–152*
Carnegie Mellon University, 74

Carpal tunnel syndrome, 209–210, *210–213*
Cars, *see* Vehicles
Carter, Jimmy (former U.S. President), 260, *260*
Caterpillar, 225
Cathode ray tube (CRT) screens
    LCD comparison, *271*
    radiation, 265–266
    reflections and glare, *266–267,* 266–270, *269*
    visual fatigue, 265
    visually exacting, 264
Ceiling impact, lighting, 58
CEN European Standard, 162
Chain of events, accidents, 331
Chains, safety, 345
Chairs
    cost-benefit analysis, 20
    design, 262–263, *263*
Checkerboard patterns, 48
Checklists
    early user testing, 127, *127*
    human factors and ergonomics, 367–371
    job aid, 281
Chess players, 73
China
    color coding, 104–105, *105*
    investigations, 29
    symbols, 113
Chunking information
    basics, 73–74
    comprehension, 117
    fault identification, 319
    memory requirements, 123
Cigars, performance improvement, *282*
Circadian rhythms, 295, *296*
Circuit fault identification, 287, *288*
Civilian American and European Surface
    Anthropometry Resource (CAESAR)
    project, 158, 160–161
Classical decision making, 79
Clean-room process, 169–170
Clippy, 124
Clothes wringing disease, *210,* 213
Clouding of vision, 57
Cockpit, *132*
Codebooks, job aid, 281
Coding, controls, 104–108
Cognitive demands table, 291, *291*
Cognitive task analysis (CTA), *see* Investigations;
    Training, skills, cognitive task analysis
Cognitive walkthrough (CW), 129–130, *132,*
    132–133, 141
Collecting and analysis, data, 38, *38–39*
Color coding
    controls design, 104–105, *105*
    fault identification, 319